# Flexible, Wearable, and Stretchable Electronics

## Devices, Circuits, and Systems
Series Editor - Krzysztof Iniewski

**Telecommunication Networks**
*Eugenio Iannone*

**Optical, Acoustic, Magnetic, and Mechanical Sensor Technologies**
*Krzysztof Iniewski*

**Biological and Medical Sensor Technologies**
*Krzysztof Iniewski*

**Graphene, Carbon Nanotubes, and Nanostuctures**
Techniques and Applications
*James E. Morris and Krzysztof Iniewski*

**Low Power Emerging Wireless Technologies**
*Reza Mahmoudi and Krzysztof Iniewski*

**High-Speed Photonics Interconnects**
*Lukas Chrostowski and Krzysztof Iniewski*

**Smart Sensors for Industrial Applications**
*Krzysztof Iniewski*

**MEMS: Fundamental Technology and Applications**
*Vikas Choudhary and Krzysztof Iniewski*

**Nanoelectronic Device Applications Handbook**
*James E. Morris and Krzysztof Iniewski*

**Novel Advances in Microsystems Technologies
and Their Applications**
*Laurent A. Francis and Krzysztof Iniewski*

**Building Sensor Networks: From Design to Applications**
*Ioanis Nikolaidis and Krzysztof Iniewski*

**Embedded and Networking Systems**
Design, Software, and Implementation
*Gul N. Khan and Krzysztof Iniewski*

**Medical Imaging**
Technology and Applications
*Troy Farncombe and Krzysztof Iniewski*

**Nanoplasmonics**
Advanced Device Applications
*James W. M. Chon and Krzysztof Iniewski*

**Testing for Small-Delay Defects in Nanoscale CMOS Integrated Circuits**
*Sandeep K. Goel and Krishnendu Chakrabarty*

**Energy Harvesting with Functional Materials
and Microsystems**
*Madhu Bhaskaran, Sharath Sriram, and Krzysztof Iniewski*

**Energy Harvesting with Functional Materials and Microsystems**
*Madhu Bhaskaran, Sharath Sriram, and Krzysztof Iniewski*

**Nanoscale Semiconductor Memories**
Technology and Applications
*Santosh K. Kurinec and Krzysztof Iniewski*

**Integrated Power Devices and TCAD Simulation**
*Yue Fu, Zhanming Li, Wai Tung Ng, and Johnny K.O. Sin*

**MIMO Power Line Communications**
Narrow and Broadband Standards, EMC, and Advanced Processing
*Lars Torsten Berger, Andreas Schwager, Pascal Pagani, and Daniel Schneider*

**Technologies for Smart Sensors and Sensor Fusion**
*Kevin Yallup and Krzysztof Iniewski*

**Smart Grids**
Clouds, Communications, Open Source, and Automation
*David Bakken*

**Microfluidics and Nanotechnology**
Biosensing to the Single Molecule Limit
*Eric Lagally*

**Nanopatterning and Nanoscale Devices for Biological
Applications**
*Šeila Selimovic´*

**High-Speed Devices and Circuits with THz Applications**
*Jung Han Choi*

**Metallic Spintronic Devices**
*Xiaobin Wang*

**Mobile Point-of-Care Monitors and Diagnostic Device Design**
*Walter Karlen*

**VLSI**
Circuits for Emerging Applications
*Tomasz Wojcicki*

**Micro- and Nanoelectronics**
Emerging Device Challenges and Solutions
*Tomasz Brozek*

**Design of 3D Integrated Circuits and Systems**
*Rohit Sharma*

**Wireless Transceiver Circuits**
System Perspectives and Design Aspects
*Woogeun Rhee*

**Soft Errors**
From Particles to Circuits
*Jean-Luc Autran and Daniela Munteanu*

**Optical Fiber Sensors**
Advanced Techniques and Applications
*Ginu Rajan*

**Laser-Based Optical Detection of Explosives**
*Paul M. Pellegrino, Ellen L. Holthoff, and Mikella E. Farrell*

**Organic Solar Cells**
Materials, Devices, Interfaces, and Modeling
*Qiquan Qiao*

**Solid-State Radiation Detectors**
Technology and Applications
*Salah Awadalla*

**CMOS**
Front-End Electronics for Radiation Sensors
*Angelo Rivetti*

**Multisensor Data Fusion**
From Algorithm and Architecture Design
to Applications
*Hassen Fourati*

**Electrostatic Discharge Protection**
Advances and Applications
*Juin J. Liou*

**Optical Imaging Devices**
New Technologies and Applications
*Ajit Khosla and Dongsoo Kim*

**Radiation Detectors for Medical Imaging**
*Jan S. Iwanczyk*

**Gallium Nitride (GaN)**
Physics, Devices, and Technology
*Farid Medjdoub*

**Mixed-Signal Circuits**
*Thomas Noulis*

**MRI**
Physics, Image Reconstruction, and Analysis
*Angshul Majumdar and Rabab Ward*

**Reconfigurable Logic**
Architecture, Tools, and Applications
*Pierre-Emmanuel Gaillardon*

**Ionizing Radiation Effects in Electronics**
From Memories to Imagers
*Marta Bagatin and Simone Gerardin*

**CMOS Time-Mode Circuits and Systems**
Fundamentals and Applications
*Fei Yuan*

**Tunable RF Components and Circuits**
Applications in Mobile Handsets
*Jeffrey L. Hilbert*

**Cell and Material Interface**
Advances in Tissue Engineering, Biosensor, Implant, and Imaging Technologies
*Nihal Engin Vrana*

**Nanomaterials**
A Guide to Fabrication and Applications
*Sivashankar Krishnamoorthy*

**Physical Design for 3D Integrated Circuits**
*Aida Todri-Sanial and Chuan Seng Tan*

**Wireless Medical Systems and Algorithms**
Design and Applications
*Pietro Salvo and Miguel Hernandez-Silveira*

**High Performance CMOS Range Imaging**
Device Technology and Systems Considerations
*Andreas Süss*

**Analog Electronics for Radiation Detection**
*Renato Turchetta*

**Power Management Integrated Circuits and Technologies**
*Mona M. Hella and Patrick Mercier*

**Circuits and Systems for Security and Privacy**
*Farhana Sheikh and Leonel Sousa*

**Sensors for Diagnostics and Monitoring**
*Kevin Yallup and Laura Basiricò*

**Biomaterials and Immune Response**
Complications, Mechanisms and Immunomodulation
*Nihal Engin Vrana*

**High-Speed and Lower Power Technologies**
Electronics and Photonics
*Jung Han Choi and Krzysztof Iniewski*

**X-Ray Diffraction Imaging**
Technology and Applications
*Joel Greenberg and Krzysztof Iniewski*

**Compressed Sensing for Engineers**
*Angshul Majumdar*

**Low Power Circuits for Emerging Applications in Communications, Computing, and Sensing**
*Krzysztof Iniewski and Fei Yuan*

**Labs on Chip**
Principles, Design, and Technology
*Eugenio Iannone*

**Energy Efficient Computing & Electronics**
Devices to Systems
*Santosh K. Kurinec and Sumeet Walia*

**Spectral, Photon Counting Computed Tomography**
Technology and Applications
*Katsuyuki Taguchi, Ira Blevis, and Krzysztof Iniewski*

**Flexible, Wearable, and Stretchable Electronics**
*Katsuyuki Sakuma*

For more information about this series, please visit: https://www.routledge.com/Devices-Circuits-and-Systems/book-series/CRCDEVCIRSYS

# Flexible, Wearable, and Stretchable Electronics

**Edited by**

KATSUYUKI SAKUMA

**Managing Editor**

KRZYSZTOF INIEWSKI

CRC Press
Taylor & Francis Group
Boca Raton London New York

CRC Press is an imprint of the
Taylor & Francis Group, an **informa** business

First edition published 2020
by CRC Press
6000 Broken Sound Parkway NW, Suite 300, Boca Raton, FL 33487-2742

and by CRC Press
2 Park Square, Milton Park, Abingdon, Oxon, OX14 4RN

---

**Library of Congress Cataloging-in-Publication Data**

---

Names: Sakuma, Katsuyuki, editor.
Title: Flexible, wearable, and stretchable electronics / edited by Katsuyuki Sakuma.
Description: First edition. | Boca Raton : CRC Press, 2020. | Series: Devices, circuits, & systems | Includes bibliographical references and index.
Identifiers: LCCN 2020030638 (print) | LCCN 2020030639 (ebook) | ISBN 9780367208905 (hardback) | ISBN 9780429263941 (ebook)
Subjects: LCSH: Flexible electronics. | Wearable technology. Classification: LCC TK7872.F54 F566 2020 (print) | LCC TK7872.F54 (ebook) | DDC 621.381--dc23
LC record available at https://lccn.loc.gov/2020030638
LC ebook record available at ttps://lccn.loc.gov/2020030639

---

ISBN: 978-0-367-20890-5 (hbk)
ISBN: 978-0-429-26394-1 (ebk)

Typeset in Palatino LT Std
by KnowledgeWorks Global Ltd.

# Contents

# *Preface*

Remarkable progress has been achieved within recent years in developing flexible, wearable, and stretchable (FWS) electronics. These electronics will play an increasingly significant role in future electronics and will open new product paradigms that conventional semiconductors are not capable of. This is because flexible electronics will allow us to build flexible circuits and devices on a substrate that can be bent, stretched, or folded without losing functionality. This revolutionary change will impact how we interact with the world around us. Future electronic devices will use flexible electronics as part of ambient intelligence and ubiquitous computing for many different applications such as consumer electronics, medical, healthcare, and security devices. Thus, it has a potential to create a huge market all over the world.

In this book, we provide a comprehensive technological review on the most state-of-the-art developments in FWS electronics. This book provides the reader a taste of what is possible with FWS electronics and how these electronics can provide unique solutions for a wide variety of applications. Each chapter includes a more detailed overview of key technology, the benefits and fundamental engineering challenges, and a hint to solve technical problems. Furthermore, the book introduces and explains new applications of flexible technology that has opened up the potential of FWS electronics. Based on these concepts, leading technology experts from industry to academia contributed each chapter.

This book will begin with an introduction of flexible electronics (Chapter 1), followed by a brief review of key materials, devices, and various applications that are used with FWS electronics. Flexible electronics in thin-film transistors (carbon nanotube-based TFTs and organic TFTs), displays, sensors, batteries, and biointegrated electronics (healthcare-monitoring devices and human-machine interfaces) will also be described.

Chapter 2 presents a comprehensive overview of stretchable conductors and devices. The use of these components is desirable in many future electronics such as in sensors, actuators, light-emitting diode arrays, energy harvesters, and storage devices. Recent performance gains of stretchable conductors including materials, structural engineering, and applications are also summarized in this chapter.

Chapter 3 describes a variety of electronic components and devices, which can be fabricated in FWS electronics. This chapter's first half is about integrated circuits, and components including conductive traces, resistors, capacitors and TFTs. The second half highlights optoelectronic devices including photovoltaics and luminescent devices.

Chapter 4 reviews the printing processes and techniques for flexible electronic applications. The printing processes have evolved to serve the fabrication of printed electronics. This chapter offers unique practical examples, methods, and tested technical solutions to face the main challenges of formulating inks, printing, and drying for industrialization and prototyping printed electronic applications. The characteristics, limitations, and advantages of the various printing processes for manufacturing (screen printing, inkjet, flexography, roll-to-roll, and gravure) are discussed.

Chapter 5 is focused on carbon nanotube (CNT)-based flexible complementary metal-oxide-semiconductor (CMOS) circuits and sensors. CNT has been widely considered as a superior candidate for flexible electronics due to its outstanding electrical and mechanical properties. The challenges and recent progress in making flexible CNT TFTs are discussed.

The application of CNT TFTs for flexible sensors, especially integrated pressure sensors, is also reviewed.

In Chapter 6, the emphasis is on flexible multifunctional sensor sheets for the application of health condition monitoring. In particular, this chapter aims to discuss a three-axis acceleration sensor using a printed piezoresistive strain sensor, a skin temperature sensor, and an electrocardiography (ECG) sensor on flexible polyethylene terephthalate (PET) film as a disposable sheet, while integrating a CNT transistor to switch functions and a UV sensor on another film as the reusable sheet. A flexible pH sensor that uses a charge-coupled device (CCD) is also discussed.

Chapter 7 explains controlled spalling for flexible and wearable sensor application. The use of traditional semiconductor-manufacturing process, which is based on bulk wafers, is very challenging to mass-produce flexible electronics. A solution for the mass production of flexible electronics is proposed. This proposed technology utilizes mechanical stress to facilitate peeling off thin film directly off a bulk silicon substrate. It is compatible with the CMOS processing technology; therefore, this process can potentially achieve high-volume manufacturability of flexible and wearable system-level electronics including sensors, signal processing, and communication units.

Chapter 8 focuses on liquid metal-based flexible and stretchable electronics. Liquid metals are extremely soft, are intrinsically stretchable, and have superior conductivity. Gallium-based liquid metal alloys have a melting point at/or near room temperature and have negligible vapor pressure and low viscosity. Unlike mercury alloys, gallium-based liquid metal alloys have shown low toxicity in addition to forming a surface oxide skin that allows them to be patterned into useful electronic components such as stretchable wires, antennas, interconnects, self-healing conductors, soft sensors, soft composites, and reconfigurable electronics.

Chapter 9 describes the fabrications and characterizations of an advanced flexible hybrid electronics (FHE). Fine-pitch interconnects are formed on large numbers of thin small dielets (50–400-μm thickness). The dielets are embedded in flexible substrates such as biomedical-grade epoxy, polydimethylsiloxane (PDMS), and hydrogel by advanced fan-out wafer-level packaging (FOWLP). This new FHE technology can be used for various wearable and biomedical applications.

Chapter 10 presents flexible electronic textiles (e-textiles) technologies using metal laminated fabric substrates and anisotropic conductive films (ACFs). Smart clothes using e-textiles are considered as one of the future wearable devices, because they combine electronic devices with the clothing that we are wearing on a daily basis. In this chapter, how the Si chip is bonded to the metal-laminated fabric substrates using nonconductive films (NCFs) lamination methods is described.

Chapter 11 describes the fabrication of high-density flexible substrates based on wafer-level redistribution and laser-assisted debonding. The systems are built up by sequential processing of polyimide layers and semi-additive structured metal layers on a temporary glass carrier wafer. Thin flexible circuit layers (only 20–50 μm) are generated with the removal of the carrier. This enables the integration of high-density routing features as well as thin active IC components into a polymer foil. An overview of flexible and stretchable systems for healthcare, mobility, and IoT are also discussed.

Chapter 12 provides a design, fabrication, and characterization of transparent antennas on thin flexible glass substrates. Among various flexible substrates, thin flexible glass has an ultra-smooth, high-temperature, and high-vacuum processable surface that makes it attractive in conventional micro/nano fabrication. Copper-based antennas were fabricated using a subtractive and a semi-additive process. Both these processes were demonstrated on sheet level and are compatible with high-throughput roll-to-roll manufacturing processing.

This book concludes with required testing and reliability characterization methods (Chapter 13). Flexible electronics in wearable applications will be subjected to bending, twisting, folding, or flexing depending on the form factor and the location of use. These actions induce mechanical stresses on flexible electronics which can lead to delamination, cracking, or shearing of the interconnects and additively printed layers. The transition from conventional device to flexible electronics mandates that there is no loss of functionality when folded, twisted, or bent.

This book is written so that the reader will have a deep understanding of the techniques, materials, fabrication, and applications of FWS electronics. I would like to acknowledge the contributing authors for their dedicated work and commitment to create this wonderful book. Without their outstanding contributions, this book would not have been possible. I am deeply grateful to all the authors for their time and effort. It is my sincere hope that this book will provide researchers and engineers with a timely comprehensive view of current flexible technology; thus, it will inspire readers to design their own novel future applications in flexible electronics.

**Katsuyuki Sakuma, Ph.D.**
*Yorktown Heights, New York*
*April 2020*

# *About the Series Editor*

**Krzysztof (Kris) Iniewski** is managing R&D at Redlen Technologies Inc., a startup company in Vancouver, Canada. Redlen's revolutionary production process for advanced semiconductor materials enables a new generation of more accurate, alldigital, radiation-based imaging solutions. Kris is also a founder of ET CMOS Inc. (http:// www.etcmos.com), an organization of high-tech events covering communications, microsystems, optoelectronics, and sensors. In his career, Dr. Iniewski held numerous faculty and management positions at University of Toronto (Toronto, Canada), University of Alberta (Edmonton, Canada), Simon Fraser University (SFU, Burnaby, Canada), and PMC-Sierra Inc (Vancouver, Canada). He has published more than 150 research papers in international journals and conferences. He holds more than 20 international patents granted in the United States, Canada, France, Germany, and Japan. He is a frequently invited speaker and has consulted for multiple organizations internationally. He has written and edited several books for CRC Press (Taylor & Francis Group), Cambridge University Press, IEEE Press, Wiley, McGraw-Hill, Artech House, and Springer. His personal goal is to contribute to healthy living and sustainability through innovative engineering solutions. In his leisurely time, Kris can be found hiking, sailing, skiing, or biking in beautiful British Columbia. He can be reached at kris.iniewski@gmail.com.

# About the Editor

**Katsuyuki Sakuma** is a research staff member at the IBM Thomas J. Watson Research Center in New York. He is also a visiting professor at the Department of Biomedical Engineering, Tohoku University, Japan, and at National Chiao Tung University, Taiwan. He has over 22 years of experience of researching three-dimensional integrated circuit technologies and performing various semiconductor packaging research and development projects. His research interests include bonding technologies, biomedical sensors, and advanced packaging.

He has published more than 90 peer-reviewed journal papers and conference proceeding papers, including four book chapters in the semiconductor and electronic packaging area. He also holds over 60 issued or pending U.S. and international patents. He has been recognized with the IBM 15th Invention Achievement Award in 2020 and an Outstanding Technical Achievement Award (OTAA) in 2015. He was also given the 2018 Exceptional Technical Achievement Award from IEEE Electronics Packaging Society and the 2017 Alumni Achievement Award from his Alma Mater, the School of Engineering at Tohoku University, for his contribution to 3D chip stack technology development in the electronics packaging industry. He was co-recipient of the IEEE CPMT Japan Society Best Presentation Award in 2012 and the IMAPS Best Paper Award of the interactive poster session in 2015.

Dr. Sakuma received his B.S. and M.S. degrees from Tohoku University, and the Ph.D. degree from Waseda University, Japan. He currently serves as an associate editor for IEEE Transactions on Components, Packaging and Manufacturing Technology (CPMT). He served as an associate editor of the Institute of Electronics, Information and Communication Engineers (IEICE, Japan) from 2003 until 2005. He has been serving as committee member of the IEEE ECTC Interconnections sub-committee since 2012 and the IEEE International Conference on 3D System Integration (IEEE 3DIC) since 2016. He has been a senior member of IEEE since 2012.

# Contributors

**Vivek T. Bharambe**
North Carolina State University
Raleigh, North Carolina

**Chloé Bois**
Printability and Graphic Communications
   Institute
Montreal, Quebec, Canada

**Zherui Cao**
Shanghai Institute of Ceramics
Chinese Academy of Sciences
Shanghai, China

**Christopher B. Cooper**
Stanford University
Palo Alto, California

**Michael D. Dickey**
North Carolina State University
Raleigh, North Carolina

**Takafumi Fukushima**
Tohoku University
Sendai, Miyagi, Japan

**Kartik Goyal**
Auburn University
Auburn, Alabama

**Neil Graddage**
National Research Council of Canada
Ottawa, Ontario, Canada

**Huan Hu**
Zhejiang University
Haining, China

**Ming-Huang Huang**
Corning Research and Development
   Corporation
Corning, New York

**Marie-Ève Huppé**
Printability and Graphic Communications
   Institute
Montreal, Quebec, Canada

**Subramanian S. Iyer**
University of California
Los Angeles, California

**Ishan D. Joshipura**
Lawrence Livermore National
   Laboratory
Livermore, California

**Erik Jung**
Fraunhofer IZM
Berlin, Germany

**Seung-Yoon Jung**
Korea Advanced Institute of Science
   and Technology
Daejeon, South Korea

**Christine Kallmayer**
Fraunhofer IZM
Berlin, Germany

**Young-Tae Kwon**
Georgia Institute of Technology
Atlanta, Georgia

**Pradeep Lall**
Auburn University
Auburn, Alabama

**Yiliang Lin**
University of Chicago
Chicago, Illinois

**Thomas Löher**
Fraunhofer IZM
Berlin, Germany

Jack P. Lombardi III
State University of New York
Binghamton, New York

Robert E. Malay
State University of New York
Binghamton, New York

Renwei Mao
Zhejiang University
Haining, China

Jinesh Narangaparambil
Auburn University
Auburn, Alabama

Kyung-Wook Paik
Korea Advanced Institute of Science
    and Technology
Daejeon, South Korea

Dishit P. Parekh
IBM Research
Albany, New York

Mark D. Poliks
State University of New York
Binghamton, New York

Scott C. Pollard
Corning Research and Development
    Corporation
Corning, New York

Michael Rozel
Printability and Graphic Communications
    Institute
Montreal, Quebec, Canada

Katsuyuki Sakuma
IBM Thomas J. Watson Research Center
Yorktown Heights, New York

James H. Schaffner
HRL Laboratories, LLC
Malibu, California

Hyok Jae Song
HRL Laboratories, LLC
Malibu, California

Jing Sun
Shanghai Institute of Ceramics
Chinese Academy of Sciences
Shanghai, China

Kuniharu Takei
Osaka Prefecture University
Sakai, Osaka, Japan

Timothy Talty
General Motors
Warren, Michigan

Jianshi Tang
Tsinghua University
Beijing, China

Ngoc Duc Trinh
Printability and Graphic Communications
    Institute
Montreal, Quebec, Canada

Ranran Wang
Shanghai Institute of Ceramics
Chinese Academy of Sciences
Shanghai, China

Darshana L. Weerawarne
State University of New York
Binghamton, New York

Woon-Hong Yeo
Georgia Institute of Technology
Atlanta, Georgia

Kai Zoschke
Fraunhofer IZM
Berlin, Germany

# 1

## Flexible, Wearable, and Stretchable Electronics

**Young-Tae Kwon**
*Georgia Institute of Technology*

**Woon-Hong Yeo**
*Georgia Institute of Technology*

**KEYWORDS:** *Wearable electronics, Flexible and stretchable electronics, Soft materials, Nanomaterials, Health monitoring, and Human-machine interfaces.*

## CONTENTS

## 1.1 Introduction

Over the last few decades, there has been an increased interest in flexible electronic devices due to their potential applications for smartphone, mobile display, and wearable healthcare system. Smart electronics, widely used today, have thin, lightweight, and wireless characteristics, which allow the employment on the human body and clothes. More recently, innovative features in curved, foldable smart devices enable enhanced mobility, user convenience, and conformal interfaces. Remarkable advances in wearable, wireless electronics have opened a new paradigm of portable, home health monitoring and therapeutics. The use of noninvasive, wearable systems for tracking biological signals provides real-time, continuous health monitoring, without the use of wired, bulky hardware in the conventional medical systems. Such biopotential signals, including electroencephalogram (EEG), electrooculogram (EOG), electrocardiogram (ECG), electromyogram (EMG), and bodily secretions, can investigate the clinical changes in brain activity, eye movement, cardiac systole-diastole cycle, and so on. Figure 1.1 illustrates some of recently developed wearable smart electronic devices and healthcare systems (Kim, Lee et al. 2016, Leleux et al. 2014, Ameri et al. 2018, Hong et al. 2019, Pickham et al. 2018).

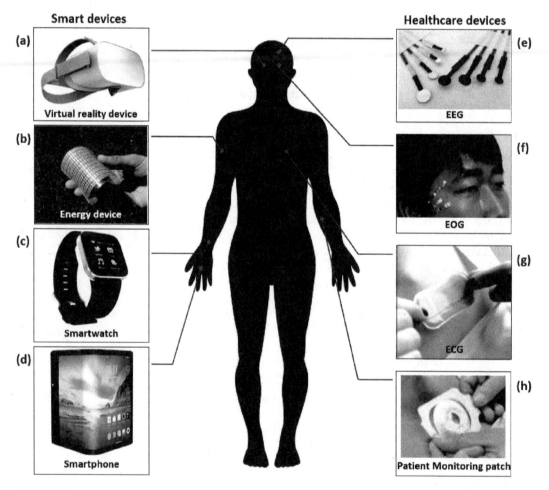

**FIGURE 1.1**
Wearable, flexible, and stretchable electronics for smart and healthcare devices. (a) Virtual reality device, (b) Flexible thin battery, (c) Wearable smart watch, and (d) Flexible smartphone. (e) EEG, (f) EOG, (g) ECG, and (h) Patient monitoring patch for human healthcare. ((b) Reproduced with permission from Kim, Lee et al. (2016), copyright 2016, ACS Publications; (e) Reproduced with permission from Leleux et al. (2014), copyright 2013, Wiley-VCH; (f) Reproduced with permission from Ameri et al. 2018), copyright 2008, Nature Publishing Group; (g) Reproduced with permission from (Hong et al. 2019), copyright 2019, Wiley-VCH; (h) Reproduced with permission from (Pickham et al. 2018), copyright 2018, Elsevier.)

Although significant progress in the field of flexible and stretchable electronics successfully achieves the commercialization in smart and healthcare applications, the currently available devices, based on rigid materials, have limited applications in biocompatibility, long-term wearability, portability, and direct integration with the skin and/or clothes. To achieve soft, stretchable electronics, capable of conforming to the dynamic surface and retaining their performance, novel approaches in advanced materials, mechanics design, and system integration are required. Once bulky and rigid materials are oriented into the downscaling of material dimensions, they easily accomplish the flexibility (Leleux et al. 2014, Lu, Suo, and Vlassak 2010). The design engineering, such as open-mesh and serpentine interconnection, is also a promising approach to endure the strain/stress, resulting in achievement of stretchability (Park et al. 2009). These concepts allow exhibiting unique electronic, optoelectronic, and bioelectronics properties that their bulk counterparts do not possess. These will be further discussed in Section 1.2.

Collectively, this chapter provides a summary of recently developed wearable, flexible, and stretchable hybrid electronics, emphasizing the advancement in materials, designs, and integration technologies. We introduce key wearable applications of flexible and stretchable electronics in thin film transistors (TFTs), displays, sensors, batteries, and biointegrated electronics. Lastly, we conclude by discussing an overview of remaining challenges and future perspectives in wearable electronics.

## 1.2 Functional Electronic Components and Devices

The strategies to integrate flexible, wearable, and stretchable devices exploit specialized materials and designs, and are classified into three main categories: supporting substrates, conducting materials, and pattern engineering (Figure 1.2). Compared to the rigid substrates, polymer-based substrate materials play a huge role in flexible and wearable technologies due to their inherent low mechanical stiffness, thermal stability, and chemical resistance. Among polymer substrates, polyimide (PI), polyethylene terephthalate (PET), and polyethylene naphthalate (PEN) exhibit great flexibility and electrical insulation (Figure 1.2a-c) (Webb et al. 2013, Wang et al. 2012, Nomura et al. 2004, Cao et al. 2016, Kaltenbrunner et al. 2013, Wang, Liu, and Zhang 2017). However, these polymer substrates with low adhesive force and high elastic modulus have been limited in the direct application to wearable and stretchable electronics (Liu, Pharr, and Salvatore 2017). Soft silicone materials, including polydimethylsiloxane (PDMS), Ecoflex, and Solaris, have the elastic modulus similar to that of skin, enabling conformal contact and stretchability without an adhesion failure (Figure 1.2d) (Kim, Lu et al. 2011, Liu, Pan, and Liou 2017). In addition, the most used paper in everyday life has also been considered as the supporting substrates for low-cost flexible electronics (Figure 1.2e) (Tobjörk and Ö sterbacka 2011, Zschieschang et al. 2011).

To successfully maintain the high electrical and mechanical performances on diverse electronics, the functional nanomaterials is applied as the conductive electrodes though either top-down or bottom-up approaches (Yu et al. 2017). The top-down approach manufacturing with e-beam evaporation and/or sputtering tools is critical to implement thin layers because the flexural rigidity is proportional to the thickness, which is called as

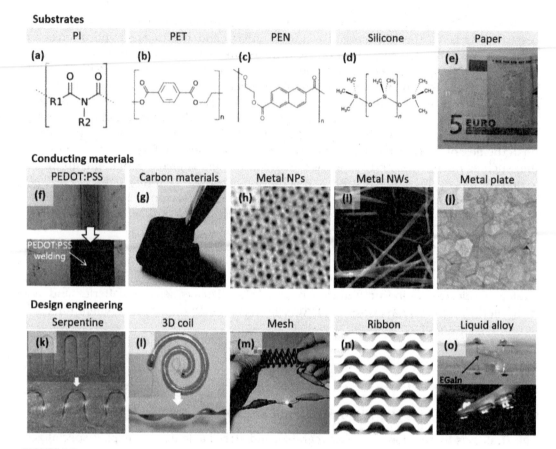

**FIGURE 1.2**

Substrates, conducting materials, and designs for flexible and stretchable electronics. Representative polymers for the electrical insulated substrates: (a) PI, (b) PET, (c) PEN, (d) Silicone elastomer, and (e) Paper. ((a-d) Reproduced with permission from Wang, Liu, and Zhang (2017), copyright 2017, Wiley-VCH. (e) Reproduced with permission from Zschieschang et al. (2011), copyright 2011, Wiley-VCH.) Representative forms of conducting nanomaterials: (f) PEDOT:PSS. (Reproduced with permission from Oh et al. (2016), copyright 2016, Wiley-VCH.) (g) Carbon materials such as CNT and graphene. (Reproduced with permission from Xu et al. (2008), copyright 2008, ACS Publications.) (h) Metal nanoparticles (NPs). (Reproduced with permission from Zhao et al. (2015), copyright 2015, ACS Publications). (i) Metal nanowires (NWs). (Reproduced with permission from Kwon et al. (2017), copyright 2017, ACS Publications.) (j) Metal plate. (Reproduced with permission from Lin et al. (2014), copyright 2014, ACS Publications.) Geometry structure designs enhance the system-level deformability: (k) Serpentine. (Reproduced with permission from Pan et al. (2017), copyright 2017, Wiley-VCH.). (l) 3D coil. (Reproduced with permission from Mohammed and Kramer (2017), copyright 2017, Wiley-VCH.). (m) Mesh. (Reproduced with permission from Guo et al. (2016), copyright 2016, ACS Publications.). (n) Ribbon. (Reproduced with permission from Wu et al. (2016), copyright 2016, Wiley-VCH.) (o) Liquid alloy. (Reproduced with permission from (Yoon et al. 2014), copyright 2014, Wiley-VCH.)

Euler-Bernoulli beam theory (Son et al. 2014, Lemaitre and Chaboche 1994). Figure 1.2f-i summarizes representative conductive materials synthesized via bottom-up approach. One of conductive polymer, poly(3,4- ethylenedioxythiophene):polystyrene sulfonate (PEDOT:PSS), with low modulus organics, can be structured with minimal degradation of electrical property under extreme deformations (Figure 1.2f) (Lipomi et al. 2012, Oh et al. 2016). Since carbon-based nanomaterials, such as carbon nanotube (CNT) and graphene, exhibit exceptionally high fracture strength and electrical performance, they enhance the mechanical robustness of electronics (Figure 1.2g) (Geim and Novoselov 2010, Kabiri Ameri et al. 2017, Sekitani et al. 2008, Xu et al. 2008). Downscaling of the metal materials (Au, Ag, and Cu) is also the most promising approach for decreasing flexural rigidity. These metal nanoparticles (NPs), nanowires (NWs), and nanoplates endure a larger amount of stress/strain strength than the bulk counterparts during extreme deformations (Figure 1.2h-j) (Kwon et al. 2018, Choi, Han et al. 2018, Lee et al. 2013, Zhao et al. 2015, Kwon et al. 2017, Lin et al. 2014).

In addition to the thinned supporting substrates and nanomaterials, the geometric layout engineering has been used to enhance the device deformation, as shown in Figure 1.2k-o.

Mechanically optimized serpentine structure design displays high system-level stretchability and low interconnect resistance, which provides a good compromise between the stress distribution and ease of device layout (Figure 1.2k) (Pan et al. 2017, Wu et al. 2016). Helical coils with sacrificial 3D feature provide a uniform distribution of deformation-induced stress by compressive bucking (Figure 1.2l) (Jang et al. 2017, Xu et al. 2015, Mohammed and Kramer 2017). Figure 1.2m captures the mesh architecture, also called kirigami shape, which can be used even under harsh operating circumstances of 100% stretchability (Guo et al. 2016). As shown in Figure 1.2n, geometry configuration with ribbon layouts contributes a route of electronics for high levels of stretchability (up to ~100%) and compressibility (up to ~25%) (Wu et al. 2016, Kim et al. 2008). An additional noteworthy way involves liquid alloy embedded in soft membranes, enabling the endurance of mechanical deformations (Figure 1.2o) (Yoon et al. 2014). Figure 1.3 displays well-assorted, deformable electronics manufactured in accordance of materials engineering described above (Wang, Hsieh, and Hwang 2011, Liu, Liu et al. 2018, Lee et al. 2011, Fukagawa et al. 2018, Kim et al. 2017, Hu et al. 2011, Wang, Guo et al. 2016, Mishra et al. 2018, Kim, Kim et al. 2016, Koo et al. 2012, Berchmans et al. 2014, Kumar et al. 2017, Barrett et al. 2014, Yeo et al. 2013, Trung et al. 2016).

## 1.3 Thin Film Transistors (TFTs)

TFT is a form of transistor fabricated by depositing thin films of an active semiconductor, dielectric, and metallic layers on supporting substrates (Lu et al. 2018). Recently, the use of nanomaterials, including CNT and organics as the active semiconductor layer, has accelerated the substrate independence in TFT (Franklin 2015). This section summarizes the materials and technologies for assembling the deformable TFT based on CNT or organics.

### 1.3.1 Carbon Nanotube-Based TFTs

One-dimensional single wall CNTs (SWCNT), which have normally p-type characteristic, exhibit extraordinary electrical, thermal, and mechanical performance due to their intrinsic molecular structure (Iijima 1991). In particular, CNTs, well dispersed in solution, are recognized as a practical means to flexible TFT electronics (Zhang, Wang, and Zhou 2012,

**FIGURE 1.3**
Flexible, wearable, and stretchable electronics for (a) thin film transistor (TFT), (b) display, (c) sensor, (d) battery, and (e) healthcare patch. ((a) Reproduced with permission from Lee et al. (2011), Wang, Hsieh, and Hwang (2011), Liu, Lui et al. (2018), copyright 2011, Wiley-VCH; copyright 2018, ACS Publications; copyright 2011, ACS Publications, respectively. (b) Reproduced with permission from Fukagawa et al. (2018), Kim et al. (2017), Hu et al. (2011), copyright 2018, Wiley-VCH; copyright 2017, Wiley-VCH; copyright 2011, Wiley-VCH, respectively. (c) Reproduced with permission from Wang, Guo et al. (2016), Mishra et al. (2018), Kim, Kim et al. (2016), copyright 2016, Wiley-VCH; copyright 2018, Elsevier; copyright 2016, Wiley-VCH, respectively. (d) Reproduced with permission from Koo et al. (2012), Berchmans et al. (2014), Kumar et al. (2017), copyright 2012, ACS Publications; copyright 2014, The Royal Society of Chemistry; copyright 2017, Wiley-VCH, respectively. (e) Reproduced with permission from Barrett et al. (2014), Yeo et al. (2013), Trung et al. (2016), copyright 2014, Elsevier; copyright 2013, Wiley-VCH; copyright 2016, Wiley-VCH, respectively.)

Miyata et al. 2011). Figure 1.4a displays directly CNT-printed TFT on elastomeric PDMS substrates (Cai et al. 2016). The fully printed TFT devices withstand stretching with tensile strain exceeding 50% while showing no significant electrical degradation. In a similar manner, the CNT TFT printed between the source (S) and drain (D) electrodes on flexible PI substrates is demonstrated to comparable performance to rigid counterparts (Figure 1.4b) (Yu et al. 2018). The TFT electronics fabricated via printing technologies can easily employ to deformable supporting membranes as well as the development of large area, scalable systems. Recent progress of CNT TFT has advanced to hybridization with functional nanomaterials to compensate the drawback of active channel consisted of only CNT. Figure 1.4c illustrates the fabrication process for the flexible CNT TFT casted with ion gel. Layers of gate (Au), gate dielectric (cross-linked poly(4-vinylphenol); cPVP), and source/drain (Au) are prepared on a flexible PEN substrate in sequence. The transferred and patterned CNT networks onto the cPVP surface are covered with ion gel using micro-dispenser printing method. The lamination of high-capacitance dielectric layer onto CNT TFT can drastically reduce the operating device voltage caused by the capacitive coupling effect between two dielectric layers sandwiching the CNT channel (Choi, Kang et al. 2018). Figure 1.4d shows the TFT and temperature sensor devices vertically integrated with hybrid n-typed InGaZnO (IGZO) and p-typed CNT transistors onto flexible PI membrane (Honda et al. 2015). Finally, a temperature sensor, which consists of CNT and PEDOT:PSS,

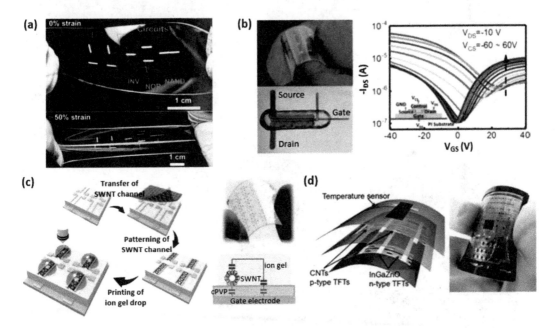

**FIGURE 1.4**
CNT-based TFT. (a) Photograph of a CNT-based TFT fabricated on a stretchable PDMS substrate. (Reproduced with permission from Cai et al. (2016), copyright 2016, ACS Publications.) (b) Optical image capturing a CNT-printed TFT on a flexible PI substrate (top-left); microscope image of the individual TFT (bottom-left); transfer curves of the CNT TFT measured under different control gate voltage on PI substrate (right). (Reproduced with permission from Yu et al. (2018), copyright 2018, ACS Publications.) (c) Illustration showing the fabrication of ion gel and CNT hybrid TFT (left); photo of the flexible ion gel and CNT hybrid TFT (top-right); Cross-sectional schematic of the TFT representing the structure of CNT, cPVP, and ion gel (bottom-left) (Reproduced with permission from Choi, Kang et al. (2018), copyright 2018, Wiley-VCH.) (d) Three-dimensional diagram of CNT and indium-gallium-zinc oxide (IGZO) hybrid TFT with temperature sensor (left). Right image shows flexible 3D TFT sensor devices. (Reproduced with permission from Honda et al. (2015), copyright 2015, Wiley-VCH.)

is laminated on the top of device. This hybrid approach of n-type and p-type nanomaterials offers not only high device yield and low power consumption comparable to the other CNT TFT electronics, but also stable temperature dependence upon a flexible condition.

## 1.3.2 Organic TFTs

Organic semiconductor materials including conjugated polymers, oligomers, or fused aromatics have a significant practical impact in many different electronic devices (Dimitrakopoulos and Malenfant 2002). Typical Organic TFT (OTFT) comprises the organic semiconductors as active layer, gate dielectrics, and three electrodes of source, drain, and gate. The voltage applied to source/drain flows through the organic semiconductor channel layer and the channel current can be modulated by the gate electrode. Since most of the organic semiconductors are soluble in organic solvents, solution-processed OTFT facilitates the employment in flexible, wearable, and stretchable membranes (Jiang et al. 2019, Hyun et al. 2015, Kelly et al. 2017). Figure 1.5a shows a diagram for flexible OTFT

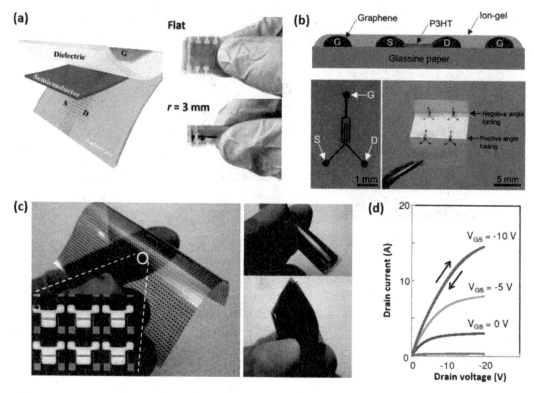

**FIGURE 1.5**
Organic TFT. (a) Schematic structure of a flexible OTFT on a PET substrate (left); example of the fabricated flat OTFT under tensile bending of 3 mm (right). (Reproduced with permission from Liu, Yin et al. (2019), copyright 2019, Wiley-VCH.) (b) Schematic illustration of OTFT fabricated on foldable glassine paper (top); optical image showing the graphene electrodes for gate, source, and drain on the glassine paper (bottom left) and manufactured OTFT devices with 3 × 4 array (bottom right). (Reproduced with permission from Hyun et al. (2015), copyright 2015, Wiley-VCH.) (c) Photo and optical microscopy image showing the paper-based flexible OTFT array (left); photographs of the bended (top-right) and folded (bottom-left) OTFT; (d) Transfer curves of the paper-based OTFT measured under different control gate voltage. ((c, d) Reproduced with permission from Fujisaki et al. (2014), copyright 2014, Wiley-VCH.)

electronics employed organic semiconducting layer between PET substrate with source/drain and the gate-deposited dielectric layer (Liu, Yin et al. 2019). The flexible OTFT has excellent bending stability under bending radii of 3 mm. Unlike the polymer membranes described previously, eco-friendly and low-cost cellulose nanofiber paper is a promising candidate as a supporting substrate for deformable electronics (Fujisaki et al. 2014). Figure 1.5b demonstrates the glassine paper provides a suitable supporting membrane for outstanding OTFT performance (Hyun et al. 2015). The device structure contains source, drain, and gate electrodes printed with graphene, organic semiconducting materials of poly(3-hexylthiophene) (P3HT), and ion-gel gate dielectric. The paper-based OTFT reveals good folding stability due to high mechanical tolerance of the paper substrate and graphene electrodes. As another example, Figure 1.5c describes that OTFT manufactured on the 20-μm thick transparent paper provides an excellent mechanical durability, including bending and folding, without tearing or fracturing. The performance exhibits a high mobility of up to 1 $cm^2V^{-1}s^{-1}$, which is similar to the OTFT arrays on the conventional plastic substrates (Figure 1.5d).

## 1.4 Displays

A major objective in the research field of next-generation displays is to develop freely deformable light-emitting diodes (LEDs) with excellent device performance, regardless of the supporting substrates (Choi, Yang, Hyeon et al. 2018). It is not an ultimate next-generation display technology; however, Samsung Electronics demonstrated newly developed, foldable smartphone with a 7.3-inch tablet-sized screen. Figure 1.6a illustrates the basic device structure of flexible/wearable LED. The multilayered LED structure consists of anode/cathode layers, hole/electron transport layers, and emission layer (Fukagawa et al. 2018, Choi, Yang, Hyeon et al. 2018). The working mechanism of LEDs is that the holes and electrons injected from each electrode take place the radiative recombination in emission layer. Here, rigid and brittle emission layers have limited the flexibility of LED. Recent advancements in flexible LEDs have been focused on two types of materials, organic LED (OLED) and quantum dot LED (QLED), due to their outstanding stability and efficiency under the deformations (Wang et al. 2011, Yokota et al. 2016, Dai et al. 2017, Kim et al. 2017, Zhao et al. 2017). The deformable OLEDs and QLEDs are introduced in this section.

OLED integrated on flexible substrates enables the electronics to be bent, rolled, and even stretched, because the organics with different bandgap energy can be solution-processed (Kalyani and Dhoble 2012, Liu, Feng et al. 2019). At present, OLED-based flexible displays are being mass produced and LG Display introduced a 77-inch ultra-high-definition (UHD) TV at the Society for Information Display 2018, which can be rolled up to 80-mm radius without compromising any of its functionality. Figure 1.6b shows the seamless, foldable OLED display, composed of two individual OLED panels, which pretend to display a single unfolded shape by a laser programmable buckling process (Kim, Kwon et al. 2011). The mechanical folding tolerance and the change in relative brightness with original brightness for the flexible OLED display are simultaneously performed under tensile stress. The device withstands bending and folding stresses with a radius of 1 mm up while exhibiting only small fluctuations of 6% in performance over 10,000 cycles. Figure 1.6c describes a flexible display with a solution-processed interlayer on PI membrane. The most important

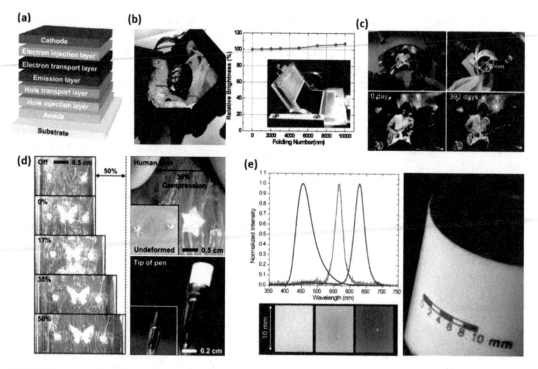

**FIGURE 1.6**

Next-generation displays. (a) Structure of a LED for flexible and stretchable displays. (b) Actual images showing the foldable OLED display (left); plot of optical brightness change in steps of 1,000 folds for a total 10,000 folds (right). Inset image is experimental setup for mechanical folding test. (Reproduced with permission from Kim, Kwon et al. (2011), copyright 2011, Wiley-VCH.) (c) Photographs capturing an operation of OLED-based flexible display as a function of storage time. ((a, c) Reproduced with permission from Fukagawa et al. (2018), copyright 2018, Wiley-VCH.) (d) Quantum dot LED (QLED) with 50% stretchability (left); QLEDs deformed to 30% compression on human skin (top-left); QLED laminated on curved substrate, tip of pen (bottom-left). (Reproduced with permission from Choi, Yang, Kim et al. (2018), copyright 2018, Wiley-VCH.) (e) Electroluminescence spectra and photograph of QD with red, green, and blue colors (left); complete QLED display on a flexible substrate. (Reproduced with permission from Wood et al. (2009), copyright 2009, Wiley-VCH.)

issue remaining for the practical application of flexible display is their short lifetimes, caused by the poor environmental stability of OLEDs (Fukagawa et al. 2018). In order to settle the issue for short lifetime of OLED, this study reports the fabrication of long-lived flexible displays by employing stable OLED using an interlayer suitable for efficient electron injection. The luminance of device maintains even though bending to a 50-mm diameter and long storage time of 392 days.

The recent progress in solution-processed patterning techniques has made it possible to integrate the flexible, full-color QLED electronics (Yang et al. 2014, Kim, Cho et al. 2011, Zhao et al. 2017). Quantum dot (QD) materials have many advantages in optical characteristics including excellent color purity, wide bandgap tunability, and high electroluminescence (EL) brightness, originating from the quantum confinement effect (Kagan et al. 2016, Gong et al. 2016, Mashford et al. 2013). Figure 1.6d shows ultrathin deformable QLED without any luminance degradation. Ultrathin nature of QLED fabricated on parylene-C membrane allows its conformal contact on various curved objects, such as human skin and pen (Choi, Yang, Kim et al. 2018). The wavy structured QLED reveals remarkable stretching up to 50%, particularly extreme 30% compression on the human skin. Figure 1.6e captures the flexible QLED display driven by the QLED arrays with three colors, red (R), green (G), and blue (B). Flexible full-color display is achieved through the inkjet printing techniques using stable QD material ink solutions, and that it contributes to high efficient and robust device architecture (Wood et al. 2009). With these technological advances, QLED can be successfully utilized to next-generation deformable display.

## 1.5 Sensors

### 1.5.1 Gases and Light Signals

Signals from surrounding environment including gases and light are crucial factors that immediately affect human health. Therefore, sensors enabling precise detection of environmental hazard have been intensively studied in the research field of deformable electronics. This section highlights flexible and wearable sensors capturing signals from gases and light. With rapid increase in hazardous gas emission, the safety with regard to human protection has become a critical issue (Wiedinmyer, Yokelson, and Gullett 2014, Perera 2016). Although conventional detection platform permits exact sensing, fast analysis, without the degradation, the integration to human skin or cloths is limited for sizes, weights, and form factor of commercialized devices. Recent advances in nanotechnologies facilitate the gas sensing based on the flexible and wearable formats. Among them, carbonaceous nanomaterials are ideal for gas molecule adsorption due to their extremely high surface-to-volume ratio (Toda, Furue, and Hayami 2015, Wang, Huang et al. 2016, Singh, Meyyappan, and Nalwa 2017, Kim et al. 2015, Choi, Kim, and Kim 2016). Furthermore, since most carbon-based nanomaterials have p-type characteristic, the charge recombination occurs in the presence of n-typed gases donating the electrons, which results in the sudden increase in resistance. Figure 1.7a illustrates the interaction of n-typed hydrogen sulfide gases with reduced graphene oxide (Choi, Kim, and Kim 2016). This strategy offers enhanced selectivity toward the gaseous target including hydrogen sulfide, ethanol, and hydrogen (Figure 1.7b). The deformation properties before and after the bending cycles are demonstrated at 20 p.p.m. concentration of the gases. Figure 1.7c shows reduced graphene oxide incorporated wearable sensor module, enabling Bluetooth-assisted wireless communication. Similarly, CNT exhibits a response to hydrogen sulfide molecules, caused by the change in conductivity (Figure 1.7d). This chapter reports that the integrating the CNT and Cu NPs selectively contributes to detecting the target gas molecules due to the modulation of the charge density (Figure 1.7e) (Asad et al. 2015). Figure 1.7f captures the fabricated flexible gas sensor with the structure in which Cu-decorated CNTs are bridged between the Al electrode fingers. Other nanomaterials including various metal oxide have been studied for flexible and wearable gas sensor (Rashid, Phan, and Chung 2013, Arena et al. 2010, Uddin et al. 2016, Zang et al. 2014, Kumaresan et al. 2018, Zheng et al. 2015).

Ultraviolet (UV) ray that reaches Earth's surface is only ~8%; however, it has the greatest impact on human health over the spectral composition of sunlight, because the excessive exposure to UVB (280–315 nm) and UVA (315–400 nm) causes a skin cancer

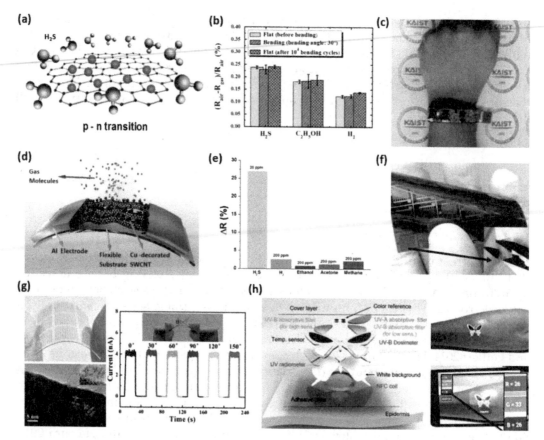

**FIGURE 1.7**

Sensors for environmental gases and light. (a) Schematic illustration showing a sensing mechanism between reduced graphene oxide and hydrogen sulfide molecules. (b) Selectivity of the sensor toward hydrogen sulfide, ethanol, and hydrogen gases during flat and bending. (c) Photograph showing wearable sensor module detecting target gases. ((a–c) Reproduced with permission from Choi, Kim, and Kim (2016), copyright 2016, Nature Publishing Group.) (d) Schematic for Cu-decorated CNT-based $H_2S$ gas sensor on a flexible substrate. (e) Plot for the sensing selectivity of $H_2S$. (f) Photo of the fabricated gas sensor devices. The inset depicts a single device. ((d–f) Reproduced with permission from Asad et al. (2015), copyright 2015, Elsevier.) (g) Optical image of flexible UV photodetectors composed of $Zn_2SnO_4$ NWs (top-left); TEM image showing the interfaces of $Zn_2SnO_4$ NWs (down-left); photo-response characteristics of the flexible device bent with different bending angles from 0° to 150° (right). (Reproduced with permission from Li, Gu et al. (2017), copyright 2017, ACS Publications.) (h) Exploded view image of multimodal, colorimetric epidermal device for capabilities in ultraviolet (UV) and temperature and in wireless operation (left); photos capturing the UV-sensing device attached on human arm and wireless communication between the device and smartphone (right). (Reproduced with permission from Araki et al. (2017), Copyright 2017, Wiley-VCH.)

(Ray et al. 2019). Such a risk has accelerated the critical development in the flexible and wearable UV-sensing platform, which can alert people to overexposure. The semiconductor nanomaterials with wide bandgap energy matching with UV wavelength are promising candidates for UV photodetector. Figure 1.7g shows flexible UV sensor manufactured by high-quality ZnO QD-decorated $Zn_2SnO_4$ NWs (Li, Gu et al. 2017). High-resolution transmission electron microscopy images demonstrate that 10-nm sized ZnO QDs are clearly decorated on the surface of the $Zn_2SnO_4$ NWs. The 0D-1D hybrid structure achieves the bending deformations without the sensing degradation. Figure 1.7h describes a wearable, epidermal UV and temperature sensor platform, taken together with near field communication and image analysis tool. The color change of sensing materials depending on the UV exposure time is wirelessly transmitted to the smartphone and the information about instantaneous UV exposure levels and skin temperature is analyzed via extraction of RGB parameters (Araki et al. 2017).

### 1.5.2 Miscellaneous Signals

Monitoring miscellaneous environmental responses including pH, temperature, and pressure is another important research field in a human being due to climate change, global warming, and acid rain caused by expanding industrialization (Li, Zhang et al. 2017, Harada et al. 2014, Ren et al. 2016, Schwartz et al. 2013). Figure 1.8a shows colorimetric, wearable pH sensor based on a micro-fluidic platform that exploits ionic liquid polymer gel. Although the colorimetric devices are less accurate than general potentiometric sensors, it is an attractive technology in terms of simplicity of material and design (Li, Askim, and Suslick 2018, Ray et al. 2019). Different four pH-sensitive dyes used in this study change color over pH range from 4.5 to 8, which cover the concentration category (pH 5–7) of human's sweat (Figure 1.8b) (Curto et al. 2012). The wearable pH-sensing system integrated into a wristband provides the potentiometric information during an exercise (Figure 1.8c). Figure 1.8d illustrates the device mechanism, enabling simultaneous monitoring temperature-pressure dual signals. Once an external pressure is loaded on a device with thermoelectric and piezo-resistive effects, the switch disconnects the one sensing module from other modules to allow concise measurement of temperature and pressure (Figure 1.8e) (Zhang et al. 2015). Figure 1.8f describes flexible and wearable sensor array worn on a prosthetic hand. The change in temperature and pressure caused by arm-wrestling corresponds to the mapping profiles of pixel signals. An additional prospective direction in flexible pressure and temperature sensing is to use field effect transistor (FET) sensor integrated with multi-stimuli responsive materials (Figure 1.8g) (Tien et al. 2014). The sensor platform comprising piezo-electric nanocomposite gate and thermos-resistive organic semiconductor allows responding to pressure and temperature simultaneously. Figure 1.8h demonstrates that the bimodal monitoring system integrated with $4 \times 4$ sensing device could identify the position and the temperature pressed with a human finger in real time.

## 1.6 Batteries

Soft battery-based energy storage technology has become one of the popular research topics for the application ranging from powering homes and electric vehicles to miniaturized electronics and medical devices (Kanamura 2017, Jaguemont, Boulon, and Dubé 2016, Gao et al. 2015, Wang et al. 2015, Montgomery and Ellis 2018). These batteries require high specific energy densities, fast recharging capabilities, and superior cycle stabilities even under various complicated deformation conditions. In this regard, rechargeable lithium-ion batteries have shown great representative power source due to their high-energy capacity and

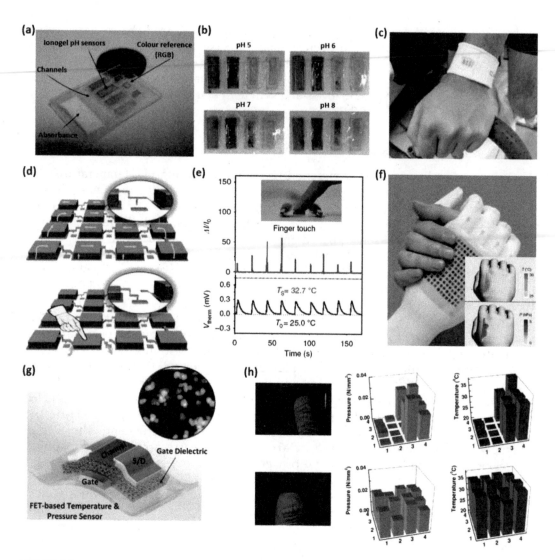

**FIGURE 1.8**
Sensors using miscellaneous signals. (a) Photo of pH sensor platform based on ionic liquid polymer gels (ionogels). (b) Demonstration of the pH sensing system tested with artificial sweat. (c) Optical image of the pH sensor system integrated into a wristband. ((a-c) Reproduced with permission from Curto et al. (2012), copyright 2012, Elsevier.) (d) Illustration showing a mechanism of wearable temperature and pressure sensor. (e) Plots for the current and output voltage responses under finger touch. (f) Photograph of a prosthetic arm-wrestling with human. Inset graph displays the temperature and pressure mapping profiles on the back of the prosthetic hand. ((d-f) Reproduced with permission from Zhang et al. (2015), copyright 2015, Nature Publishing Group.) (g) Schematic of the flexible device for simultaneous sensing of pressure and temperature. (h) Read-out pressure-temperature bimodal signals of a pressed sensor array by human thumb. ((g-i) Reproduced with permission from Tien et al. (2014), copyright 2014, Wiley-VCH.)

long-term cyclability (Ren et al. 2014, Koo et al. 2012, Jeong et al. 2011). Figure 1.9a shows an example of a flexible lithium-ion battery made from graphene form, a three-dimensional interconnected network, loaded with Li4Ti5O12 and LiFePO4 as current collectors (Li et al. 2012). Owing to the graphene form with both a highly conductive pathway for electrons/lithium ions and thin structure, the lithium-ion battery provides sufficient power to turn on the commercial LED under flexible stress (Figure 1.9b). Figure 1.9c demonstrates

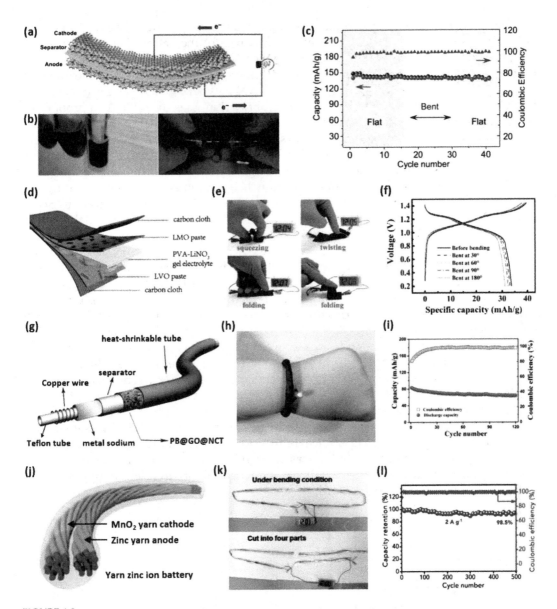

**FIGURE 1.9**
Flexible batteries. (a) Illustration, (b) photo, and (c) cyclic performance of flexible lithium-ion battery made from three-dimensional graphene form. ((a-c) Reproduced with permission from (Li et al. 2012), copyright 2012, National Academy of Sciences. (d) Schematic diagram displaying the textile lithium-ion battery. (e) Demonstration for powering a digital watch upon squeezing, twisting, and folding. (f) Galvanostatic charging/discharging graph under different bending angles. ((d-f) Reproduced with permission from (Liu, Li et al. 2018), copyright 2018, Elsevier.) (g) Schematic illustration for the tube-type flexible sodium ion battery. (h) Demonstration of an LED lighting under wrist condition. (i) Graph for cyclic performance of flexible sodium ion battery. ((g-i) Reproduced with permission from Zhu et al. (2017), copyright 2017, Wiley-VCH.) (j) Illustration showing the double-helix yarn zinc-ion battery enabling the stretching and bending. (k) Tailoring test of zinc-ion battery. Each battery cut into four parts can power the digital watch. (l) Long-term cycling performance of yarn zinc-ion battery. ((j-l) Reproduced with permission from Li et al. (2018), copyright 2018, ACS Publications.)

the effect of bending on the performance of the flexible battery. After 20 bends to a radius of 5 mm, the change in electrochemical performances of the battery is negligibly small compared to the original flat battery. Fabric-type membrane represents a great potential for the flexible and wearable energy storage electronics. Figure 1.9d illustrates textile flexible lithium-ion battery, integrating with LiMn2O4 cathode, $LiNO_3$ gel electrolyte, and LiV3O8 anode materials between two carbon fabrics (Liu, Li et al. 2018). As a proof of its superior flexibility, the battery could continually power a digital watch even in the extreme deformations (Figure 1.9e). Figure 1.9f shows an additional demonstration for exceptional bendability of the lithium-ion battery. The galvanostatic charge/discharge curves exhibit a consistent performance regardless of the several cycles of bending at 30°, 60°, 90°, and 180°.

Although lithium-based batteries are currently being explored, increased consumption and rarity of lithium resources represent a drawback for large-scale application (Slater et al. 2013, Liu, Sun et al. 2019). Thus, the battery system based on the wide availability and low cost, including sodium, zinc, and biofuel, is heavily considered. In particular, since sodium resource is the fourth most abundant element in the Earth crust, sodium-based batteries have been recently studied as an alternative chemistry to lithium-ion battery (Slater et al. 2013). Figure 1.9g describes the structure of textile tube-type flexible sodium ion battery. The battery structure consists of a hollow Teflon tube as a supporting substrate, a copper wire anode, a sodium foil for electrolyte, a flexible separator, a Prussian blue graphene composite (PB@GO@NTC) cathode, and a shrinkable tube (Zhu et al. 2017). To prove its potential application to wearable electronics, the bracelet-type battery is utilized to power a commercial red LED (Figure 1.9h). This type of battery exhibits high mechanical flexibility, good electronic conductivity, and superior electrochemical stability (Figure 1.9i). Figure 1.9j displays a representative example of stretchable, textile, zinc ion battery, which is produced by dipping and electrodepositing two CNT yarns into the $MnO_2$ solution and zinc electrolyte solution, respectively (Li et al. 2018). One remarkable point of the zinc-ion battery is its stretchability and tailorability. The battery can power a commercial digit watch even though it was cut into four parts (Figure 1.9k). Furthermore, the zinc-ion battery demonstrates excellent long-term stability to retain high capacity of 98.5% even after 500 cycles (Figure 1.9l).

## 1.7 Bio-Integrated Electronics

### 1.7.1 Healthcare-Monitoring Devices

Soft, skin-interfaced sensors and electronics enable the measurements of noninvasive biophysical and biochemical signals associated with activity of the heart, brain, eyes, muscles, and metabolite for healthcare monitoring. Compared to commercially developed classes of healthcare-monitoring device with rigid and bulky characteristics, flexible and wearable electronics mounted on various human body parts allow long-term integration, continuous and real-time health monitoring, and accurate feedback, without the user's inconvenience. Figure 1.10 summarizes recent reported skin-integrated electronics for healthcare monitoring using the electrophysiological (EP) and metabolite signals. Electrophysiological signals such as ECG, EEG, and EMG provide important diagonal information for cardiology, neural disorder, nerve, and muscle health (Chen et al. 2017, Liu, Pharr, and Salvatore 2017, Kwon et al. 2018). In a conventional clinical setting, EP measurements have been conducted by

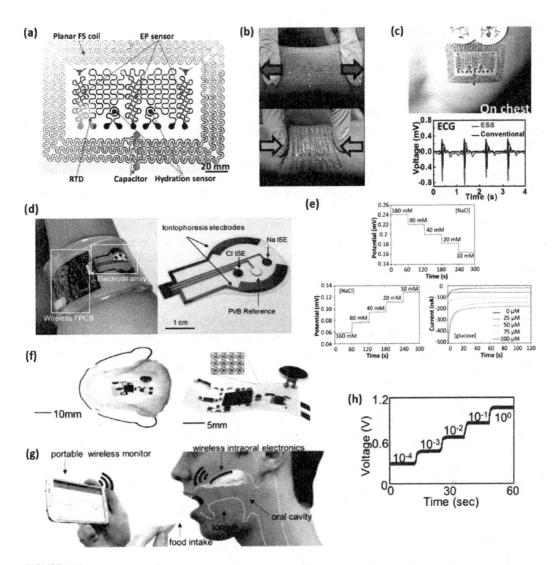

**FIGURE 1.10**

Wearable devices for human health monitoring. (a) Photograph showing an epidermal electronic system (EES) which incorporates three electrophysiological (EP) electrodes, two coaxial dot-ring impedance sensor, a resistance temperature detector (RTD), and hydration sensor in planar filamentary serpentine (FS) layouts. (b) EES on human skin demonstrating robust deformability during stretch (top) and compression (bottom). (c) ECG simultaneously measured by EES and conventional electrodes. ((a-c) Reproduced with permission from Yang et al. (2015), Copyright 2015, Wiley-VCH.) (d) Images showing the wireless, wearable sweat-sensing platform (left) and iontophoresis/sweat sensor electrodes for Na⁺ and Cl⁻ monitoring. (e) Open-circuit potential responses of the Na⁺, Cl⁻, and glucose. ((d, e) Reproduced with permission from Emaminejad et al. (2017), copyright 2017, National Academy of Sciences.) (f) Illustration and (g) overview displaying wearable intraoral electronics for quantification of sodium intake via wireless, real-time monitoring. (h) Plot of voltage change from the intraoral sensor according to the sodium concentration. ((f-h) Reproduced with permission from Lee et al. (2018), copyright 2018, National Academy of Sciences.)

employing at least two gel electrodes in direct contact with skin around the targeted tissues. However, the use of gels and sticky tapes causes critical issues, which are skin irritation and allergic reaction (Mishra et al. 2017). Figure 1.10a-c describes wearable epidermal electronic system (EES) (Yang et al. 2015). Multifunctional EES integrated on the transparent adhesive tattoo paper includes three EP electrodes, two dot-ring impedance sensor, a resistance temperature detector (RTD), and hydration sensor (Figure 1.10a). Since all sensors are connected with Au-based filamentary serpentine structure, the EES exhibits excellent deformable characteristics on human skin (Figure 1.10b).

Figure 1.10c displays representative ECG measurement from human chest using EES and conventional gel electrodes. Overall, the measured quality of biopotentials from the EES and conventional sensors agrees qualitatively well, supported by a similar level of ECG signal.

Metabolites in sweat, saliva, and tears are essential information for health state of human body. Figure 1.10d captures the noninvasive, wearable sweat-monitoring system integrated with electrochemically enhanced iontophoresis interface that allows measuring real-time $Na^+$, $Cl^-$, glucose in the sweat (Emaminejad et al. 2017). The sweat-sensing platform containing electrode array and wireless flexible printed circuit board (FPCB) consists of sodium/chloride ion-selective electrodes (ISE) for monitoring of $Na^+$/$Cl^-$ ions and iontophoresis electrodes for detecting the glucose. Figure 1.10e displays the voltage and chronoamperometric responses from the wearable autonomous sweat sensor, enabling the monitoring in 10–160 mM NaCl contents and glucose concentration range from 0 to 100 µM. As another example, Figure 1.10f-h presents stretchable sodium-sensor platform integrated in the oral cavity using biochemical assessments of saliva (Lee et al. 2018). The soft membrane electronics with serpentine interconnects are laminated on the surface of a custom-made porous retainer (Figure 1.10f), resulting in high consistent functionalities during the deformations. This novel class of intraoral device, which can easily be connected with a personal smartphone, enables a long-range wireless, real-time quantification of sodium intake (Figure 1.10g). Figure 1.10h demonstrates that the fast responsive ion-selective sodium electrodes allow the accurate change in time-dynamic voltage according to the various concentrations of sodium solution.

### 1.7.2 Human-Machine Interfaces

From the perspective of expanding the application space of epidermal electronics, human-machine interfaces represent another important research study in the area of humanoid robotics, prosthetic hand, drone, display interface, and machine-assisted living (Liu, Pharr, and Salvatore 2017, Xu et al. 2016, Herbert et al. 2018). The physiological signals, measured from the skin-mounted flexible/wearable electronics, can be translated to electrical signals for controlling the machine or providing feedback (Huang et al. 2019). The biopotentials include EEG, EOG, and EMG, as described in Section 1.7.1. To achieve sensitive and accurate interaction between human and machines, it is important to comply with seamless, conformal contact of devices to the human skin. Figure 1.11 captures representative recent reports for human-machine interfaces, controlled by the biopotentials, come from skin electronics.

Figure 1.11a shows eight tripolar concentric ring and capacitive electrodes capable of high-fidelity EEG recording (Norton et al. 2015). The ultrathin and wearable neural platform, including Au-PI-Au layers on a 3-µm thick elastomer, makes conformal contact to uneven skin surfaces, such as auricle and mastoid (Figure 1.11b). The combined area of auricle and mastoid offers unique electrical isolation from other regions of the scalp,

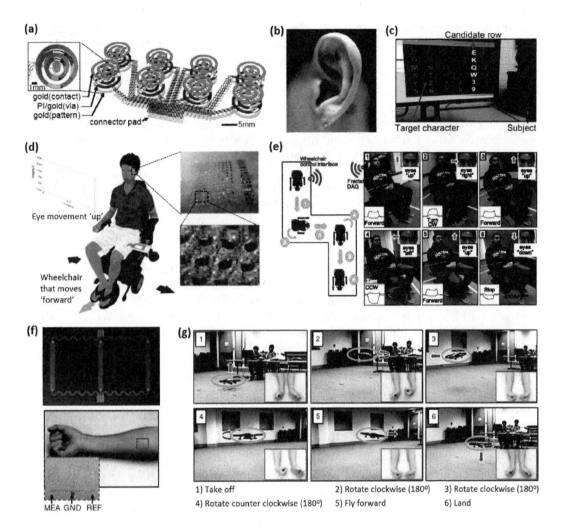

**FIGURE 1.11**
Examples of human-machine interfaces. (a-c) Brain-computer interfaces. (a) Schematic illustration and (b) EEG electrodes on the auricle and mastoid. (c) Images showing text speller with brain-computer interfaces. ((a-c) Reproduced with permission from Norton et al. (2015), copyright 2015, National Academy of Sciences.) (d, e) Human-wheelchair interface. (d) Illustration of a subject wearing electrodes to control the wheelchair via EOG signals. Inset photos capturing the soft gold electrodes. (e) A set of photos showing a real time, wireless control of the wheelchair through eye movements. There are four control commands depending on eye movements including up, right, left, and down. Inset graphs show the EOG signal generated by different eye movements. ((d, e) Reproduced with permission from Mishra et al. (2017), copyright 2017, Elsevier.) (f, g) Human-quadrotor interfaces. (f) Gold electrodes mounted on the forearm. (g) Images of quadrotor control via EMG signal. Insets capture the control gestures. ((f, g) Reproduced with permission from Jeong et al. (2013), copyright 2013, Wiley-VCH.)

enhancing the EEG signals. The neuronal EEG activity recorded from brain enables direct communication with external computers for improving our understanding of language comprehension. In this chapter, examples in text spellers with steady state visually evoked potentials and event-related potentials demonstrate advanced progresses in brain-computer interface technologies (Figure 1.11c). A subject is provided with two flickering classification windows at a unique frequency on a computer screen. When the user chooses and stares a desired character, the classifier selects the representative group of letters (Norton et al. 2015). This assistive technology to help patients with paralyzed limbs is demonstrated via noninvasive, wearable electronics for EEG monitoring.

EOG signals, driven from an eye movement, can be utilized to control a wheelchair. Figure 1.11d presents skin-like bioelectrodes with mechanical compliance in both stretchability and bendability (up to 180°) (Mishra et al. 2017). Two sets of Au/PI electrodes integrated on the elastomeric supporting membrane are placed on the outer canthus of each eye for horizontal eye movements and upper eyebrow/lower eyelid for vertical eye movements. A Bluetooth-based system incorporating the soft electronics demonstrates a precise, hand-free operation of a robotic wheelchair via four different commands (Figure 1.11e): 'eye up' to go forward, 'eye down' to stop movement, 'eye left' to turn counterclockwise, and 'eye right' to turn clockwise (Mishra et al. 2017). These ergonomic human-wheelchair interfaces exhibit the real-time control to navigate the designed pathway from the position 1 to position 6. Figure 1.11f captures a wearable skin electronics on the forearm, capable of high-fidelity EMG recording for human-drone interfaces (Jeong et al. 2013). Geometrical serpentine Au electrodes as a measurement, ground, and reference reveal the robust stretchable property of 30% and conformal contact that occurs when the adhesive energy is greater than the sum of bending and elastic energy (Jeong et al. 2013). EMG signals from four different bimanual gestures prove 91.1% classification accuracy with distinct commands of quadrotor (Figure 1.11g): 'Out' to land and take off, 'In' to fly forward, 'Left' to rotate counterclockwise, and 'Right' to rotate clockwise.

## 1.8 Conclusions and Future Outlook

The evolution in soft materials, nanoengineering, and system integration, summarized in this chapter, offers the advancement of flexible and stretchable electronics for various wearable applications, including transistors, LEDs, sensors, batteries, and human healthcare devices. Ultrathin silicone elastomers, organic NPs, and integration of functional chips and soft membranes have accelerated the development of soft, deformable electronics, improving device functionality and compatibility with humans. Downscaling of the material dimensions, including carbon materials, NWs, and nanoplates, is another promising strategy to reduce the flexural rigidity, thus enduring external strain. Additionally, the study of stretchable mechanics and engineering of geometric design offer enhanced mechanical reliability in dynamic, cyclic use of the electronics.

Nevertheless, several challenges remain in the development of real-world applicable, wearable electronics. Recently, new substrate like smart textiles is widely studied to replace polymer membranes due to the advantage of compatibility with the existing clothes and reliability in a long-term use (Zang et al. 2016). Integration of a display with wearable electronics requires a thorough investigation of thin encapsulation materials to replace the current high-modulus layer, which can endure the stretchable stress (Koo et al. 2018). From the perspective of attaching various functional sensors directly to human body, biocompatibility and longevity of key materials should be further investigated (Han et al. 2017). Although the advances in the materials for cathode, anode, and electrolyte realize the deformable batteries, the use of metal oxide with brittle and flammable characteristics in battery components remains an urgent problem to be solved (Tan et al. 2017, Liu et al. 2017). Lastly, the electronics for healthcare monitoring must be further developed to distinguish the biophysical or biochemical signals from noisy background during human movements in daily activities (Wang, Liu, and Zhang 2017). We anticipate that future commitment to these remaining challenges will promise a reliable, real-world application of fully functional, low-profile, wireless, wearable electronics.

## Acknowledgments

We acknowledge the support of the Alzheimer's Association (2019-AARGD-NTF-643460) and the National Institutes of Health (NIH R21AG064309). The content is solely the responsibility of the authors and does not necessarily represent the official views of the NIH.

## References

Ameri, Shideh Kabiri, Myungsoo Kim, Irene Agnes Kuang, Withanage K Perera, Mohammed Alshiekh, Hyoyoung Jeong, Ufuk Topcu, Deji Akinwande, and Nanshu Lu. 2018. Imperceptible electrooculography graphene sensor system for human–robot interface. *NPJ 2D Materials and Applications* 2(1):19.

Araki, Hitoshi, Jeonghyun Kim, Shaoning Zhang, Anthony Banks, Kaitlyn E Crawford, Xing Sheng, Philipp Gutruf, Yunzhou Shi, Rafal M Pielak, and John A Rogers. 2017. Materials and device designs for an epidermal UV colorimetric dosimeter with near field communication capabilities. *Advanced Functional Materials* 27(2):1604465.

Arena, A, N Donato, G Saitta, A Bonavita, G Rizzo, and G Neri. 2010. Flexible ethanol sensors on glossy paper substrates operating at room temperature. *Sensors and Actuators B: Chemical* 145(1):488–494.

Asad, Mohsen, Mohammad Hossein Sheikhi, Mahdi Pourfath, and Mahmood Moradi. 2015. High sensitive and selective flexible $H_2S$ gas sensors based on Cu nanoparticle decorated SWCNTs. *Sensors and Actuators B: Chemical* 210:1–8.

Barrett, Paddy M, Ravi Komatireddy, Sharon Haaser, Sarah Topol, Judith Sheard, Jackie Encinas, Angela J Fought, and Eric J Topol. 2014. Comparison of 24-hour Holter monitoring with 14-day novel adhesive patch electrocardiographic monitoring. *The American Journal of Medicine* 127(1):95. e11–17.

Berchmans, Sheela, Amay J Bandodkar, Wenzhao Jia, Julian Ramírez, Ying S Meng, and Joseph Wang. 2014. An epidermal alkaline rechargeable Ag–Zn printable tattoo battery for wearable electronics. *Journal of Materials Chemistry A* 2(38):15788–15795.

Cai, Le, Suoming Zhang, Jinshui Miao, Zhibin Yu, and Chuan Wang. 2016. Fully printed stretchable thin-film transistors and integrated logic circuits. *ACS Nano* 10(12):11459–11468.

Cao, Xuan, Christian Lau, Yihang Liu, Fanqi Wu, Hui Gui, Qingzhou Liu, Yuqiang Ma, Haochuan Wan, Moh R Amer, and Chongwu Zhou. 2016. Fully screen-printed, large-area, and flexible active-matrix electrochromic displays using carbon nanotube thin-film transistors. *ACS Nano* 10(11):9816–9822.

Chen, Yang, Zhong-yi Wang, Gang Yuan, and Lan Huang. 2017. An overview of online based platforms for sharing and analyzing electrophysiology data from big data perspective. *Wiley Interdisciplinary Reviews: Data Mining and Knowledge Discovery* 7(4):e1206.

Choi, Moon Kee, Jiwoong Yang, Taeghwan Hyeon, and Dae-Hyeong Kim. 2018. Flexible quantum dot light-emitting diodes for next-generation displays. *NPJ Flexible Electronics* 2(1):10.

Choi, Moon Kee, Jiwoong Yang, Dong Chan Kim, Zhaohe Dai, Junhee Kim, Hyojin Seung, Vinayak S Kale, Sae Jin Sung, Chong Rae Park, and Nanshu Lu. 2018. Extremely vivid, highly transparent, and ultrathin quantum dot light-emitting diodes. *Advanced Materials* 30(1):1703279.

Choi, Seon-Jin, Sang-Joon Kim, and Il-Doo Kim. 2016. Ultrafast optical reduction of graphene oxide sheets on colorless polyimide film for wearable chemical sensors. *NPG Asia Materials* 8(9):e315.

Choi, Suji, Sang Ihn Han, Dongjun Jung, Hye Jin Hwang, Chaehong Lim, Soochan Bae, Ok Kyu Park, Cory M Tschabrunn, Mincheol Lee, and Sun Youn Bae. 2018. Highly conductive, stretchable and biocompatible Ag–Au core–sheath nanowire composite for wearable and implantable bioelectronics. *Nature Nanotechnology* 13(11):1048–1056.

Choi, Yongsuk, Joohoon Kang, Ethan B Secor, Jia Sun, Hyoungjun Kim, Jung Ah Lim, Moon Sung Kang, Mark C Hersam, and Jeong Ho Cho. 2018. Capacitively coupled hybrid ion gel and carbon nanotube thin-film transistors for low voltage flexible logic circuits. *Advanced Functional Materials* 28(34):1802610.

Curto, Vincenzo F, Cormac Fay, Shirley Coyle, Robert Byrne, Corinne O'Toole, Caroline Barry, Sarah Hughes, Niall Moyna, Dermot Diamond, and Fernando Benito-Lopez. 2012. Real-time sweat pH monitoring based on a wearable chemical barcode micro-fluidic platform incorporating ionic liquids. *Sensors and Actuators B: Chemical* 171:1327–1334.

Dai, Xingliang, Yunzhou Deng, Xiaogang Peng, and Yizheng Jin. 2017. Quantum-dot light-emitting diodes for large-area displays: towards the dawn of commercialization. *Advanced Materials* 29(14):1607022.

Dimitrakopoulos, Christos D, and Patrick RL Malenfant. 2002. Organic thin film transistors for large area electronics. *Advanced Materials* 14(2):99–117.

Emaminejad, Sam, Wei Gao, Eric Wu, Zoe A Davies, Hnin Yin Yin Nyein, Samyuktha Challa, Sean P Ryan, Hossain M Fahad, Kevin Chen, and Ziba Shahpar. 2017. Autonomous sweat extraction and analysis applied to cystic fibrosis and glucose monitoring using a fully integrated wearable platform. *Proceedings of the National Academy of Sciences* 114(18):4625–4630.

Franklin, Aaron D. 2015. Nanomaterials in transistors: from high-performance to thin-film applications. *Science* 349(6249):aab2750.

Fujisaki, Yoshihide, Hirotaka Koga, Yoshiki Nakajima, Mitsuru Nakata, Hiroshi Tsuji, Toshihiro Yamamoto, Taiichiro Kurita, Masaya Nogi, and Naoki Shimidzu. 2014. Transparent nanopaper-based flexible organic thin-film transistor array. *Advanced Functional Materials* 24(12): 1657–1663.

Fukagawa, Hirohiko, Tsubasa Sasaki, Toshimitsu Tsuzuki, Yoshiki Nakajima, Tatsuya Takei, Genichi Motomura, Munehiro Hasegawa, Katsuyuki Morii, and Takahisa Shimizu. 2018. Long-lived flexible displays employing efficient and stable inverted organic light-emitting diodes. *Advanced Materials* 30(28):1706768.

Gao, Xian-Zhong, Zhong-Xi Hou, Zheng Guo, and Xiao-Qian Chen. 2015. Reviews of methods to extract and store energy for solar-powered aircraft. *Renewable and Sustainable Energy Reviews* 44:96–108.

Geim, Andre K, and Konstantin S Novoselov. 2010. The rise of graphene. In *Nanoscience and Technology: A Collection of Reviews from Nature Journals*, 11–19. World Scientific.

Gong, Xiwen, Zhenyu Yang, Grant Walters, Riccardo Comin, Zhijun Ning, Eric Beauregard, Valerio Adinolfi, Oleksandr Voznyy, and Edward H Sargent. 2016. Highly efficient quantum dot near-infrared light-emitting diodes. *Nature Photonics* 10(4):253.

Guo, Hengyu, Min-Hsin Yeh, Ying-Chih Lai, Yunlong Zi, Changsheng Wu, Zhen Wen, Chenguo Hu, and Zhong Lin Wang. 2016. All-in-one shape-adaptive self-charging power package for wearable electronics. *ACS Nano* 10(11):10580–10588.

Han, Su-Ting, Haiyan Peng, Qijun Sun, Shishir Venkatesh, Kam-Sing Chung, Siu Chuen Lau, Ye Zhou, and VAL Roy. 2017. An overview of the development of flexible sensors. *Advanced Materials* 29(33):1700375.

Harada, Shingo, Kenichiro Kanao, Yuki Yamamoto, Takayuki Arie, Seiji Akita, and Kuniharu Takei. 2014. Fully printed flexible fingerprint-like three-axis tactile and slip force and temperature sensors for artificial skin. *ACS Nano* 8(12):12851–12857.

Herbert, Robert, Jong-Hoon Kim, Yun Kim, Hye Lee, and Woon-Hong Yeo. 2018. Soft material-enabled, flexible hybrid electronics for medicine, healthcare, and human-machine interfaces. *Materials* 11(2):187.

Honda, Wataru, Shingo Harada, Shohei Ishida, Takayuki Arie, Seiji Akita, and Kuniharu Takei. 2015. High-performance, mechanically flexible, and vertically integrated 3D carbon nanotube and InGaZnO complementary circuits with a temperature sensor. *Advanced Materials* 27(32):4674–4680.

Hong, Yongseok Joseph, Hyoyoung Jeong, Kyoung Won Cho, Nanshu Lu, and Dae-Hyeong Kim. 2019. Wearable and implantable devices for cardiovascular healthcare: from monitoring to therapy based on flexible and stretchable electronics. *Advanced Functional Materials* 29(19):1808247. https://doi.org/10.1002/adfm.201808247

Hu, Xiaolong, Peter Krull, Bassel De Graff, Kevin Dowling, John A Rogers, and William J Arora. 2011. Stretchable inorganic-semiconductor electronic systems. *Advanced Materials* 23(26):2933–2936.

Huang, Yan, Xiangyu Fan, Shih-Chi Chen, and Ni Zhao. 2019. Emerging technologies of flexible pressure sensors: materials, modeling, devices, and manufacturing. *Advanced Functional Materials* 29(12):1808509.

Hyun, Woo Jin, Ethan B Secor, Geoffrey A Rojas, Mark C Hersam, Lorraine F Francis, and C Daniel Frisbie. 2015. All-printed, foldable organic thin-film transistors on glassine paper. *Advanced Materials* 27(44):7058–7064.

Iijima, Sumio. 1991. Helical microtubules of graphitic carbon. *Nature* 354(6348):56.

Jaguemont, J, L Boulon, and Y Dubé. 2016. A comprehensive review of lithium-ion batteries used in hybrid and electric vehicles at cold temperatures. *Applied Energy* 164:99–114.

Jang, Kyung-In, Kan Li, Ha Uk Chung, Sheng Xu, Han Na Jung, Yiyuan Yang, Jean Won Kwak, Han Hee Jung, Juwon Song, and Ce Yang. 2017. Self-assembled three dimensional network designs for soft electronics. *Nature Communications* 8:15894.

Jeong, Goojin, Young-Ugk Kim, Hansu Kim, Young-Jun Kim, and Hun-Joon Sohn. 2011. Prospective materials and applications for Li secondary batteries. *Energy & Environmental Science* 4(6):1986–2002.

Jeong, Jae-Woong, Woon-Hong Yeo, Aadeel Akhtar, James JS Norton, Young-Jin Kwack, Shuo Li, Sung-Young Jung, Yewang Su, Woosik Lee, and Jing Xia. 2013. Materials and optimized designs for human-machine interfaces via epidermal electronics. *Advanced Materials* 25(47):6839–6846.

Jiang, Chen, Hyung Woo Choi, Xiang Cheng, Hanbin Ma, David Hasko, and Arokia Nathan. 2019. Printed subthreshold organic transistors operating at high gain and ultralow power. *Science* 363(6428):719–723.

Kabiri Ameri, Shideh, Rebecca Ho, Hongwoo Jang, Li Tao, Youhua Wang, Liu Wang, David M Schnyer, Deji Akinwande, and Nanshu Lu. 2017. Graphene electronic tattoo sensors. *ACS Nano* 11(8):7634–7641.

Kagan, Cherie R, Efrat Lifshitz, Edward H Sargent, and Dmitri V Talapin. 2016. Building devices from colloidal quantum dots. *Science* 353(6302):aac5523.

Kaltenbrunner, Martin, Tsuyoshi Sekitani, Jonathan Reeder, Tomoyuki Yokota, Kazunori Kuribara, Takeyoshi Tokuhara, Michael Drack, Reinhard Schwödiauer, Ingrid Graz, and Simona Bauer-Gogonea. 2013. An ultra-lightweight design for imperceptible plastic electronics. *Nature* 499(7459):458.

Kalyani, N Thejo, and SJ Dhoble. 2012. Organic light emitting diodes: energy saving lighting technology—A review. *Renewable and Sustainable Energy Reviews* 16(5):2696–2723.

Kanamura, Kiyoshi. 2017. Large-scale batteries for green energy society. In *Electrochemical Science for a Sustainable Society*, 175–195. Springer.

Kelly, Adam G, Toby Hallam, Claudia Backes, Andrew Harvey, Amir Sajad Esmaeily, Ian Godwin, João Coelho, Valeria Nicolosi, Jannika Lauth, and Aditya Kulkarni. 2017. All-printed thin-film transistors from networks of liquid-exfoliated nanosheets. *Science* 356(6333):69–73.

Kim, Dae-Hyeong, Jizhou Song, Won Mook Choi, Hoon-Sik Kim, Rak-Hwan Kim, Zhuangjian Liu, Yonggang Y Huang, Keh-Chih Hwang, Yong-wei Zhang, and John A Rogers. 2008. Materials and noncoplanar mesh designs for integrated circuits with linear elastic responses to extreme mechanical deformations. *Proceedings of the National Academy of Sciences* 105(48):18675–18680.

Kim, Daeil, Doyeon Kim, Hyunkyu Lee, Yu Ra Jeong, Seung-Jung Lee, Gwangseok Yang, Hyoungjun Kim, Geumbee Lee, Sanggeun Jeon, and Goangseup Zi. 2016. Body-attachable and stretchable multisensors integrated with wirelessly rechargeable energy storage devices. *Advanced Materials* 28(4):748–756.

Kim, Jaemin, Hyung Joon Shim, Jiwoong Yang, Moon Kee Choi, Dong Chan Kim, Junhee Kim, Taeghwan Hyeon, and Dae-Hyeong Kim. 2017. Ultrathin quantum dot display integrated with wearable electronics. *Advanced Materials* 29(38):1700217.

Kim, Sun Jin, Han Eol Lee, Hyeongdo Choi, Yongjun Kim, Ju Hyung We, Ji Seon Shin, Keon Jae Lee, and Byung Jin Cho. 2016. High-performance flexible thermoelectric power generator using laser multiscanning lift-off process. *ACS Nano* 10(12):10851–10857.

Kim, Sunkook, Hyuk-Jun Kwon, Sunghun Lee, Hongshik Shim, Youngtea Chun, Woong Choi, Jinho Kwack, Dongwon Han, MyoungSeop Song, and Sungchul Kim. 2011. Low-power flexible organic light-emitting diode display device. *Advanced Materials* 23(31):3511–3516.

Kim, Tae-Ho, Kyung-Sang Cho, Eun Kyung Lee, Sang Jin Lee, Jungseok Chae, Jung Woo Kim, Do Hwan Kim, Jang-Yeon Kwon, Gehan Amaratunga, and Sang Yoon Lee. 2011. Full-colour quantum dot displays fabricated by transfer printing. *Nature Photonics* 5(3):176.

Kim, Yeon Hoo, Sang Jin Kim, Yong-Jin Kim, Yeong-Seok Shim, Soo Young Kim, Byung Hee Hong, and Ho Won Jang. 2015. Self-activated transparent all-graphene gas sensor with endurance to humidity and mechanical bending. *ACS Nano* 9(10):10453–10460.

Kim, Dae-Hyeong, Nanshu Lu, Rui Ma, Yun-Soung Kim, Rak-Hwan Kim, Shuodao Wang, Jian Wu, Sang Min Won, Hu Tao, Ahmad Islam, Ki Jun Yu, Tae-il Kim, Raeed Chowdhury, Ming Ying, Lizhi Xu, Ming Li, Hyun-Joong Chung, Hohyun Keum, Martin McCormick, Ping Liu, Yong-Wei Zhang, Fiorenzo G. Omenetto, Yonggang Huang, Todd Coleman, and John A. Rogers. 2011. Epidermal electronics. *Science* 333(6044):838.

Koo, Ja Hoon, Dong Chan Kim, Hyung Joon Shim, Tae-Ho Kim, and Dae-Hyeong Kim. 2018. Flexible and stretchable smart display: materials, fabrication, device design, and system integration. *Advanced Functional Materials* 28(35):1801834.

Koo, Min, Kwi-Il Park, Seung Hyun Lee, Minwon Suh, Duk Young Jeon, Jang Wook Choi, Kisuk Kang, and Keon Jae Lee. 2012. Bendable inorganic thin-film battery for fully flexible electronic systems. *Nano Letters* 12(9):4810–4816.

Kumar, Rajan, Jaewook Shin, Lu Yin, Jung-Min You, Ying Shirley Meng, and Joseph Wang. 2017. All-printed, stretchable Zn-Ag$_2$O rechargeable battery via hyperelastic binder for self-powering wearable electronics. *Advanced Energy Materials* 7(8):1602096.

Kumaresan, Yogeenth, Ryeri Lee, Namsoo Lim, Yusin Pak, Hyeonghun Kim, Woochul Kim, and Gun-Young Jung. 2018. Extremely flexible indium-gallium-zinc oxide (igzo) based electronic devices placed on an ultrathin poly (methyl methacrylate)(pmma) substrate. *Advanced Electronic Materials* 4(7):1800167.

Kwon, Young-Tae, Yun-Soung Kim, Yongkuk Lee, Shinjae Kwon, Minseob Lim, Yoseb Song, Yong-Ho Choa, and Woon-Hong Yeo. 2018. Ultrahigh conductivity and superior interfacial adhesion of a nanostructured, photonic-sintered copper membrane for printed flexible hybrid electronics. *ACS Applied Materials & Interfaces* 10(50):44071–44079.

Kwon, Young-Tae, Jong Woon Moon, Yo-Min Choi, Seil Kim, Seung Han Ryu, and Yong-Ho Choa. 2017. Novel concept for fabricating a flexible transparent electrode using the simple and scalable self-assembly of Ag nanowires. *The Journal of Physical Chemistry C* 121(10):5740–5746.

Lee, Seoung-Ki, Beom Joon Kim, Houk Jang, Sung Cheol Yoon, Changjin Lee, Byung Hee Hong, John A Rogers, Jeong Ho Cho, and Jong-Hyun Ahn. 2011. Stretchable graphene transistors with printed dielectrics and gate electrodes. *Nano Letters* 11(11):4642–4646.

Lee, Yongkuk, Connor Howe, Saswat Mishra, Dong Sup Lee, Musa Mahmood, Matthew Piper, Youngbin Kim, Katie Tieu, Hun-Soo Byun, and James P Coffey. 2018. Wireless, intraoral hybrid electronics for real-time quantification of sodium intake toward hypertension management. *Proceedings of the National Academy of Sciences* 115(21):5377–5382.

Lee, Young-In, Seil Kim, Seung-Boo Jung, Nosang V Myung, and Yong-Ho Choa. 2013. Enhanced electrical and mechanical properties of silver nanoplatelet-based conductive features direct printed on a flexible substrate. *ACS Applied Materials & Interfaces* 5(13):5908–5913.

Leleux, Pierre, Jean-Michel Badier, Jonathan Rivnay, Christian Bénar, Thierry Hervé, Patrick Chauvel, and George G Malliaras. 2014. Conducting polymer electrodes for electroencephalography. *Advanced Healthcare Materials* 3(4):490–493.

Lemaitre, Jean, and Jean-Louis Chaboche. 1994. *Mechanics of Solid Materials*. Cambridge University Press.

Li, Hongfei, Zhuoxin Liu, Guojin Liang, Yang Huang, Yan Huang, Minshen Zhu, Zengxia Pei, Qi Xue, Zijie Tang, and Yukun Wang. 2018. Waterproof and tailorable elastic rechargeable yarn zinc ion batteries by a cross-linked polyacrylamide electrolyte. *ACS Nano* 12(4):3140–3148.

Li, Ludong, Leilei Gu, Zheng Lou, Zhiyong Fan, and Guozhen Shen. 2017. ZnO quantum dot decorated $Zn_2SnO_4$ nanowire heterojunction photodetectors with drastic performance enhancement and flexible ultraviolet image sensors. *ACS Nano* 11(4):4067–4076.

Li, Na, Zongping Chen, Wencai Ren, Feng Li, and Hui-Ming Cheng. 2012. Flexible graphene-based lithium ion batteries with ultrafast charge and discharge rates. *Proceedings of the National Academy of Sciences* 109(43):17360–17365.

Li, Qiao, Li-Na Zhang, Xiao-Ming Tao, and Xin Ding. 2017. Review of flexible temperature sensing networks for wearable physiological monitoring. *Advanced Healthcare Materials* 6(12):1601371.

Li, Zheng, Jon R Askim, and Kenneth S Suslick. 2018. The optoelectronic nose: colorimetric and fluorometric sensor arrays. *Chemical Reviews* 119(1):231–292.

Lin, Zhaoyang, Yu Chen, Anxiang Yin, Qiyuan He, Xiaoqing Huang, Yuxi Xu, Yuan Liu, Xing Zhong, Yu Huang, and Xiangfeng Duan. 2014. Solution processable colloidal nanoplates as building blocks for high-performance electronic thin films on flexible substrates. *Nano Letters* 14(11):6547–6553.

Lipomi, Darren J, Jennifer A Lee, Michael Vosgueritchian, Benjamin C-K Tee, John A Bolander, and Zhenan Bao. 2012. Electronic properties of transparent conductive films of PEDOT: PSS on stretchable substrates. *Chemistry of Materials* 24(2):373–382.

Liu, Huan-Shen, Bo-Cheng Pan, and Guey-Sheng Liou. 2017. Highly transparent AgNW/PDMS stretchable electrodes for elastomeric electrochromic devices. *Nanoscale* 9(7):2633–2639.

Liu, Qingzhou, Yihang Liu, Fanqi Wu, Xuan Cao, Zhen Li, Mervat Alharbi, Ahmad N Abbas, Moh R Amer, and Chongwu Zhou. 2018. Highly sensitive and wearable $In_2O_3$ nanoribbon transistor biosensors with integrated on-chip gate for glucose monitoring in body fluids. *ACS Nano* 12(2):1170–1178.

Liu, Wei, Min-Sang Song, Biao Kong, and Yi Cui. 2017. Flexible and stretchable energy storage: recent advances and future perspectives. *Advanced Materials* 29(1):1603436.

Liu, Yang, Zehang Sun, Ke Tan, Dienguila Kionga Denis, Jinfeng Sun, Longwei Liang, Linrui Hou, and Changzhou Yuan. 2019. Recent progress in flexible non-lithium based rechargeable batteries. *Journal of Materials Chemistry A* 7(9):4353–4382.

Liu, Yue-Feng, Jing Feng, Yan-Gang Bi, Da Yin, and Hong-Bo Sun. 2019. Recent developments in flexible organic light-emitting devices. *Advanced Materials Technologies* 4(1):1800371.

Liu, Yuhao, Matt Pharr, and Giovanni Antonio Salvatore. 2017. Lab-on-skin: a review of flexible and stretchable electronics for wearable health monitoring. *ACS Nano* 11(10):9614–9635.

Liu, Zhuoxin, Hongfei Li, Minshen Zhu, Yan Huang, Zijie Tang, Zengxia Pei, Zifeng Wang, Zhicong Shi, Jun Liu, and Yang Huang. 2018. Towards wearable electronic devices: a quasi-solid-state aqueous lithium-ion battery with outstanding stability, flexibility, safety and breathability. *Nano Energy* 44:164–173.

Liu, Ziyang, Zhigang Yin, Jianbin Wang, and Qingdong Zheng. 2019. Polyelectrolyte dielectrics for flexible low-voltage organic thin-film transistors in highly sensitive pressure sensing. *Advanced Functional Materials* 29(1):1806092.

Lu, Nanshu, Zhigang Suo, and Joost J Vlassak. 2010. The effect of film thickness on the failure strain of polymer-supported metal films. *Acta Materialia* 58(5):1679–1687.

Lu, Nianduan, Wenfeng Jiang, Quantan Wu, Di Geng, Ling Li, and Ming Liu. 2018. A review for compact model of thin-film transistors (TFTs). *Micromachines* 9(11):599.

Mashford, Benjamin S, Matthew Stevenson, Zoran Popovic, Charles Hamilton, Zhaoqun Zhou, Craig Breen, Jonathan Steckel, Vladimir Bulović, Moungi Bawendi, and Seth Coe-Sullivan. 2013. High-efficiency quantum-dot light-emitting devices with enhanced charge injection. *Nature Photonics* 7(5):407.

Mishra, Rupesh K, Aida Martin, Tatsuo Nakagawa, Abbas Barfidokht, Xialong Lu, Juliane R Sempionatto, Kay Mengjia Lyu, Aleksandar Karajic, Mustafa M Musameh, and Ilias L Kyratzis. 2018. Detection of vapor-phase organophosphate threats using wearable conformable integrated epidermal and textile wireless biosensor systems. *Biosensors and Bioelectronics* 101:227–234.

Mishra, Saswat, James JS Norton, Yongkuk Lee, Dong Sup Lee, Nicolas Agee, Yanfei Chen, Youngjae Chun, and Woon-Hong Yeo. 2017. Soft, conformal bioelectronics for a wireless human-wheel-chair interface. *Biosensors and Bioelectronics* 91:796–803.

Miyata, Yasumitsu, Kazunari Shiozawa, Yuki Asada, Yutaka Ohno, Ryo Kitaura, Takashi Mizutani, and Hisanori Shinohara. 2011. Length-sorted semiconducting carbon nanotubes for high-mobility thin film transistors. *Nano Research* 4(10):963–970.

Mohammed, Mohammed G, and Rebecca Kramer. 2017. All-printed flexible and stretchable electronics. *Advanced Materials* 29(19):1604965.

Montgomery, Jay A, and Christopher R Ellis. 2018. Longevity of cardiovascular implantable electronic devices. *Cardiac Electrophysiology Clinics* 10(1):1–9.

Nomura, Kenji, Hiromichi Ohta, Akihiro Takagi, Toshio Kamiya, Masahiro Hirano, and Hideo Hosono. 2004. Room-temperature fabrication of transparent flexible thin-film transistors using amorphous oxide semiconductors. *Nature* 432(7016):488.

Norton, James JS, Dong Sup Lee, Jung Woo Lee, Woosik Lee, Ohjin Kwon, Phillip Won, Sung-Young Jung, Huanyu Cheng, Jae-Woong Jeong, and Abdullah Akce. 2015. Soft, curved electrode systems capable of integration on the auricle as a persistent brain–computer interface. *Proceedings of the National Academy of Sciences* 112(13):3920–3925.

Oh, Jin Young, Sunghee Kim, Hong-Koo Baik, and Unyong Jeong. 2016. Conducting polymer dough for deformable electronics. *Advanced Materials* 28(22):4455–4461.

Pan, Taisong, Matt Pharr, Yinji Ma, Rui Ning, Zheng Yan, Renxiao Xu, Xue Feng, Yonggang Huang, and John A Rogers. 2017. Experimental and theoretical studies of serpentine interconnects on ultrathin elastomers for stretchable electronics. *Advanced Functional Materials* 27(37):1702589.

Park, Sang-Il, Yujie Xiong, Rak-Hwan Kim, Paulius Elvikis, Matthew Meitl, Dae-Hyeong Kim, Jian Wu, Jongseung Yoon, Chang-Jae Yu, and Zhuangjian Liu. 2009. Printed assemblies of inorganic light-emitting diodes for deformable and semitransparent displays. *Science* 325(5943):977–981.

Perera, Frederica P. 2016. Multiple threats to child health from fossil fuel combustion: impacts of air pollution and climate change. *Environmental Health Perspectives* 125(2):141–148.

Pickham, David, Nic Berte, Mike Pihulic, Andre Valdez, Barbara Mayer, and Manisha Desai. 2018. Effect of a wearable patient sensor on care delivery for preventing pressure injuries in acutely ill adults: a pragmatic randomized clinical trial (LS-HAPI study). *International Journal of Nursing Studies* 80:12–19.

Rashid, Tonny-Roksana, Duy-Thach Phan, and Gwiy-Sang Chung. 2013. A flexible hydrogen sensor based on Pd nanoparticles decorated ZnO nanorods grown on polyimide tape. *Sensors and Actuators B: Chemical* 185:777–784.

Ray, Tyler R, Jungil Choi, Amay J Bandodkar, Siddharth Krishnan, Philipp Gutruf, Limei Tian, Roozbeh Ghaffari, and John A Rogers. 2019. Bio-integrated wearable systems: a comprehensive review. *Chemical Reviews* 119(8):5461–5533.

Ren, Jing, Ye Zhang, Wenyu Bai, Xuli Chen, Zhitao Zhang, Xin Fang, Wei Weng, Yonggang Wang, and Huisheng Peng. 2014. Elastic and wearable wire-shaped lithium-ion battery with high electrochemical performance. *Angewandte Chemie International Edition* 53(30):7864–7869.

Ren, Xiaochen, Ke Pei, Boyu Peng, Zhichao Zhang, Zongrong Wang, Xinyu Wang, and Paddy KL Chan. 2016. A low-operating-power and flexible active-matrix organic-transistor temperature-sensor array. *Advanced Materials* 28(24):4832–4838.

Schwartz, Gregor, Benjamin C-K Tee, Jianguo Mei, Anthony L Appleton, Do Hwan Kim, Huiliang Wang, and Zhenan Bao. 2013. Flexible polymer transistors with high pressure sensitivity for application in electronic skin and health monitoring. *Nature Communications* 4:1859.

Sekitani, Tsuyoshi, Yoshiaki Noguchi, Kenji Hata, Takanori Fukushima, Takuzo Aida, and Takao Someya. 2008. A rubberlike stretchable active matrix using elastic conductors. *Science* 321(5895):1468–1472.

Singh, Eric, M Meyyappan, and Hari Singh Nalwa. 2017. Flexible graphene-based wearable gas and chemical sensors. *ACS Applied Materials & Interfaces* 9(40):34544–34586.

Slater, Michael D, Donghan Kim, Eungje Lee, and Christopher S Johnson. 2013. Sodium-ion batteries. *Advanced Functional Materials* 23(8):947–958.

Son, Donghee, Jongha Lee, Shutao Qiao, Roozbeh Ghaffari, Jaemin Kim, Ji Eun Lee, Changyeong Song, Seok Joo Kim, Dong Jun Lee, and Samuel Woojoo Jun. 2014. Multifunctional wearable devices for diagnosis and therapy of movement disorders. *Nature Nanotechnology* 9(5):397.

Tan, Peng, Bin Chen, Haoran Xu, Houcheng Zhang, Weizi Cai, Meng Ni, Meilin Liu, and Zongping Shao. 2017. Flexible Zn–and Li–air batteries: recent advances, challenges, and future perspectives. *Energy & Environmental Science* 10(10):2056–2080.

Tien, Nguyen Thanh, Sanghun Jeon, Do-Il Kim, Tran Quang Trung, Mi Jang, Byeong-Ung Hwang, Kyung-Eun Byun, Jihyun Bae, Eunha Lee, and Jeffrey B-H Tok. 2014. A flexible bimodal sensor array for simultaneous sensing of pressure and temperature. *Advanced Materials* 26(5):796–804.

Tobjörk, Daniel, and Ronald Österbacka. 2011. Paper electronics. *Advanced Materials* 23(17):1935–1961.

Toda, Kei, Ryo Furue, and Shinya Hayami. 2015. Recent progress in applications of graphene oxide for gas sensing: a review. *Analytica Chimica Acta* 878:43–53.

Trung, Tran Quang, Subramaniyan Ramasundaram, Byeong-Ung Hwang, and Nae-Eung Lee. 2016. An all-elastomeric transparent and stretchable temperature sensor for body-attachable wearable electronics. *Advanced Materials* 28(3):502–509.

Uddin, ASM Iftekhar, Usman Yaqoob, Duy-Thach Phan, and Gwiy-Sang Chung. 2016. A novel flexible acetylene gas sensor based on PI/PTFE-supported Ag-loaded vertical ZnO nanorods array. *Sensors and Actuators B: Chemical* 222:536–543.

Wang, Chuan, Jun-Chau Chien, Kuniharu Takei, Toshitake Takahashi, Junghyo Nah, Ali M Niknejad, and Ali Javey. 2012. Extremely bendable, high-performance integrated circuits using semiconducting carbon nanotube networks for digital, analog, and radio-frequency applications. *Nano Letters* 12(3):1527–1533.

Wang, Chung-Hwa, Chao-Ying Hsieh, and Jenn-Chang Hwang. 2011. Flexible organic thin-film transistors with silk fibroin as the gate dielectric. *Advanced Materials* 23(14):1630–1634.

Wang, Tao, Da Huang, Zhi Yang, Shusheng Xu, Guili He, Xiaolin Li, Nantao Hu, Guilin Yin, Dannong He, and Liying Zhang. 2016. A review on graphene-based gas/vapor sensors with unique properties and potential applications. *Nano-Micro Letters* 8(2):95–119.

Wang, Ting, Yunlong Guo, Pengbo Wan, Han Zhang, Xiaodong Chen, and Xiaoming Sun. 2016. Flexible transparent electronic gas sensors. *Small* 12(28):3748–3756.

Wang, Xuewen, Zheng Liu, and Ting Zhang. 2017. Flexible sensing electronics for wearable/attachable health monitoring. *Small* 13(25):1602790.

Wang, Yuxing, Bo Liu, Qiuyan Li, Samuel Cartmell, Seth Ferrara, Zhiqun Daniel Deng, and Jie Xiao. 2015. Lithium and lithium ion batteries for applications in microelectronic devices: a review. *Journal of Power Sources* 286:330–345.

Wang, ZB, MG Helander, J Qiu, DP Puzzo, MT Greiner, ZM Hudson, S Wang, ZW Liu, and ZH Lu. 2011. Unlocking the full potential of organic light-emitting diodes on flexible plastic. *Nature Photonics* 5(12):753.

Webb, R Chad, Andrew P Bonifas, Alex Behnaz, Yihui Zhang, Ki Jun Yu, Huanyu Cheng, Mingxing Shi, Zuguang Bian, Zhuangjian Liu, and Yun-Soung Kim. 2013. Ultrathin conformal devices for precise and continuous thermal characterization of human skin. *Nature Materials* 12(10):938.

Wiedinmyer, Christine, Robert J Yokelson, and Brian K Gullett. 2014. Global emissions of trace gases, particulate matter, and hazardous air pollutants from open burning of domestic waste. *Environmental Science & Technology* 48(16):9523–9530.

Wood, Vanessa, Matthew J Panzer, Jianglong Chen, Michael S Bradley, Jonathan E Halpert, Moungi G Bawendi, and Vladimir Bulović. 2009. Inkjet-printed quantum dot–polymer composites for full-color ac-driven displays. *Advanced Materials* 21(21):2151–2155.

Wu, Hao, YongAn Huang, Feng Xu, Yongqing Duan, and Zhouping Yin. 2016. Energy harvesters for wearable and stretchable electronics: from flexibility to stretchability. *Advanced Materials* 28(45):9881–9919.

Xu, Baoxing, Aadeel Akhtar, Yuhao Liu, Hang Chen, Woon-Hong Yeo, Sung Il Park, Brandon Boyce, Hyunjin Kim, Jiwoo Yu, and Hsin-Yen Lai. 2016. An epidermal stimulation and sensing platform for sensorimotor prosthetic control, management of lower back exertion, and electrical muscle activation. *Advanced Materials* 28(22):4462–4471.

Xu, Sheng, Zheng Yan, Kyung-In Jang, Wen Huang, Haoran Fu, Jeonghyun Kim, Zijun Wei, Matthew Flavin, Joselle McCracken, and Renhan Wang. 2015. Assembly of micro/nanomaterials into complex, three-dimensional architectures by compressive buckling. *Science* 347(6218):154–159.

Xu, Yuxi, Hua Bai, Gewu Lu, Chun Li, and Gaoquan Shi. 2008. Flexible graphene films via the filtration of water-soluble noncovalent functionalized graphene sheets. *Journal of the American Chemical Society* 130(18):5856–5857.

Yang, Shixuan, Ying-Chen Chen, Luke Nicolini, Praveenkumar Pasupathy, Jacob Sacks, Becky Su, Russell Yang, Daniel Sanchez, Yao-Feng Chang, and Pulin Wang. 2015. "Cut-and- paste" manufacture of multiparametric epidermal sensor systems. *Advanced Materials* 27(41):6423–6430.

Yang, Xuyong, Evren Mutlugun, Cuong Dang, Kapil Dev, Yuan Gao, Swee Tiam Tan, Xiao Wei Sun, and Hilmi Volkan Demir. 2014. Highly flexible, electrically driven, top-emitting, quantum dot light-emitting stickers. *ACS Nano* 8(8):8224–8231.

Yeo, Woon-Hong, Yun-Soung Kim, Jongwoo Lee, Abid Ameen, Luke Shi, Ming Li, Shuodao Wang, Rui Ma, Sung Hun Jin, and Zhan Kang. 2013. Multifunctional epidermal electronics printed directly onto the skin. *Advanced Materials* 25(20):2773–2778.

Yokota, Tomoyuki, Peter Zalar, Martin Kaltenbrunner, Hiroaki Jinno, Naoji Matsuhisa, Hiroki Kitanosako, Yutaro Tachibana, Wakako Yukita, Mari Koizumi, and Takao Someya. 2016. Ultraflexible organic photonic skin. *Science Advances* 2(4):e1501856.

Yoon, Jangyeol, Soo Yeong Hong, Yein Lim, Seung-Jung Lee, Goangseup Zi, and Jeong Sook Ha. 2014. Design and fabrication of novel stretchable device arrays on a deformable polymer substrate with embedded liquid-metal interconnections. *Advanced Materials* 26(38):6580–6586.

Yu, Ki Jun, Zheng Yan, Mengdi Han, and John A Rogers. 2017. Inorganic semiconducting materials for flexible and stretchable electronics. *NPJ Flexible Electronics* 1(1):4.

Yu, Min, Haochuan Wan, Le Cai, Jinshui Miao, Suoming Zhang, and Chuan Wang. 2018. Fully printed flexible dual-gate carbon nanotube thin-film transistors with tunable ambipolar characteristics for complementary logic circuits. *ACS Nano* 12(11):11572–11578.

Zang, Weili, Yuxin Nie, Dan Zhu, Ping Deng, Lili Xing, and Xinyu Xue. 2014. Core–shell $In_2O_3/ZnO$ nanoarray nanogenerator as a self-powered active gas sensor with high $H_2S$ sensitivity and selectivity at room temperature. *The Journal of Physical Chemistry C* 118(17):9209–9216.

Zang, Yaping, Dazhen Huang, Chong-an Di, and Daoben Zhu. 2016. Device engineered organic transistors for flexible sensing applications. *Advanced Materials* 28(22):4549–4555.

Zhang, Fengjiao, Yaping Zang, Dazhen Huang, Chong-an Di, and Daoben Zhu. 2015. Flexible and self-powered temperature–pressure dual-parameter sensors using microstructure-frame-supported organic thermoelectric materials. *Nature Communications* 6:8356.

Zhang, Jialu, Chuan Wang, and Chongwu Zhou. 2012. Rigid/flexible transparent electronics based on separated carbon nanotube thin-film transistors and their application in display electronics. *ACS Nano* 6(8):7412–7419.

Zhao, Fangchao, Dustin Chen, Shuai Chang, Hailong Huang, Kwing Tong, Changtao Xiao, Shuyu Chou, Haizheng Zhong, and Qibing Pei. 2017. Highly flexible organometal halide perovskite quantum dot based light-emitting diodes on a silver nanowire–polymer composite electrode. *Journal of Materials Chemistry C* 5(3):531–538.

Zhao, Wei, Thomas Rovere, Darshana Weerawarne, Gavin Osterhoudt, Ning Kang, Pharrah Joseph, Jin Luo, Bonggu Shim, Mark Poliks, and Chuan-Jian Zhong. 2015. Nanoalloy printed and pulse-laser sintered flexible sensor devices with enhanced stability and materials compatibility. *ACS Nano* 9(6):6168–6177.

Zheng, ZQ, JD Yao, B Wang, and GW Yang. 2015. Light-controlling, flexible and transparent ethanol gas sensor based on ZnO nanoparticles for wearable devices. *Scientific Reports* 5:11070.

Zhu, Yun-hai, Shuang Yuan, Di Bao, Yan-bin Yin, Hai-xia Zhong, Xin-bo Zhang, Jun-min Yan, and Qing Jiang. 2017. Decorating waste cloth via industrial wastewater for tube-type flexible and wearable sodium-ion batteries. *Advanced Materials* 29(16):1603719.

Zschieschang, Ute, Tatsuya Yamamoto, Kazuo Takimiya, Hirokazu Kuwabara, Masaaki Ikeda, Tsuyoshi Sekitani, Takao Someya, and Hagen Klauk. 2011. Organic electronics on banknotes. *Advanced Materials* 23(5):654–658.

# 2

---

# *Stretchable Conductor*

**Zherui Cao**
*Shanghai Institute of Ceramics*
*Chinese Academy of Sciences*

**Ranran Wang**
*Shanghai Institute of Ceramics*
*Chinese Academy of Sciences*

**Jing Sun**
*Shanghai Institute of Ceramics*
*Chinese Academy of Sciences*

## CONTENTS

## 2.1 Introduction

Highly conductive and intrinsically stretchable conductors are pivotal components for soft electronics such as stretchable light-emitting diode arrays, sensors, actuators, energy harvesting, and storage devices.[1-8] In the past few years, rapid developments in conductive nanomaterials, new concepts in structural design, fabrication techniques, and applications have contributed to significant progress in the achievement of stretchable conductors. These advances have provided ever-increasing performance such as higher conductivity and stretchability, lower resistance variation under tensile strain, longer cycle life, and so on. Table 2.1 summarizes the main performance and applications of stretchable conductors reported in very recent years. Notably, the performance of stretchable conductors shows great dependence on materials and structure designs. In this chapter, the most widely investigated materials and structure designs to construct high-performance stretchable conductors will be reviewed and potential applications will be summarized.

**TABLE 2.1**

Electromechanical Performance and Applications of the Stretchable Conductors

| Materials | Structure | Conductivity | Stretchability (%) | Changes in Electrical Performance under Strains | Application |
|---|---|---|---|---|---|
| AgNP/PU[9] | Percolation network | 17,460 S cm$^{-1}$ | 490 | 236 S cm$^{-1}$ at 490% strain | Interconnection |
| AgNP/SBS[10] | Percolation network | 5400 S cm$^{-1}$ | 100 | 2200 S cm$^{-1}$ at 100% strain | Interconnection |
| AgNW-AgNP/ SBS[11] | Percolation network | 2450 S cm$^{-1}$ | 220 | $\sigma/\sigma_0 = 4.4\%$ at 100% strain | Sensor |
| AgNW/PDMS[12] | Percolation network | 8130 S cm$^{-1}$ | 80 | $\Delta R/R_0 = 4.5$ at 50% strain | Interconnection, sensor |
| AgNW/PDMS[13] | Percolation network | 10 $\Omega$ sq$^{-1}$ | 160 | 100–1000 $\Omega$ sq$^{-1}$ at 140% strain | Actuator |
| AgNW/Ecoflex[1] | Buckle | 9 $\Omega$ sq$^{-1}$ | 460 | Resistance maintained up to 460% strain | Interconnection |
| AgNW/SBS[14] | Serpentine | 12,000 S cm$^{-1}$ | 100 | Conductivity maintained up to 100% strain | Heater |
| AgNW/fiber[15] | Helix | 4018 S cm$^{-1}$ | 500 | 688 S cm$^{-1}$ at 500% strain | Interconnection |
| AgNW/fiber[5] | Helix | 0.5 $\Omega$ cm$^{-1}$ | 2000 | $\Delta R/R_0 = 9.2\%$ at 2000% strain | Interconnection, actuator, sensor |
| AgNW/nanofiber | Nanomesh | 8.2 $\Omega$ sq$^{-1}$ | 50 | $\Delta R/R_0 = 1.3$ at 50% strain | Interconnection, sensor |
| AgNW-GO/ PUA[16] | Percolation network | 14 $\Omega$ sq$^{-1}$ | 80 | $\Delta R/R_0 = 9.6$ at 80% strain | Interconnection |
| Ag flakes/silicone adhesive[17] | Percolation network | 15,100 S cm$^{-1}$ | 240 | 1110 S cm$^{-1}$ at 240% strain | Interconnection |

*(Continued)*

**TABLE 2.1** (*Continued*)

Electromechanical Performance and Applications of the Stretchable Conductors

| Materials | Structure | Conductivity | Stretchability (%) | Changes in Electrical Performance under Strains | Application |
|-----------|-----------|--------------|--------------------|-----------------------------------------------|-------------|
| Ag flake/fluoroelastomer[18] | Percolation network | 6168 S cm$^{-1}$ | 400 | 935 S cm$^{-1}$ at 400% strain | Interconnection |
| Ag flake-Ecoflex/hydrogel[19] | Percolation network | 542 S cm$^{-1}$ | 1780 | $\Delta R/R_0 = 154$ at 1780% strain | Interconnection |
| Ag flake-EGaInPs/EVA[20] | Percolation network | 8331 S cm$^{-1}$ | 1000 | $\Delta R/R_0 = 6.2$ at 1000% strain | Interconnection, sensor |
| Ag flake-PVDF nanofiber/fluoroelastomer[3] | Percolation network | 3667 S cm$^{-1}$ | 450 | 1018 S cm$^{-1}$ at 450% strain | Sensor |
| AuNP/PU[21] | Percolation network | 11,000 S cm$^{-1}$ (LBL) 1800 S cm$^{-1}$ (VAF) | 115 (LBL) 486 (VAF) | 2400 S cm$^{-1}$ at 110% strain (LBL) 94 S cm$^{-1}$ at 110% strain (VAF) | Interconnection |
| Au-AgNW/SBS[22] | Serpentine | 41,850 S cm$^{-1}$ | 266 | $\sigma/\sigma_0 = 10\%$ at 266% strain | Heater, sensor |
| Au nanosheet/Ecoflex[23] | Percolation network | 1667 S cm$^{-1}$ | 200 | Resistance maintained at 60% (longitudinal), 140% (transverse) strain | Interconnection |
| Au/PDMS[24] | Percolation network | 47.5 Ω cm$^{-1}$ | 100 | $\Delta R/R_0 = 50\%$ at 50% strain | Interconnection |
| Au/PDMS[25] | Nanomesh | 8 Ω sq$^{-1}$ | 50 | $\Delta R/R_0 = 40\%$ at 50% strain | Interconnection |
| Au/PDMS[26] | Nanomesh | 21 Ω sq$^{-1}$ | 160 | 67 Ω sq$^{-1}$ at 160% strain | Interconnection |
| Au/PVA fiber[27] | Nanomesh | $5.3 \times 10^{-5}$ Ω cm | 40 | Conductivity maintained up to 15% strain | Interconnection, sensor |
| Au/silk[28] | Buckle | 7 Ω sq$^{-1}$ | 100 | $\Delta R/R_0 = 1.45$ at 40% strain | Interconnection |
| CuNW/fiber[29] | Helix | 2.5 Ω cm$^{-1}$ | 100 | $\Delta R/R_0 = 50\%$ at 50% strain | Heater |
| Cu-AgNW/PDMS[2] | Percolation network | 2040 S cm$^{-1}$ | 250–350 | $\Delta R/R_0 = 1.46$ at 150% strain | Interconnection |
| Cu/PDMS[30] | Nanomesh | 0.9 Ω sq$^{-1}$ | 70 | $\Delta R/R_0 = 4$ at 70% strain | Interconnection |
| EGaIn/SEBS[31] | Percolation network | $3 \times 10^{-5}$ Ω cm | 700–800 | 0.08 Ω cm$^{-1}$ at 500% strain | Interconnection |
| CNT/PDMS[32] | Percolation network | 108 S cm$^{-1}$ | 100 | $\Delta R/R_0 = 0.1$ at 100% strain | Interconnection |
| CNT/PDMS[33] | Nanomesh | 2000 S cm$^{-1}$ | 60 | $\Delta R/R_0 = 1.25$ at 60% strain | Interconnection |
| CNT/PDMS[34] | Buckle | 211 S cm$^{-1}$ | 100 | $\Delta R/R_0 = 4\%$ at 100% strain | Interconnection |

(*Continued*)

**TABLE 2.1 (*Continued*)**

Electromechanical Performance and Applications of the Stretchable Conductors

| Materials | Structure | Conductivity | Stretchability (%) | Changes in Electrical Performance under Strains | Application |
|---|---|---|---|---|---|
| CNT/fluoroelastomer[35] | Percolation network | 57 S cm$^{-1}$ | 134 | 8 S cm$^{-1}$ at 134% strain | Interconnection |
| CNT/SEBS[6] | Buckle | 26–2100 S cm$^{-1}$ | 1320 | $\Delta R/R_0 < 5\%$ at 1000% strain | Interconnection, actuator, sensor |
| CNT/PANI[36] | Buckle | 2.75 $\Omega$ sq$^{-1}$ | 200 | $\Delta R/R_0 < 3\%$ at 200% strain | Supercapacitor |
| Graphene/PDMS[37] | Percolation network | 10 S cm$^{-1}$ | 90 | $\Delta R/R_0 = 1.8$ at 90% strain | Interconnection |
| Graphene/PDMS[38] | Buckle | 600 $\Omega$ cm$^{-1}$ | 40 | $\Delta R/R_0 = 25$ at 40% strain | Supercapacitor |
| PEDOT:PSS-rGO/PDMS[39] | Percolation network | 1010 S cm$^{-1}$ | 20 | Conductivity maintained up to 15% strain | Interconnection |
| PEDOT:PSS-ionic compounds/SEBS[40] | Percolation network | 3100 S cm$^{-1}$ | 800 | 56 S cm$^{-1}$ at 800% strain | Interconnection |
| PEDOT:PSS/PAAC[41] | Percolation network | 0.127 S cm$^{-1}$ | 399 | $\Delta R/R_0 = 10\%$ at 80% strain | Interconnection |
| PEDOT-PU/hydrogel[42] | Percolation network | 40 S cm$^{-1}$ | 100 | 120 S cm$^{-1}$ at 100% strain | Interconnection |

## 2.2 Materials

### 2.2.1 Substrate Materials

The mechanical properties of the elastic substrates play a vital role in the performance of the stretchable conductors. The substrates can be mainly divided into three categories: silicone rubbers, and hydrogels and textiles. Silicone rubbers, such as polydimethylsiloxane (PDMS) and Ecoflex are widely used for the substrates of stretchable conductors because of their easy processability and chemical durability. The cross-linking reactions of these rubbers are often carried out with the aid of cross-linking agents. Therefore, the mechanical properties of these rubbers can be controlled by adjusting the weight ratios of the cross-linking agents or the reaction conditions (e.g., temperature and time).[43, 44] Furthermore, the surface characters of these rubbers can be effectively modified by various physical or chemical treatments, which can tune the bonding between the conductive materials and substrates and enhance the performance of the stretchable conductors readily.[45–47] Compared with silicone rubbers, hydrogels consist of long-chain polymer networks infiltrated with water instead of cross-linkers. Due to the specific network structures, hydrogels possess light weight, high stretchability, and favorable biocompatibility.[42, 48–50] In addition, hydrogels can be readily prepared with self-healing abilities by adding healing agents or engineering dynamic polymer networks, which mimic the properties of human skin.[51, 52] Apart from polymeric materials, textiles are another substrates for adsorbing conductive materials and providing deformability.[53-55] The textiles can be stretchable fibers or 2D planar fabrics woven by intertwined fibers. More importantly, the textile-based stretchable conductors are convenient for large-scale production and integration into clothes,[56] which can greatly broaden the application scope of the stretchable conductors.

## 2.2.2 Conductive Materials

Conductive materials of the stretchable conductors should adapt to the deformation of the structures under tensile strain. As the bulk rigid materials lack flexibility and stretchability, novel nanomaterials have been widely studied to provide the desirable electrical performance and flexibility. The frequently used conductive materials for stretchable conductors can be mainly divided into three categories: metal materials, carbon materials, and conducting polymers. All kinds of materials possess different mechanical and electrical properties and suitable materials should be chosen according to the practical requirements.

### 2.2.2.1 Silver Nanomaterials

Silver exhibits the highest bulk electrical conductivity ($6.3 \times 10^5$ S cm$^{-1}$) of all metals and excellent oxidation resistance, which is usually used for electrical circuits.[57] Silver-based nanomaterials, including 0D silver nanoparticles (AgNPs), 1D silver nanowires (AgNWs), and 2D silver nanoflakes, have been confirmed as appropriate conductive materials of the stretchable conductors in the literature.[9, 10, 12, 19, 58-60] Matsuhisa et al. fabricated printable stretchable conductors by in situ formation of silver nanoparticles from silver flakes (Figure 2.1a).[18] Ag flakes, fluorine rubbers, fluorine surfactant, and methyl isobutyl ketone (MIBK) were mixed with a magnetic stirrer for 12 h in a weight ratio of 3:1:0.1:1. Then the ink was printed on the substrate, followed by drying at 80°C for 1 h and 120°C for 1 h. Figure 2.1b shows that the AgNPs were formed around the Ag flakes with the aid of surfactant. Figure 2.1c compares the conductivities versus strain of the stretchable conductors with or without surfactant and with or without thermal treatment at 120°C. The stretchable conductors with surfactant exhibited a high conductivity of 4919 S cm$^{-1}$ without strain and 719 S cm$^{-1}$ at a strain of 300%. For the stretchable conductors without surfactant, the conductivities were 3727 S cm$^{-1}$ and 297 S cm$^{-1}$ at strains of 0% and 300%, respectively. On the contrary, stretchable conductors with surfactant prepared at low temperature exhibited much larger resistance at strains of 0% and 300%. Therefore, the surfactant and thermal treatment are essential for the generation of AgNPs in the composites, leading to high electrical performance over a wide range of applied strain.

### 2.2.2.2 Gold Nanomaterials

Gold is a noble metal with high conductivity ($4.1 \times 10^5$ S cm$^{-1}$) and resistance to corrosion and oxidization.[61] Similar to silver, gold nanostructures include 0D AuNPs, 1D AuNWs, and 2D Au nanoflakes.[24, 28, 62, 63] Kim et al. developed stretchable conductors of polyurethane (PU) containing Au nanoparticles prepared by either layer-by-layer (LBL) assembly or vacuum-assisted flocculation (VAF).[21] Figures 2.1d and e compare the morphologies of the AuNP/PU composites prepared by the methods of LBL or VAF. Nanoparticles in the LBL composites dispersed more homogeneously, which resulted in more efficient conducting pathways than in the VAF composites, in which nanoparticles aggregated more severely. On the contrast, the presence of larger PU domains led to a lower conductivity and higher stretchability of the VAF composites. The conductivities of the 5 × LBL and

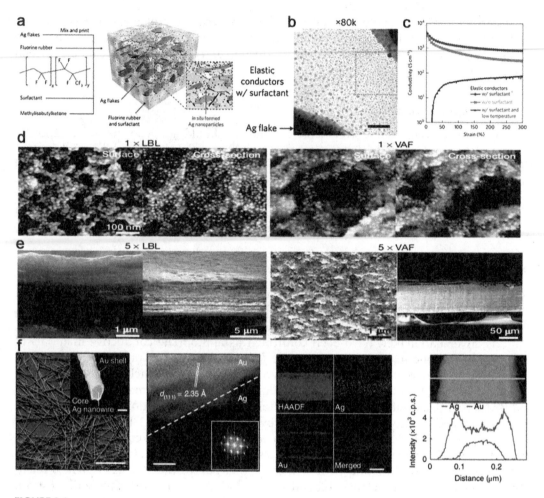

**FIGURE 2.1**

(a) Chemical components of the Ag precursor ink. (b) TEM image of the composite fabricated with the use of a surfactant. Scale bar: 100 nm. (c) Conductivities versus strain of the stretchable conductors with or without surfactant and with or without high-temperature treatment. (d) SEM images of 1 × LBL and 1 × VAF. (e) Cross-sectional SEM images of 5 × LBL and 5 × VAF. (f) SEM, TEM, and EDS characterization of the Ag-Au nanowires.

5 × VAF composites were 11,000 S cm$^{-1}$ and 1800 S cm$^{-1}$, while the stretching ranges of 5 × LBL and 5 × VAF composites were 115% and 486%, respectively.

Recently, Choi et al. prepared Ag-Au sheath-core nanowire composite for wearable and implantable bioelectronics.[22] Figure 2.1f presents the SEM and TEM characterization of the Ag-AuNWs, which demonstrated the sheath-core structure of the alloy nanowires. The atomic intensity profile in a line scan of the Ag-AuNWs shows that the mean diameter of the AgNW and thickness of the Au shell were 140 and 35 nm, respectively. After mixing the Ag-AuNWs with SBS/toluene solution under appropriate ratio, the solution was cast into a mold and dried at 25–85°C. The conductivity of the composite reached 41,850 S cm$^{-1}$ before stretch and retained 104 S cm$^{-1}$ at the strain of about 150%.

### 2.2.2.3 Copper Nanomaterials

Owing to the abundance of copper over other noble metals, as well as its excellent bulk conductivity ($5.96 \times 10^5$ S cm$^{-1}$), copper-based nanomaterials may offer cost benefits in the applications of stretchable conductors.[64] Up to now, CuNWs have been the most investigated copper nanomaterial because of their facile synthesis and excellent mechanical properties.[65–68] Won et al. reported a highly stretchable, helical CuNW conductor.[69] The as-synthesized CuNWs had an average diameter of 66 ± 17 nm and an average length of 450 μm (Figure 2.2a). The nanowires were dispersed in isopropyl alcohol (IPA) at a concentration of 203.8 mg ml$^{-1}$ and subsequently chemically treated with lactic acid to completely

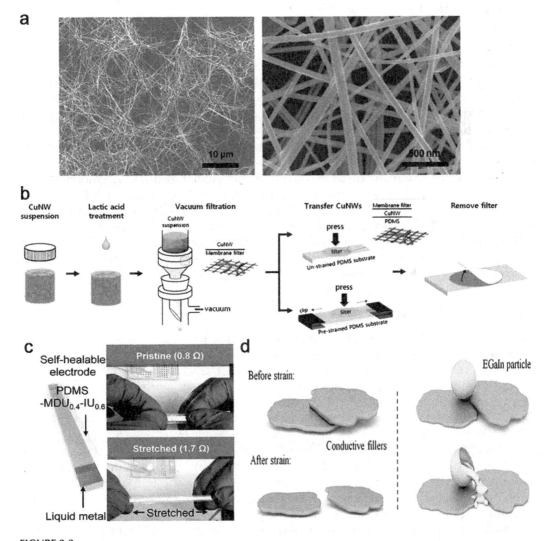

**FIGURE 2.2**
(a) SEM images of the CuNWs after chemical treatment. (b) Schematic illustration representation of the preparation of the stretchable CuNW conductors. (c) Schematic of the self-healing conductors (left). Optical images of the conductor equipped with the LED lamp before and after stretching (right). (d) Schematic of the stretchable conductors without (left) and with EGaInPs (right).

remove the surface oxide/hydroxide, which influenced the electrical conductivity of the CuNW network films. As shown in Figure 2.2b, the CuNW films were prepared by vacuum filtration, followed by transferring onto a PDMS substrate.

Compared with AgNWs, the main disadvantage of CuNWs is their low resistance to oxidization.[70] Song et al. developed transparent Cu@Cu$_4$Ni nanowire composites against oxidation, bending, stretching, and twisting for flexible optoelectronics.[71] CuNWs with Cu$_4$Ni alloying shells were synthesized and embedded into the PDMS substrates. The fluctuation amplitude of resistance is within $2\,\Omega\,sq^{-1}$ within 30 days, meaning that at $\Delta R/R_0=1$, the actual lifetime is estimated to be more than 1200 days, indicating the good oxidization resistance in the natural environment.

### 2.2.2.4 Liquid Metal

The shape self-adaptation of liquid metal can ensure the stable connection between the conductive pathways of the stretchable conductors during deformation.[31,72,73] Furthermore, the fluidity of liquid metal is appropriate for the self-healing stretchable conductors. Kang et al. prepared a self-healing elastomer (PDMS-MPU0.4-IU0.6) to support and encapsulate the liquid metal EGaIn (Figure 2.2c).[74] This composite exhibited high stretchability with stable and low resistance. After being cut into two pieces, the stretchable conductor could self-heal at room temperature even under water.

The liquid metal can also be used to fill the cracks in the conductive network to reduce the loss of conductivity. Wang et al. prepared EGaIn particles (EGaInPs) through sonication in acetone for 30 min, followed by mixing it with Ag flakes and EVA/chlorobenzene solution.[20] Due to the fast evaporation rate of chlorobenzene, the conductive ink was dried to maintain the printed structures after being printed onto the substrate under room temperature. Figure 2.2d presents the schematic of the stretchable conductors without and with EGaInPs. The conductive fillers slid with strain while the Ga$_2$O$_3$ shells of EGaInPs broke under deformation. The released EGaIn bridged between the conductive fillers, which kept the electrical performance stable.

### 2.2.2.5 Carbon Nanotubes

Carbon nanotubes (CNTs) are tubular carbon molecules, which are formed by one or several rolled-up carbon layers around nano-sized central cylindrical hollow cores.[75–77] Owing to the high conductivity, elastic modulus, and low cost, CNTs have been widely used in the fabrication of stretchable conductors to enhance the electromechanical properties.[33,78] Zhu's group fabricated transparent stretchable conductors with continuous CNT ribbons.[79] The CNT ribbons were directly drawn from vertically grown multi-walled carbon nanotubes (MWNT) forests and then transferred onto PDMS substrates. Figure 2.3a shows that CNTs became wavy in the lateral direction after a stretching-releasing cycle at the strain of 80%. Due to the mismatch of the mechanical properties between the conductive materials and substrates, the CNTs could not recover to the initial state and formed the wavy structures. The resistance of the conductor increased under stretching and kept stable during releasing (Figure 2.3b). Figure 2.3c presents the periodic wavy structures of the conductor prepared by the prestrain-release-buckling strategy.[34] As shown in Figure 2.3d, the wavy CNT ribbons were able to accommodate large stretching up to the prestrain level (100%) with an increase in resistance of about 4.1%. At a strain larger than the prestrain, the CNT ribbons showed a strain dependence of the resistance because of the sliding between the CNTs, similar to the structure change in Figures 2.3a and b.

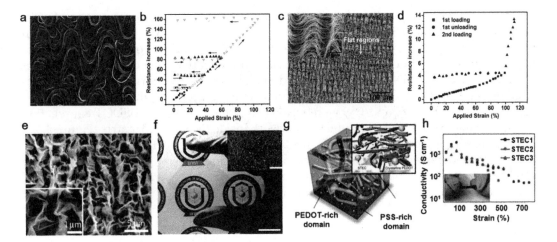

**FIGURE 2.3**
(a) SEM image showing the lateral buckling of CNTs induced by stretching-release buckling. (b) Resistance variation versus strain of the CNT/PDMS composite. (c) SEM image of a buckled CNT ribbon induced by releasing the prestrain of PDMS. (d) Resistance variation versus strain of the CNT/PDMS composite. (e) Crumpled graphene film formed by sequentially relaxing the substrate along two prestretched direction. (f) Photograph of graphene-AgNW hybrid film on a PET substrate. The inset shows a SEM image of this composite. Scale: 2 cm. (g) Schematic illustration of the structure change of the PEDOT:PSS polymer induced by the addition of ionic compounds. (h) Conductivities of the PEDOT:PSS composites with different additives as functions of applied strain.

### 2.2.2.6 Graphene

As a novel 2D material, graphene has attracted extensive academic and industrial interests. Graphene exhibits excellent electron mobility, good chemical and thermal stability, high strength, and large specific surface area due to the unique 2D structure.[80–82] Similar with CNTs, graphene and graphene-based hybrid composites have been confirmed as suitable conductive materials for stretchable conductors.[37, 83, 84] Zang et al. investigated the multifunctionality and control of the crumpling and unfolding of graphene.[85] Few-layer graphene films grown on nickel films by chemical vapor deposition were attached onto a PDMS stamp. Then the graphene/PDMS sample is stamped on a biaxially prestretched elastomer film of VHB acrylic 4905 to transfer the graphene to the substrate. As shown in Figure 2.3e, crumpled hierarchical structures are formed after the release of the prestrain. The crumpled graphene is superhydrophobic and can be stretched to 450%. The reversible crumpling-unfolding process can be used to tune the wettability and optical transmittance of the graphene. Furthermore, Lee et al. reported stretchable conductors based on 2D graphene and 1D AgNWs.[86] Figure 2.3f presents a photo and SEM image of the graphene/AgNW film on a PET substrate. With the hybrid nanostructures, the stretchable conductor possessed high mechanical flexibility as well as good optical transparency (94% in visible range) and electric conductivity (33 $\Omega$ sq$^{-1}$).

### 2.2.2.7 Conducting Polymers

Conducting polymers are a class of organic polymers that conduct electricity. Although their charge-transport performance and stability in ambient conditions are relatively lower than other metal and carbon materials, the conducting polymers possess better compatibility with the elastic substrates.[87, 88] Poly(3,4-ethylenedioxythiophene):poly(styre-nesulfonate) (PEDOT:PSS) is a well-known conducting polymer, while it cannot maintain its electrical properties over 5% strain in the absence of any additive or posttreatment due to the rigidity. Therefore, Wang et al. reported a highly stretchable conducting polymer composite based on PEDOT:PSS with the addition of ionic compounds to improve the conductivity and stretchability.[40] With the aid of the ionic compounds, the electrostatic interactions between PEDOT and PSS were weakened, which facilitated PEDOT aggregation and the formation of a highly conductive network inside the soft PSS matrix (Figure 2.3g). Thin films were processed onto SEBS substrates by spin coating, followed by an annealing and post-deposition treatment. As shown in Figure 2.3h, the resulting stretchable conductor exhibited enhanced electromechanical performance, with the conductivity of 4100 S cm$^{-1}$ under 100% strain while maintaining 100 S cm$^{-1}$ at 600% strain.

## 2.3 Stretchable Structures

The stretchable conductors are fabricated by incorporating the conductive materials with the elastomeric substrates. The structures of the conductive networks change under various deformation, showing a significant impact on the performance. The recently developed stretchable conductors mainly adopt five categories of structures, i.e., percolation networks, buckled structures, nanomeshes, serpentine structures, and helix structures.

### 2.3.1 Percolation Networks

As the most common structure in the stretchable conductors, the percolation network has attracted wide interests because of their easy processability. In the percolation networks, the conductive materials are dispersed on the surface or inside the elastic substrates depending on the composite-formation process. Conducting elements stack to form the conductive network when their content is above the threshold, which endows the composite materials with tunable electromechanical properties.

Under tensile strain, the relatively rigid conductive materials will slide with the deformation of the elastic substrates. The structure change of in the percolation networks is readily observed in literature.[16, 17, 89, 90] Moon et al. fabricated highly stretchable patterned gold conductors and observed the stretchability of the Au nanosheet film on an Ecoflex substrate.[23] As shown in Figures 2.4a-c, the overlapped Au nanosheets separated with each other at the strain of 30% and recovered after releasing. Figure 2.4d shows the resistivity variation of the Au nanosheet film at high strains ($\varepsilon$ = 80%, 100%, and 120%) during 1000 stretching cycles. The Au nanosheet film showed excellent cycle performance up to $\varepsilon$ = 100% and the resistance variation began to increase gradually at $\varepsilon$ = 120%. Reinforcing the interface combination is a common method to improve the electromechanical performance of the stretchable conductors. Matsuhisa et al. fabricated printable elastic conductors by in situ formation of silver nanoparticles from silver flakes.[18] The conductivity and stretchability were enhanced by the regulation of the surfactant, heating processes, and molecular weight of the elastomer. 1D conductive nanomaterials, such as CNTs and metal nanowires, have also been applied to prepare the stretchable conductors. Lee et al. reported the multiscale nanocomposite combining the enhanced mechanical compliance, electrical conductivity, and optical transparency due to the hybrid AgNW and CNT percolation networks.[91]

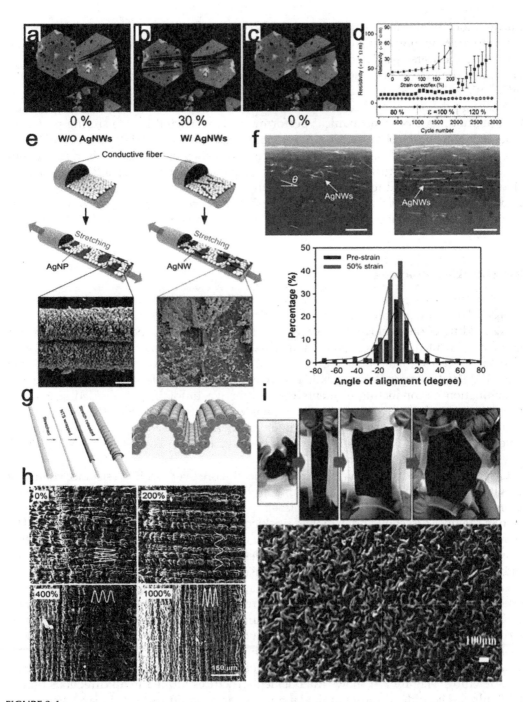

**FIGURE 2.4**

(a-c) Optical images of the structure change in the nanosheet film during stretching. (d) Resistivity variation of the Au nanosheet film as a function of the stretching cycle. The inset shows the resistivity variation versus strain. (e) Comparison of the structure changes of the composite fibers with or without AgNWs. Scale bar: 30 μm. (f) Sliding and angular distribution of AgNWs under stretching. Scale bars: 20 μm. (g) Schematic illustration of the fabrication and structure of the buckled sheath-core fibers. (h) SEM images showing the evolution of the buckled sheath structure of a composite fiber during stretch from 0 to 1000% strain. (i) Photos of the buckled CNT film at various stretching states (top). SEM images of the buckled structures in the CNT films (bottom).

Furthermore, 1D conductive materials were also introduced in the 2D materials to improve the properties by integrating the sliding and bridging effects. As shown in Figure 2.4e and f, Lee et al. fabricated AgNW reinforced highly stretchable conductive fibers.[11] The AgNWs slid to the preferred orientations at high strain to keep the conductive paths between the adjacent AgNPs.

Although the percolation networks are widely studied because of the simple preparation method, its electromechanical performance is restricted by the sliding-preferred orientation macromechanics. First, sufficient conductive materials are essential for keeping conductive paths under large deformation whereas the relatively rigid materials will increase the stiffness of stretchable conductors. Second, the sliding of conductive materials will lead to the resistance increase and the failure to realize large strain and small resistance variation simultaneously. Furthermore, the conducting elements cannot recover completely after stress release, which is unfavorable for the cycling stability.

### 2.3.2 Buckled Structures

Constructing buckled structures is another effective route to prepare stretchable conductors.[92-95] The stress on the conductive materials under tensile strain can be relieved by the unfolding of the buckled conductive network, guaranteeing the stable resistance. For stretching buckled structures on elastomeric substrates, macromechanics can be divided into three stages: (i) the buckled structures exhibit gradual decrease in the buckling amplitude and increase in the buckling wavelength at low tensile strain; (ii) breaking and delamination of conducting elements occur after the unfolding of buckled structures; (iii) separation of conducting elements happens with further increase in the tensile strain, resulting in the failure of stretchable conductors. Therefore, the conductivity of the stretchable conductors with buckled structures can keep stable under large strain through adjusting the magnitude of prestrain.

Liu et al. fabricated hierarchically buckled sheath-core fibers with a resistance change of less than 5% for a 1000% stretch.[6] As shown in Figure 2.4g, CNT sheets were wrapped on a prestretched SEBS fiber core. After releasing, the CNT film was subjected to compression in the fiber direction and exhibited buckled structure on the surface of the fiber. The morphology evolution of the composite fiber at different stretching stage from 0 to 1000% was observed in Figure 2.4h. The axial buckles gradually flattened with the appearing of belt-direction buckles because of the increase in length and decrease in diameter. No breaking or delamination of conducting elements occurred during stretching, leading to the stable electrometrical properties. For the stretchable conductors with plane substrates, unidirectional and omnidirectional prestretching can be introduced to construct buckled structures. Stretchable conductors based on buckled networks of long AgNW were fabricated by Lee et al.[1] The conductor could maintain low resistance up to 460% strain. Yu et al. reported isotropic buckled CNT films for supercapacitors by the omnidirectional prestretching.[36] Figure 2.4i shows the buckled structures formed in all directions on the CNT films. The films could be uniaxial, biaxial, and omnidirectionally stretched up to 200% strain without a notable resistance increase, which ensures the stable performance of the supercapacitors.

From literatures on stretchable conductors with buckled structures, unfolding of buckled structures has been proved to be an efficient mechanics to reduce the negative influence of the stress on the conducting elements. However, the out-of-plane buckled structures increase the roughness of the surface of the conductors, which would affect the applications in the transparent stretchable electrodes. Moreover, compared with CNTs and graphene, unfolding of buckled metal nanowires is difficult due to the stiffness of metal, which results in larger resistance variation during stretching. It is still a challenge to realize high conductivity, large strain, and low resistance change simultaneously for researchers.

### 2.3.3 Nanomeshes

Conductive nanomeshes with interconnected networks and large pores are very promising to prepare stretchable conductors.[27, 30, 84, 96, 97] Under tensile strain, the fully interconnected networks can avoid contact resistance change and the large pores provide enough distortion space to reduce the effect of deformation on the conductive materials meanwhile increase the transparency.

Wu et al. prepared stretchable conductors by depositing metal on electrospun nanofibers.[25] A fully interconnected nanomesh of metal nanotrough was obtained after removing the nanofiber templates. The conductors based on Au nanotrough network possessed high transmittance and conductivity, as well as stable resistance upon tensile strain of up to 50%. In addition, Wu et al. reported the preparation of transparent and flexible 2D silver meshes using recrystallized ice crystals as templates.[98] The size and line width of silver meshes obtained after freeze-drying can be easily adjusted, which in turn varied the transmittance and conductivity of the obtained 2D silver film. Besides unit cell distortion, distributed rupture of nanomesh upon tensile strain was studied in detail by Guo et al.[26] The distributed rupture can be directly evidenced in the Au nanomesh at high tensile strain (Figures 2.5a and b). At high tensile strain, aligned large unit cell with ruptured ligaments got distributed in the nanomesh, indicating the significant contribution of this mechanics for the stretchability of the nanomesh. Therefore, the Au nanomesh/PDMS conductor exhibited very low resistance and stable cyclic performance even at a tensile strain of 100%, as shown in Figure 2.5c. The function of the distributed rupture is further illustrated in Figures 2.5d-f, showing that when part of ligaments in the mesh is cut before stretching, the mesh can be elongated out-of-plane to a large extent without breaking into two halves. Recently, Wang et al. reported the giant Poisson's effect for wrinkle-free stretchable transparent electrodes, which focused on the study of the effect of the nanomesh structure on the performance.[99] Due to the nearly constant volumes, homogeneous materials generally possess a small Poisson's ratio below 0.5, which readily wrinkle upon deformation, affecting the smoothness and transparency. On the contrast, the nanomesh structure exhibits a giant Poisson's ratio above 2 because of the grid space, which is found to be under biaxial tension without wrinkles after the stretching-releasing process.

Constructing nanomesh structures can be considered an appropriate strategy to achieve highly conductive and transparent stretchable conductors. How to simplify the processing on the basis of precise control of the mesh structure is the key problem to solve.

### 2.3.4 Serpentine Structures

In recent years, Roger's group has been focusing on the study of the stretchable conductors with serpentine structures. Benefited from the unique structures, conductive materials can bear large applied strain and maintain the electrical performance stable. Related works about the serpentine structures have been widely reported in the literatures, indicating their potentials for the stretchable conductors.[100–102]

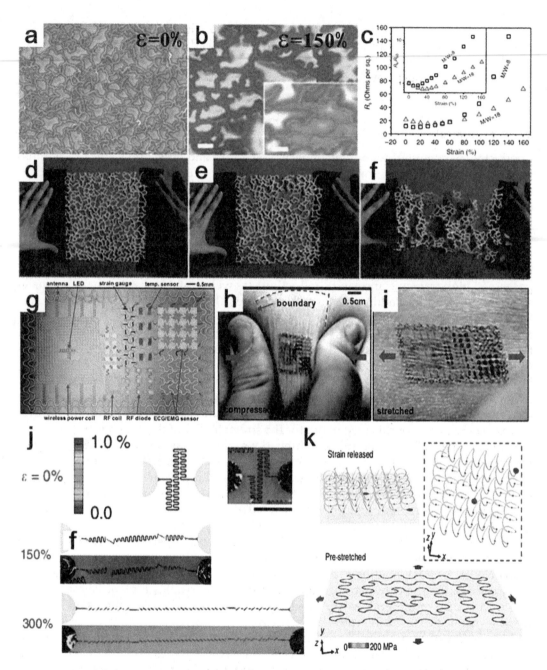

**FIGURE 2.5**
SEM images of the Au nanomesh before (a) and after (b) stretching at tensile strain of 150%. Scale bar is 1 μm and that in the inset is 500 nm. (c) Resistance variations versus strain of the Au nanomeshes. (d-f) Schematic illustration for deformation of an irregular mesh using a laser-cut sheet of paper as an example. (g) Image of a demonstration platform for multifunctional electronics. Images of the platform on skin when compressed (h) and stretched (i). (j) Experimental and computational results to show the mechanism of the hierarchical serpentine layout upon stretching. (k) FEA of the process for the 3D serpentine structure.

As shown in Figure 2.5g, Kim et al. reported multifunctional sensors and other electronic elements on an elastomeric sheet using serpentine interconnects as stretchable circuits.[103] The compliant integrated platform could be adhered tightly to skin via van der Waals forces alone, exhibiting the promising applications in the epidermal electronics (Figures 2.5h and i). The serpentine interconnects were bent and straightened under compressing and stretching to reduce the impact of the deformation on the conductive materials. Furthermore, hierarchical and out-of-plane geometries were designed to improve the stretchability. Xu et al. fabricated stretchable batteries with hierarchical serpentine interconnects.[104] Figure 2.5j shows the structure changes in the finite element analysis (FEA) and experiment, which are highly consistent. The motions in the first level started when the second level was almost fully extended, corresponding to a strain of 150%, resulting in a reversible tensile range of 300% in total. As the complex serpentine structure in plane increases the area of the conductor, which brings difficulties to the practical use, Jang et al. transferred the conductive network with 2D serpentine structure onto a biaxially prestretched elastic substrate. 3D serpentine structure was obtained after the release of prestrain, which reduced the area and increased the stretchability of the device (Figure 2.5k).[105]

The layout of the serpentine structures contributes to the precise control of the electromechanical performance of the stretchable conductors under deformation. Nevertheless, it is still a difficult challenge to realize large-scale production, which is limited by the complicated manufacturing.

### 2.3.5 Helix Structures

Compared with the previous structures, 3D helix structure can provide larger deformation space and accommodate higher strain.[69] Conductive fibers have been an active area of research because of the significant potential for developing lightweight, flexible, and stretchable electronic devices on textile products.[106, 107] Combining the advantages of the helix structure and fiber substrate, our group has done a series work of fiber-based stretchable conductors with helix structures.

Cheng et al. reported highly conductive and stretchable electric circuits made from covered yarns and AgNWs (Figure 2.6a).[15] The double-spiral yarns can not only adsorb the AgNW but also possess the helix structure to relieve the negative influence of large strain on the conductive networks. As shown in Figures 2.6b and c, the composite fiber exhibited a high conductivity up to 4018 S cm$^{-1}$, which remained as high as 688 S cm$^{-1}$ at the tensile strain of 500%. In addition, the conductivity of the composite fiber remained perfectly stable after 1000 cycles of bending and leveled off at 183 S cm$^{-1}$ after 1000 cyclic stretching events of 200% strain. The method is facile and appropriate for scalable production. However, there still exists the problem of conductivity decay at high tensile strain. The coil yarns around the fiber slid during stretching, which caused the gaps between yarns emerge and broaden with strain. Therefore, the connection bridged across adjacent yarns by AgNWs was disrupted, reducing the conductive pathways on the film. To improve the conductivity retention capability against stretch deformation, we proposed a novel hierarchically helix structure inspired by the adaptive tendril coiling of climbing plants.[5] Figure 2.6d shows the design and fabrication of the biomimetic conductive tendril. The key concept for the design lies in the incorporation of a helical conductive yarn into an elastic polymer layer, which is coated on another prestretched elastic substrate. After the release of prestretch, the strain mismatch at the interface gives the conductive tendril an intrinsic curvature and it self-shapes into the coiled

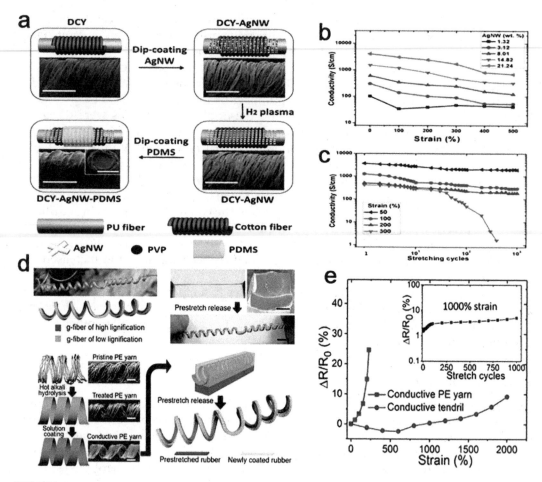

**FIGURE 2.6**

(a) Schematic illustration of the fabrication process of the composite fiber. (b) Variation of electrical conductivity versus strain of DCY-AgNW-PDMS with different AgNW mass fractions. (c) Conductivity of DCY-AgNW-PDMS after increasing stretching times at strains of 50%, 100%, 200%, and 300%, respectively. (d) Design and fabrication of the biomimetic conductive tendril. (e) Relative resistance variations versus strain of the conductive PE yarn and the conductive tendril. Inset is the relative resistance variation under cyclic stretching test at strain of 1000% up to 1000 cycles.

structure featuring perversion in between oppositely handed helices with equal numbers. Figure 2.6e displays the relative resistance variation versus tensile strain of both the conductive PE yarn and the corresponding conductive tendril. The conductive PE yarn exhibited a stretch range of 225% owing to its helix structure and a sharp increase of resistance by 24.6% due to the generation of micro cracks in the conductive network. In comparison, the conductive tendril indicated a markedly enlarged stretch range up to 2000% and an appreciably smaller resistance variation: 1.5% increase at the strain of 1000% and 9.2% at the strain of 2000%. Further cyclic stretching test (inset of Figure 2.6e) revealed a reliable durability of the conductive tendril: only 5.1% increase of resistance after 1000 times of stretching tests at the strain of 1000%.

This significantly improved performance is attributed to the elastic buffering connection effect of the helix structure, which greatly mitigates the actual strain on the conductive materials by reversible structural adaptation. The properties of the helix structure could be further explored due to the potential in the utilization of the stretchable conductors for flexible electronics.

## 2.4 Applications

In Sections 2.1–2.3, we have reviewed recent progress in developing new materials and structures for stretchable conductors. Benefited from the excellent mechanical compliance and high electric conductivity, a variety of stretchable electronic devices, such as interconnects, sensors, energy storage devices, actuators, heaters, and their integrated systems can be developed.

### 2.4.1 Interconnects

A straightforward application of stretchable conductors is stretchable interconnects that can be used to connect rigid device components.[2, 31] The stretchable interconnects have been widely used in stretchable LED systems where stretchable conductors connect and illuminate the LED elements.

Lee et al. developed flexible LED arrays on a paper and highly stretchable LED circuits on an elastic polymer substrate with elastic electrical interconnects based on the patterned AgNW-Ecoflex composites (Figure 2.7a).[1] The highly conductive and stretchable conductors essentially absorbed most of the strain during folding and crumpling, while LED devices experienced little strain. Thus, the electrical performance of the system was maintained, which could be utilized in the flexible displays. In addition, Matsuhisa et al. fabricated fully printed elastic sensor networks laminated on the textiles substrate, in which pressure and temperature sensors on rigid islands of photoresist and polyurethane were wired using Ag/rubber conductors (Figure 2.7b).[18] In this way, the sensing elements could be isolated from strain when the whole system was stretched. The interconnect could be stretched by 120% and guaranteed the stability of the sensors, indicating the promising applications for human-machine interaction or stretchable robotics.

### 2.4.2 Sensors

Two conductors with a dielectric layer sandwiched in between form a parallel-plate capacitor whose capacitance changes as a result of the dimensional change, change in the dielectric constant or change in the surrounding electric field, which is the basis of many multifunctional sensors. A number of pressure or strain capacitive sensors based on stretchable conductors have been reported in the literature.[108–111]

Lee et al. fabricated a capacitive type of highly sensitive textile-based pressure sensor by coating the PDMS on the surface of the AgNP/SBS conductive fiber and stacking the PDMS-coated fibers perpendicularly to each other as shown in Figure 2.7c.[4] When the cross point of the fibers was pressed, the increase in the contact area as well as the reduction in the distance between separated electrodes induced a significant increase in the capacitance (Figure 2.7d). The sensitivity of 0.21 kPa$^{-1}$ was obtained in the low-pressure range under 2 kPa and there was a reduction in sensitivity to 0.064 kPa$^{-1}$ at the pressure region above 2 kPa (Figure 2.7e). Similarly, capacitive sensors can respond to an applied

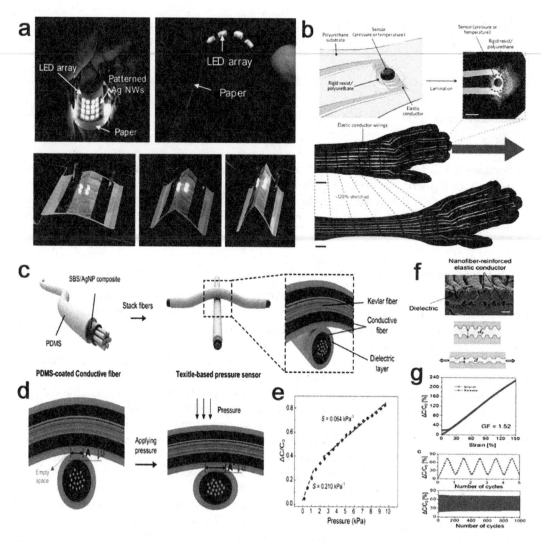

**FIGURE 2.7**
(a) Images of a flexible paper display with LEDs using AgNW electrodes. (b) Stretchable and fully printed sensor networks for stretchable robotics. (c) Schematic illustration of the fabrication of the pressure sensor. (d) Schematic illustration showing structural change of the pressure sensor under applied pressure. (e) Relative capacitance variation versus pressure of the capacitive sensor. (f) SEM cross-sectional image (top) and schematic of the capacitive strain sensor (bottom). Scale bar: 100 μm. (g) Relative capacitance variation and cycling performance of the strain sensor during stretching and releasing.

tensile strain due to the decreased distance between the two electrodes, also resulting in an increase in capacitance. Recently, a highly durable stretchable conductor was fabricated by Jin et al. through embedding PVDF nanofibers into Ag/fluoroelastomer to reinforce the toughness and suppress the crack growth.[3] To fabricate the capacitive-type strain sensor, the conductive films and stretchable dielectric layer were stacked together as a sandwich structure on a prestretched elastomer and released as shown in Figure 2.7f. Upon stretching, the thickness of the dielectric layer changed so that the capacitance increased. The relationship between the capacitance and strain was highly linear ($R^2 = 0.995$) with a gauge factor of 1.52 and excellent cyclic stability and reproducibility were obtained during the cyclic stretching test under the tensile strain of 50% (Figure 2.7g).

The capacitive sensors can function well with highly linear response because of the stable performance of the stretchable conductors under deformation, which is a good choice for artificial skins for robots, touch panels, and wearable smart devices.

### 2.4.3 Energy-Storage Devices

Stretchable energy-storage and conversion devices that can accommodate large strains represent another indispensable part of stretchable electronics. Many related works have been reported including stretchable supercapacitors, batteries, and solar cells.[38, 112, 113] As an important class of energy-storage devices, supercapacitors possess high power density, fast charging/discharging, and long cycle life. Due to the advantages in wearable electronics, the elastic fiber structures have been widely used for the stretchable supercapacitors in the recent literatures.[7, 8, 114–116]

Xu et al. fabricated stretchable wire-shaped supercapacitors through the prestrain-release-buckling strategy.[117] As shown in Figure 2.8a, the supercapacitor is composed of

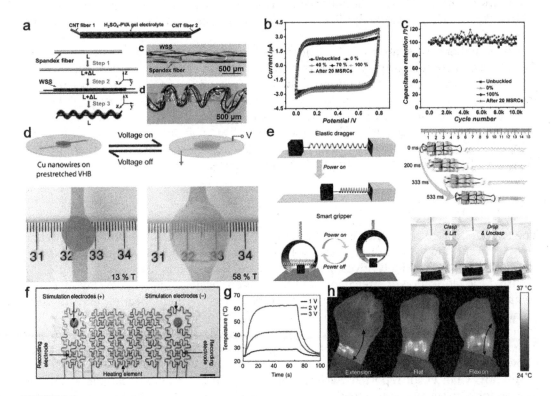

**FIGURE 2.8**

(a) Schematics of the structure and fabrication of the stretchable wire-shaped supercapacitors. (b) CV curves of stretchable supercapacitor at different applied strains. (c) Variations in capacitance retention with cycle number. (d) Schematic and photos of a transparent copper nanowire actuator. (e) The conductive tendrils as actuators in applications of a dragger and a smart gripper. (f) Photo of a multifunctional wearable electronic patch consisting of bipolar stimulation electrodes, electrophysiological-signal-recording electrodes, and a heater. Scale bar: 1 cm. (g) Temperature responses of the heater in the wearable device with applied voltages of 1, 2, and 3 V. (h) Infrared camera images showing reliable heating performance of the wearable device on a wrist.

a CNT fiber electrode, a spandex fiber substrate and $H_2SO_4$-poly (vinyl alcohol) (PVA) gel as the solid electrolyte. With the buckled structure of the electrodes, the supercapacitor exhibited better energy-storage capabilities than flat structure, which possessed stretchability up to 100% without much deterioration in capacitance, cycling stability, power, and energy density (Figures 2.8b and c). In addition, Zhang et al. developed an elastic fiber-shaped supercapacitor using two buckled aligned CNT/polyaniline (PANI) composite sheets as electrodes, which could be stretched by over 400%.[118] The stretchable supercapacitors can be easily woven into textiles with stable electrochemical performances under dynamic stretching or bending.

### 2.4.4 Actuators

Stretchable conductors can be used in actuators as well.[85, 119–121] Actuators can exhibit volumetric change in response to voltage stimulation and thus have broad applications in artificial muscles.

Wu et al. reported a stretchable transparent actuator by transferring CuNWs to both the top and the bottom sides of a prestretched VHB film.[89] Under voltage stimulation, the dielectric elastomer contracted in thickness and expanded in area (Figure 2.8d). The actuator exhibited area strain of larger than 200% when the applied voltage was 4.8 kV.

A fiber actuator was fabricated by Cheng et al. using a hierarchically helix structure.[5] The PE fibers possess a negative linear thermal expansion coefficient. Joule heating of the conductive network generates internal contractive force in the conductive PE yarn along the helix line direction, which resultantly causes a contraction in the actuator. As shown in Figure 2.8e, the actuators can not only work independently as an elastic dragger but also carry out programmable and sophisticated motility in an integrative level as a smart gripper. The excellent actuation performance of the actuators empowered their further deployment as artificial muscles in practical applications.

### 2.4.5 Heaters

The temperature of the stretchable conductors will increase following Joule's law when direct current flows through it. Resistive-type stretchable heaters based on the thermal response have been widely investigated for wearable electronics.[14, 29, 122, 123] Hong et al. reported the fabrication of stretchable and transparent heaters with AgNW/PDMS nanocomposites.[124] A constant bias was applied to the device for joule heating and the temperature was increased to 200°C until the device failed at the bias voltage of 10 V. When stretched, the heater exhibited a stable operation with reduced temperature, since the heat generated by Joule heating at a fixed bias voltage is inversely proportional to the resistance of overall network of AgNWs.

Recently, Choi et al. reported the fabrication of multifunctional wearable electronic devices for both sensing and actuating applications on human skin, using nanocomposites of Ag-Au NWs and SBS elastomers.[22] Owing to the intrinsic softness of composite electrodes as well as a serpentine-shaped pattern of electrodes, the stretchable conductor was able to keep stable heating capability under deformation (Figure 2.8f). Figure 2.8g shows temperature-time profiles with various input voltages. The softness of the patch ensures firm contact with the skin, allowing reliable heat transfer even when the wrist is flexed or extended (Figure 2.8h).

# References

1. Lee, P.; Lee, J.; Lee, H.; Yeo, J.; Hong, S.; Nam, K. H.; Lee, D.; Lee, S. S.; Ko, S. H. Highly stretchable and highly conductive metal electrode by very long metal nanowire percolation network. *Adv. Mater.* **2012,** *24* (25), 3326–3332.

2. Catenacci, M. J.; Reyes, C.; Cruz, M. A.; Wiley, B. J. Stretchable conductive composites from Cu-Ag nanowire felt. *ACS Nano* **2018,** *12* (4), 3689–3698.

3. Jin, H.; Nayeem, M. O. G.; Lee, S.; Matsuhisa, N.; Inoue, D.; Yokota, T.; Hashizume, D.; Someya, T. Highly durable nanofiber-reinforced elastic conductors for skin-tight electronic textiles. *ACS Nano* **2019,** *13* (7), 7905–7912.

4. Lee, J.; Kwon, H.; Seo, J.; Shin, S.; Koo, J. H.; Pang, C.; Son, S.; Kim, J. H.; Jang, Y. H.; Kim, D. E.; Lee, T. Conductive fiber-based ultrasensitive textile pressure sensor for wearable electronics. *Adv. Mater.* **2015,** *27* (15), 2433–2439.

5. Cheng, Y.; Wang, R.; Chan, K. H.; Lu, X.; Sun, J.; Ho, G. W. A biomimetic conductive tendril for ultrastretchable and integratable electronics, muscles, and sensors. *ACS Nano* **2018,** *12* (4), 3898–3907.

6. Liu, Z. F.; Fang, S.; Moura, F. A.; Ding, J. N.; Jiang, N.; Di, J.; Zhang, M.; Lepro, X.; Galvao, D. S.; Haines, C. S.; Yuan, N. Y.; Yin, S. G.; Lee, D. W.; Wang, R.; Wang, H. Y.; Lv, W.; Dong, C.; Zhang, R. C.; Chen, M. J.; Yin, Q.; Chong, Y. T.; Zhang, R.; Wang, X.; Lima, M. D.; Ovalle-Robles, R.; Qian, D.; Lu, H.; Baughman, R. H. Hierarchically buckled sheath-core fibers for superelastic electronics, sensors, and muscles. *Science* **2015,** *349* (6246), 400–404.

7. Qu, G.; Cheng, J.; Li, X.; Yuan, D.; Chen, P.; Chen, X.; Wang, B.; Peng, H. A fiber supercapacitor with high energy density based on hollow graphene/conducting polymer fiber electrode. *Adv. Mater.* **2016,** *28* (19), 3646–3652.

8. Liu, M.; Cong, Z.; Pu, X.; Guo, W.; Liu, T.; Li, M.; Zhang, Y.; Hu, W.; Wang, Z. L. High-energy asymmetric supercapacitor yarns for self-charging power textiles. *Adv. Funct. Mater.* **2019,** 1806298.

9. Ma, R.; Lee, J.; Choi, D.; Moon, H.; Baik, S. Knitted fabrics made from highly conductive stretchable fibers. *Nano Lett.* **2014,** *14* (4), 1944–1951.

10. Park, M.; Im, J.; Shin, M.; Min, Y.; Park, J.; Cho, H.; Park, S.; Shim, M. B.; Jeon, S.; Chung, D. Y.; Bae, J.; Park, J.; Jeong, U.; Kim, K. Highly stretchable electric circuits from a composite material of silver nanoparticles and elastomeric fibres. *Nat. Nanotechnol.* **2012,** *7* (12), 803–809.

11. Lee, S.; Shin, S.; Lee, S.; Seo, J.; Lee, J.; Son, S.; Cho, H. J.; Algadi, H.; Al-Sayari, S.; Kim, D. E.; Lee, T. Ag nanowire reinforced highly stretchable conductive fibers for wearable electronics. *Adv. Funct. Mater.* **2015,** *25* (21), 3114–3121.

12. Xu, F.; Zhu, Y. Highly conductive and stretchable silver nanowire conductors. *Adv. Mater.* **2012,** *24* (37), 5117–5122.

13. Yun, S.; Niu, X.; Yu, Z.; Hu, W.; Brochu, P.; Pei, Q. Compliant silver nanowire-polymer composite electrodes for bistable large strain actuation. *Adv. Mater.* **2012,** *24* (10), 1321–1327.

14. Choi, S.; Park, J.; Hyun, W.; Kim, J.; Kim, J.; Lee, Y. B.; Song, C.; Hwang, H. J.; Kim, J. H.; Hyeon, T.; Kim, D. H. Stretchable heater using ligand-exchanged silver nanowire nanocomposite for wearable articular thermotherapy. *ACS Nano* **2015,** *9* (6), 6626–6633.

15. Cheng, Y.; Wang, R. R.; Sun, J.; Gao, L. Highly conductive and ultrastretchable electric circuits from covered yarns and silver nanowires. *ACS Nano* **2015,** *9* (4), 3887–3895.

16. Liang, J.; Li, L.; Tong, K.; Ren, Z.; Hu, W.; Niu, X.; Chen, Y.; Pei, Q. Silver nanowire percolation network soldered with graphene oxide at room temperature and its application for fully stretchable polymer light-emitting diodes. *ACS Nano* **2014,** *8* (2), 1590–1600.

17. Li, Z.; Le, T.; Wu, Z.; Yao, Y.; Li, L.; Tentzeris, M.; Moon, K. S.; Wong, C. P. Rational design of a printable, highly conductive silicone-based electrically conductive adhesive for stretchable radio-frequency antennas. *Adv. Funct. Mater.* **2015,** *25* (3), 464–470.

18. Matsuhisa, N.; Inoue, D.; Zalar, P.; Jin, H.; Matsuba, Y.; Itoh, A.; Yokota, T.; Hashizume, D.; Someya, T. Printable elastic conductors by in situ formation of silver nanoparticles from silver flakes. *Nat. Mater.* **2017,** *16* (8), 834–840.

19. Kim, S. H.; Jung, S.; Yoon, I. S.; Lee, C.; Oh, Y.; Hong, J. M. Ultrastretchable conductor fabricated on skin-like hydrogel-elastomer hybrid substrates for skin electronics. *Adv. Mater.* **2018**, e1800109.

20. Wang, J.; Cai, G.; Li, S.; Gao, D.; Xiong, J.; Lee, P. S. Printable superelastic conductors with extreme stretchability and robust cycling endurance enabled by liquid-metal particles. *Adv. Mater.* **2018**, 1706157.

21. Kim, Y.; Zhu, J.; Yeom, B.; Di Prima, M.; Su, X.; Kim, J. G.; Yoo, S. J.; Uher, C.; Kotov, N. A. Stretchable nanoparticle conductors with self-organized conductive pathways. *Nature* **2013**, *500* (7460), 59–63.

22. Choi, S.; Han, S. I.; Jung, D.; Hwang, H. J.; Lim, C.; Bae, S.; Park, O. K.; Tschabrunn, C. M.; Lee, M.; Bae, S. Y.; Yu, J. W.; Ryu, J. H.; Lee, S. W.; Park, K.; Kang, P. M.; Lee, W. B.; Nezafat, R.; Hyeon, T.; Kim, D. H. Highly conductive, stretchable and biocompatible Ag-Au core-sheath nanowire composite for wearable and implantable bioelectronics. *Nat. Nanotechnol.* **2018**, *13*, 1048–1056.

23. Moon, G. D.; Lim, G. H.; Song, J. H.; Shin, M.; Yu, T.; Lim, B.; Jeong, U. Highly stretchable patterned gold electrodes made of Au nanosheets. *Adv. Mater.* **2013**, *25* (19), 2707–2712.

24. Zhang, B.; Lei, J.; Qi, D.; Liu, Z.; Wang, Y.; Xiao, G.; Wu, J.; Zhang, W.; Huo, F.; Chen, X. Stretchable conductive fibers based on a cracking control strategy for wearable electronics. *Adv. Funct. Mater.* **2018**, *28* (29), 1801683.

25. Wu, H.; Kong, D.; Ruan, Z.; Hsu, P. C.; Wang, S.; Yu, Z.; Carney, T. J.; Hu, L.; Fan, S.; Cui, Y. A transparent electrode based on a metal nanotrough network. *Nat. Nanotechnol.* **2013**, *8* (6), 421–425.

26. Guo, C. F.; Sun, T.; Liu, Q.; Suo, Z.; Ren, Z. Highly stretchable and transparent nanomesh electrodes made by grain boundary lithography. *Nat. Commun.* **2014**, *5*, 3121.

27. Miyamoto, A.; Lee, S.; Cooray, N. F.; Lee, S.; Mori, M.; Matsuhisa, N.; Jin, H.; Yoda, L.; Yokota, T.; Itoh, A.; Sekino, M.; Kawasaki, H.; Ebihara, T.; Amagai, M.; Someya, T. Inflammation-free, gas-permeable, lightweight, stretchable on-skin electronics with nanomeshes. *Nat. Nanotechnol.* **2017**, *12* (9), 907–913.

28. Chen, G.; Matsuhisa, N.; Liu, Z.; Qi, D.; Cai, P.; Jiang, Y.; Wan, C.; Cui, Y.; Leow, W. R.; Liu, Z.; Gong, S.; Zhang, K. Q.; Cheng, Y.; Chen, X. Plasticizing silk protein for on-skin stretchable electrodes. *Adv. Mater.* **2018**, *30* (21), e1800129.

29. Cheng, Y.; Zhang, H.; Wang, R.; Wang, X.; Zhai, H.; Wang, T.; Jin, Q.; Sun, J. Highly stretchable and conductive copper nanowire based fibers with hierarchical structure for wearable heaters. *ACS Appl. Mater. Interfaces* **2016**, *8* (48), 32925–32933.

30. Yu, Y.; Zhang, Y.; Li, K.; Yan, C.; Zheng, Z. Bio-inspired chemical fabrication of stretchable transparent electrodes. *Small* **2015**, *11* (28), 3444–3449.

31. Zhu, S.; So, J. H.; Mays, R.; Desai, S.; Barnes, W. R.; Pourdeyhimi, B.; Dickey, M. D. Ultrastretchable fibers with metallic conductivity using a liquid metal alloy core. *Adv. Funct. Mater.* **2013**, *23* (18), 2308–2314.

32. Kim, K. H.; Vural, M.; Islam, M. F. Single-walled carbon nanotube aerogel-based elastic conductors. *Adv. Mater.* **2011**, *23* (25), 2865–2869.

33. Cai, L.; Li, J.; Luan, P.; Dong, H.; Zhao, D.; Zhang, Q.; Zhang, X.; Tu, M.; Zeng, Q.; Zhou, W.; Xie, S. Highly transparent and conductive stretchable conductors based on hierarchical reticulate single-walled carbon nanotube architecture. *Adv. Funct. Mater.* **2012**, *22* (24), 5238–5244.

34. Xu, F.; Wang, X.; Zhu, Y.; Zhu, Y. Wavy ribbons of carbon nanotubes for stretchable conductors. *Adv. Funct. Mater.* **2012**, *22* (6), 1279–1283.

35. Sekitani, T.; Noguchi, Y.; Hata, K.; Fukushima, T.; Aida, T.; Someya, T. A rubberlike stretchable active matrix using elastic conductors. *Science* **2008**, *321* (5895), 1468–1472.

36. Yu, J.; Lu, W.; Pei, S.; Gong, K.; Wang, L.; Meng, L.; Huang, Y.; Smith, J. P.; Booksh, K. S.; Li, Q.; Byun, J. H.; Oh, Y.; Yan, Y.; Chou, T. W. Omnidirectionally stretchable high-performance supercapacitor based on isotropic buckled carbon nanotube films. *ACS Nano* **2016**, *10* (5), 5204–5211.

37. Chen, Z.; Ren, W.; Gao, L.; Liu, B.; Pei, S.; Cheng, H. M. Three-dimensional flexible and conductive interconnected graphene networks grown by chemical vapour deposition. *Nat. Mater.* **2011**, *10* (6), 424–428.

38. Chen, T.; Xue, Y. H.; Roy, A. K.; Dai, L. M. Transparent and stretchable high-performance supercapacitors based on wrinkled graphene electrodes. *ACS Nano* **2014**, *8* (1), 1039–1046.

39. Seol, Y. G.; Trung, T. Q.; Yoon, O. J.; Sohn, I. Y.; Lee, N. E. Nanocomposites of reduced graphene oxide nanosheets and conducting polymer for stretchable transparent conducting electrodes. *J. Mater. Chem.* **2012**, *22* (45), 23759.

40. Wang, Y.; Zhu, C. X.; Pfattner, R.; Yan, H. P.; Jin, L. H.; Chen, S. C.; Molina-Lopez, F.; Lissel, F.; Liu, J.; Rabiah, N. I.; Chen, Z.; Chung, J. W.; Linder, C.; Toney, M. F.; Murmann, B.; Bao, Z. A highly stretchable, transparent, and conductive polymer. *Sci. Adv.* **2017**, *3* (3), e1602076.

41. Feig, V. R.; Tran, H.; Lee, M.; Bao, Z. Mechanically tunable conductive interpenetrating network hydrogels that mimic the elastic moduli of biological tissue. *Nat. Commun.* **2018**, *9* (1), 2740.

42. Sasaki, M.; Karikkineth, B. C.; Nagamine, K.; Kaji, H.; Torimitsu, K.; Nishizawa, M. Highly conductive stretchable and biocompatible electrode-hydrogel hybrids for advanced tissue engineering. *Adv. Healthc. Mater.* **2014**, *3* (11), 1919–1927.

43. Das, A.; Banerji, A.; Mukherjee, R. Programming feature size in the thermal wrinkling of metal polymer bilayer by modulating substrate viscoelasticity. *ACS Appl. Mater. Interfaces* **2017**, *9* (27), 23255–23262.

44. Wang, T.; Wang, R.; Cheng, Y.; Sun, J. Quasi in situ polymerization to fabricate copper nanowire-based stretchable conductor and its applications. *ACS Appl. Mater. Interfaces* **2016**, *8* (14), 9297–9304.

45. Lee, H.; Lee, K.; Park, J. T.; Kim, W. C.; Lee, H. Well-ordered and high density coordination-type bonding to strengthen contact of silver nanowires on highly stretchable polydimethylsiloxane. *Adv. Funct. Mater.* **2014**, *24* (21), 3276–3283.

46. Akter, T.; Kim, W. S. Reversibly stretchable transparent conductive coatings of spray-deposited silver nanowires. *ACS Appl. Mater. Interfaces* **2012**, *4* (4), 1855–1859.

47. Li, Y.; Cui, P.; Wang, L.; Lee, H.; Lee, K.; Lee, H. Highly bendable, conductive, and transparent film by an enhanced adhesion of silver nanowires. *ACS Appl. Mater. Interfaces* **2013**, *5* (18), 9155–9160.

48. Zhou, Y.; Wan, C.; Yang, Y.; Yang, H.; Wang, S.; Dai, Z.; Ji, K.; Jiang, H.; Chen, X.; Long, Y. Highly stretchable, elastic, and ionic conductive hydrogel for artificial soft electronics. *Adv. Funct. Mater.* **2019**, *29* (1), 1806220.

49. Zhao, X.; Chen, F.; Li, Y.; Lu, H.; Zhang, N.; Ma, M. Bioinspired ultra-stretchable and antifreezing conductive hydrogel fibers with ordered and reversible polymer chain alignment. *Nat. Commun.* **2018**, *9* (1), 3579.

50. Yao, B.; Wang, H.; Zhou, Q.; Wu, M.; Zhang, M.; Li, C.; Shi, G. Ultrahigh-conductivity polymer hydrogels with arbitrary structures. *Adv. Mater.* **2017**, *29* (28).

51. Cai, G.; Wang, J.; Qian, K.; Chen, J.; Li, S.; Lee, P. S. Extremely stretchable strain sensors based on conductive self-healing dynamic cross-links hydrogels for human-motion detection. *Adv. Sci. (Weinh.)* **2017**, *4* (2), 1600190.

52. Liao, M.; Wan, P.; Wen, J.; Gong, M.; Wu, X.; Wang, Y.; Shi, R.; Zhang, L. Wearable, healable, and adhesive epidermal sensors assembled from mussel-inspired conductive hybrid hydrogel framework. *Adv. Funct. Mater.* **2017**, *27* (48), 1703852.

53. Jin, H.; Matsuhisa, N.; Lee, S.; Abbas, M.; Yokota, T.; Someya, T. Enhancing the performance of stretchable conductors for e-textiles by controlled ink permeation. *Adv. Mater.* **2017**, 1605848.

54. Wang, C.; Li, X.; Gao, E.; Jian, M.; Xia, K.; Wang, Q.; Xu, Z.; Ren, T.; Zhang, Y. Carbonized silk fabric for ultrastretchable, highly sensitive, and wearable strain sensors. *Adv. Mater.* **2016**, *28* (31), 6640–6648.

55. Ge, J.; Sun, L.; Zhang, F. R.; Zhang, Y.; Shi, L. A.; Zhao, H. Y.; Zhu, H. W.; Jiang, H. L.; Yu, S. H. A stretchable electronic fabric artificial skin with pressure-, lateral strain-, and flexion-sensitive properties. *Adv. Mater.* **2016**, *28* (4), 722–728.

56. Karim, N.; Afroj, S.; Tan, S.; He, P.; Fernando, A.; Carr, C.; Novoselov, K. S. Scalable production of graphene-based wearable e-textiles. *ACS Nano* **2017**, *11* (12), 12266–12275.

57. Liu, C. H.; Yu, X. Silver nanowire-based transparent, flexible, and conductive thin film. *Nanoscale Res. Lett.* **2011**, *6*, 75.

58. Wang, S.; Gong, L.; Shang, Z.; Ding, L.; Yin, G.; Jiang, W.; Gong, X.; Xuan, S. Novel safeguarding tactile e-skins for monitoring human motion based on SST/PDMS-AgNW-PET hybrid structures. *Adv. Funct. Mater.* **2018**, *28* (18), 1707538.

59. Song, P.; Qin, H.; Gao, H. L.; Cong, H. P.; Yu, S. H. Self-healing and superstretchable conductors from hierarchical nanowire assemblies. *Nat. Commun.* **2018**, *9* (1), 2786.

60. Matsuhisa, N.; Kaltenbrunner, M.; Yokota, T.; Jinno, H.; Kuribara, K.; Sekitani, T.; Someya, T. Printable elastic conductors with a high conductivity for electronic textile applications. *Nat. Commun.* **2015**, *6*, 7461.

61. Zhu, B. W.; Gong, S.; Cheng, W. L. Softening gold for elastronics. *Chem. Soc. Rev.* **2019**, *48* (6), 1668–1711.

62. Liu, Z.; Wang, H.; Huang, P.; Huang, J.; Zhang, Y.; Wang, Y.; Yu, M.; Chen, S.; Qi, D.; Wang, T.; Jiang, Y.; Chen, G.; Hu, G.; Li, W.; Yu, J.; Luo, Y.; Loh, X. J.; Liedberg, B.; Li, G.; Chen, X. Highly stable and stretchable conductive films through-thermal-radiation-assisted metal encapsulation. *Adv. Mater.* **2019**, 1901360.

63. Lee, S.; Sasaki, D.; Kim, D.; Mori, M.; Yokota, T.; Lee, H.; Park, S.; Fukuda, K.; Sekino, M.; Matsuura, K.; Shimizu, T.; Someya, T. Ultrasoft electronics to monitor dynamically pulsing cardiomyocytes. *Nat. Nanotechnol.* **2018**, *14*, 156–160.

64. Zhao, S. F.; Han, F.; Li, J. H.; Meng, X. Y.; Huang, W. P.; Cao, D. X.; Zhang, G. P.; Sun, R.; Wong, C. P. advancements in copper nanowires: synthesis, purification, assemblies, surface modification, and applications. *Small* **2018**, *14* (26), 1800047.

65. Yang, X.; Hu, X. T.; Wang, Q. X.; Xiong, J.; Yang, H. J.; Meng, X. C.; Tan, L. C.; Chen, L.; Chen, Y. W. Large-scale stretchable semiembedded copper nanowire transparent conductive films by an electrospinning template. *ACS Appl. Mater. Interfaces* **2017**, *9* (31), 26468–26475.

66. Im, H. G.; Jung, S. H.; Jin, J.; Lee, D.; Lee, J.; Lee, D.; Lee, J. Y.; Kim, I. D.; Bae, B. S. Flexible transparent conducting hybrid film using a surface-embedded copper nanowire network: a highly oxidation-resistant copper nanowire electrode for flexible optoelectronics. *ACS Nano* **2014**, *8* (10), 10973–10979.

67. Han, S.; Hong, S.; Ham, J.; Yeo, J.; Lee, J.; Kang, B.; Lee, P.; Kwon, J.; Lee, S. S.; Yang, M. Y.; Ko, S. H. Fast plasmonic laser nanowelding for a Cu-nanowire percolation network for flexible transparent conductors and stretchable electronics. *Adv. Mater.* **2014**, *26* (33), 5808–5814.

68. Yin, Z.; Song, S. K.; You, D. J.; Ko, Y.; Cho, S.; Yoo, J.; Park, S. Y.; Piao, Y.; Chang, S. T.; Kim, Y. S. Novel synthesis, coating, and networking of curved copper nanowires for flexible transparent conductive electrodes. *Small* **2015**, *11* (35), 4576–4583.

69. Won, Y.; Kim, A.; Yang, W.; Jeong, S.; Moon, J. A highly stretchable, helical copper nanowire conductor exhibiting a stretchability of 700%. *NPG Asia Mater.* **2014**, *6* (9), e132.

70. Hsu, P. C.; Wu, H.; Carney, T. J.; McDowell, M. T.; Yang, Y.; Garnett, E. C.; Li, M.; Hu, L. B.; Cui, Y. Passivation coating on electrospun copper nanofibers for stable transparent electrodes. *ACS Nano* **2012**, *6* (6), 5150–5156.

71. Song, J.; Li, J.; Xu, J.; Zeng, H. Superstable transparent conductive Cu@Cu₄Ni nanowire elastomer composites against oxidation, bending, stretching, and twisting for flexible and stretchable optoelectronics. *Nano Lett.* **2014**, *14* (11), 6298–6305.

72. Lu, T.; Finkenauer, L.; Wissman, J.; Majidi, C. Rapid prototyping for soft-matter electronics. *Adv. Funct. Mater.* **2014**, *24* (22), 3351–3356.

73. Ma, B.; Xu, C.; Chi, J.; Chen, J.; Zhao, C.; Liu, H. A versatile approach for direct patterning of liquid metal using magnetic field. *Adv. Funct. Mater.* **2019**, 1901370.

74. Kang, J.; Son, D.; Wang, G. N.; Liu, Y.; Lopez, J.; Kim, Y.; Oh, J. Y.; Katsumata, T.; Mun, J.; Lee, Y.; Jin, L.; Tok, J. B.; Bao, Z. Tough and water-insensitive self-healing elastomer for robust electronic skin. *Adv. Mater.* **2018**, *30* (13), e1706846.

75. Calvaresi, M.; Quintana, M.; Rudolf, P.; Zerbetto, F.; Prato, M. Rolling up a Graphene sheet. *ChemPhysChem* **2013**, *14* (15), 3447–3453.

76. Baughman, R. H.; Zakhidov, A. A.; de Heer, W. A. Carbon nanotubes - the route toward applications. *Science* **2002,** *297* (5582), 787–792.

77. Cao, Q.; Rogers, J. A. Ultrathin films of single-walled carbon nanotubes for electronics and sensors: a review of fundamental and applied aspects. *Adv. Mater.* **2009,** *21* (1), 29–53.

78. Zhang, Y.; Sheehan, C. J.; Zhai, J.; Zou, G.; Luo, H.; Xiong, J.; Zhu, Y. T.; Jia, Q. X. Polymer-embedded carbon nanotube ribbons for stretchable conductors. *Adv. Mater.* **2010,** *22* (28), 3027–3031.

79. Zhu, Y.; Xu, F. Buckling of aligned carbon nanotubes as stretchable conductors: a new manufacturing strategy. *Adv. Mater.* **2012,** *24* (8), 1073–1077.

80. Huang, X.; Zeng, Z. Y.; Fan, Z. X.; Liu, J. Q.; Zhang, H. Graphene-based electrodes. *Adv. Mater.* **2012,** *24* (45), 5979–6004.

81. Pang, S. P.; Hernandez, Y.; Feng, X. L.; Mullen, K. Graphene as transparent electrode material for organic electronics. *Adv. Mater.* **2011,** *23* (25), 2779–2795.

82. Zhu, Y. W.; Murali, S.; Cai, W. W.; Li, X. S.; Suk, J. W.; Potts, J. R.; Ruoff, R. S. Graphene and graphene oxide: synthesis, properties, and applications. *Adv. Mater.* **2010,** *22* (35), 3906–3924.

83. Liu, N.; Chortos, A.; Lei, T.; Jin, L. H.; Kim, T. R.; Bae, W. G.; Zhu, C. X.; Wang, S. H.; Pfattner, R.; Chen, X. Y.; Sinclair, R.; Bao, Z. A. Ultratransparent and stretchable graphene electrodes. *Sci. Adv.* **2017,** *3* (9), e1700159.

84. Han, J.; Lee, J. Y.; Lee, J.; Yeo, J. S. Highly stretchable and reliable, transparent and conductive entangled graphene mesh networks. *Adv. Mater.* **2018,** *30* (3), 1704626.

85. Zang, J.; Ryu, S.; Pugno, N.; Wang, Q.; Tu, Q.; Buehler, M. J.; Zhao, X. Multifunctionality and control of the crumpling and unfolding of large-area graphene. *Nat. Mater.* **2013,** *12* (4), 321–325.

86. Lee, M. S.; Lee, K.; Kim, S. Y.; Lee, H.; Park, J.; Choi, K. H.; Kim, H. K.; Kim, D. G.; Lee, D. Y.; Nam, S.; Park, J. U. High-performance, transparent, and stretchable electrodes using graphene-metal nanowire hybrid structures. *Nano Lett.* **2013,** *13* (6), 2814–2821.

87. Lima, R. M. A. P.; Alcaraz-Espinoza, J. J.; da Silva, F. A. G.; de Oliveira, H. P. Multifunctional wearable electronic textiles using cotton fibers with polypyrrole and carbon nanotubes. *ACS Appl. Mater. Interfaces* **2018,** *10* (16), 13783–13795.

88. Lee, Y. Y.; Kang, H. Y.; Gwon, S. H.; Choi, G. M.; Lim, S. M.; Sun, J. Y.; Joo, Y. C. A strain-insensitive stretchable electronic conductor: PEDOT:PSS/Acrylamide Organogels. *Adv. Mater.* **2016,** *28* (8), 1636–1643.

89. Wu, J.; Zang, J.; Rathmell, A. R.; Zhao, X.; Wiley, B. J. Reversible sliding in networks of nanowires. *Nano Lett.* **2013,** *13* (6), 2381–2386.

90. Ryu, S.; Lee, P.; Chou, J. B.; Xu, R. Z.; Zhao, R.; Hart, A. J.; Kim, S. G. Extremely elastic wearable carbon nanotube fiber strain sensor for monitoring of human motion. *ACS Nano* **2015,** *9* (6), 5929–5936.

91. Lee, P.; Ham, J.; Lee, J.; Hong, S.; Han, S.; Suh, Y. D.; Lee, S. E.; Yeo, J.; Lee, S. S.; Lee, D.; Ko, S. H. Highly stretchable or transparent conductor fabrication by a hierarchical multiscale hybrid nanocomposite. *Adv. Funct. Mater.* **2014,** *24* (36), 5671–5678.

92. Jiang, Z.; Fukuda, K.; Xu, X.; Park, S.; Inoue, D.; Jin, H.; Saito, M.; Osaka, I.; Takimiya, K.; Someya, T. Reverse-offset printed ultrathin ag mesh for robust conformal transparent electrodes for high-performance organic photovoltaics. *Adv. Mater.* **2018,** *30* (26), e1707526.

93. Jiang, T.; Huang, R.; Zhu, Y. Interfacial sliding and buckling of monolayer graphene on a stretchable substrate. *Adv. Funct. Mater.* **2014,** *24* (3), 396–402.

94. Jiang, S. J.; Zhang, H. B.; Song, S. Q.; Ma, Y. W.; Li, J. H.; Lee, G. H.; Han, Q. W.; Liu, J. Highly stretchable conductive fibers from few-walled carbon nanotubes coated on poly(m-phenylene isophthalamide) polymer core/shell structures. *ACS Nano* **2015,** *9* (10), 10252–10257.

95. Shin, U. H.; Jeong, D. W.; Kim, S. H.; Lee, H. W.; Kim, J. M. Elastomer-infiltrated vertically aligned carbon nanotube film-based wavy-configured stretchable conductors. *ACS Appl. Mater. Interfaces* **2014,** *6* (15), 12909–12914.

96. Fan, Y. J.; Li, X.; Kuang, S. Y.; Zhang, L.; Chen, Y. H.; Liu, L.; Zhang, K.; Ma, S. W.; Liang, F.; Wu, T.; Wang, Z. L.; Zhu, G. Highly robust, transparent, and breathable epidermal electrode. *ACS Nano* **2018,** *12* (9), 9326–9332.

97. Jang, H. Y.; Lee, S. K.; Cho, S. H.; Ahn, J. H.; Park, S. Fabrication of metallic nanomesh: Pt nanomesh as a proof of concept for stretchable and transparent electrodes. *Chem. Mater.* **2013**, *25* (17), 3535–3538.

98. Wu, S.; Li, L.; Xue, H.; Liu, K.; Fan, Q.; Bai, G.; Wang, J. Size controllable, transparent, and flexible 2d silver meshes using recrystallized ice crystals as templates. *ACS Nano* **2017**, *11* (10), 9898–9905.

99. Wang, Y.; Liu, Q.; Zhang, J.; Hong, T.; Sun, W.; Tang, L.; Arnold, E.; Suo, Z.; Hong, W.; Ren, Z.; Guo, C. F. Giant Poisson's effect for wrinkle-free stretchable transparent electrodes. *Adv. Mater.* **2019**, *31* (35), 1902955.

100. Yeo, W. H.; Kim, Y. S.; Lee, J.; Ameen, A.; Shi, L. K.; Li, M.; Wang, S. D.; Ma, R.; Jin, S. H.; Kang, Z.; Huang, Y. G.; Rogers, J. A. Multifunctional epidermal electronics printed directly onto the skin. *Adv. Mater.* **2013**, *25* (20), 2773–2778.

101. Kim, J.; Lee, M.; Shim, H. J.; Ghaffari, R.; Cho, H. R.; Son, D.; Jung, Y. H.; Soh, M.; Choi, C.; Jung, S.; Chu, K.; Jeon, D.; Lee, S. T.; Kim, J. H.; Choi, S. H.; Hyeon, T.; Kim, D. H. Stretchable silicon nanoribbon electronics for skin prosthesis. *Nat. Commun.* **2014**, *5*, 5747.

102. Lee, H.; Choi, T. K.; Lee, Y. B.; Cho, H. R.; Ghaffari, R.; Wang, L.; Choi, H. J.; Chung, T. D.; Lu, N.; Hyeon, T.; Choi, S. H.; Kim, D.-H. A graphene-based electrochemical device with thermoresponsive microneedles for diabetes monitoring and therapy. *Nat. Nanotechnol.* **2016**, *11* (6), 566–572.

103. Kim, D. H.; Lu, N.; Ma, R.; Kim, Y. S.; Kim, R. H.; Wang, S.; Wu, J.; Won, S. M.; Tao, H.; Islam, A.; Yu, K. J.; Kim, T. I.; Chowdhury, R.; Ying, M.; Xu, L.; Li, M.; Chung, H. J.; Keum, H.; McCormick, M.; Liu, P.; Zhang, Y. W.; Omenetto, F. G.; Huang, Y.; Coleman, T.; Rogers, J. A. Epidermal electronics. *Science* **2011**, *333* (6044), 838–843.

104. Xu, S.; Zhang, Y.; Cho, J.; Lee, J.; Huang, X.; Jia, L.; Fan, J. A.; Su, Y.; Su, J.; Zhang, H.; Cheng, H.; Lu, B.; Yu, C.; Chuang, C.; Kim, T. I.; Song, T.; Shigeta, K.; Kang, S.; Dagdeviren, C.; Petrov, I.; Braun, P. V.; Huang, Y.; Paik, U.; Rogers, J. A. Stretchable batteries with self-similar serpentine interconnects and integrated wireless recharging systems. *Nat. Commun.* **2013**, *4*, 1543.

105. Jang, K. I.; Li, K.; Chung, H. U.; Xu, S.; Jung, H. N.; Yang, Y.; Kwak, J. W.; Jung, H. H.; Song, J.; Yang, C.; Wang, A.; Liu, Z.; Lee, J. Y.; Kim, B. H.; Kim, J. H.; Lee, J.; Yu, Y.; Kim, B. J.; Jang, H.; Yu, K. J.; Kim, J.; Lee, J. W.; Jeong, J. W.; Song, Y. M.; Huang, Y.; Zhang, Y.; Rogers, J. A. Self-assembled three dimensional network designs for soft electronics. *Nat. Commun.* **2017**, *8*, 15894.

106. Ma, R.; Kang, B.; Cho, S.; Choi, M.; Baik, S. Extraordinarily high conductivity of stretchable fibers of polyurethane and silver nanoflowers. *ACS Nano* **2015**, *9* (11), 10876–10886.

107. Wei, Y.; Chen, S.; Yuan, X.; Wang, P.; Liu, L. Multiscale wrinkled microstructures for piezoresistive fibers. *Adv. Funct. Mater.* **2016**, *26* (28), 5078–5085.

108. Lipomi, D. J.; Vosgueritchian, M.; Tee, B. C. K.; Hellstrom, S. L.; Lee, J. A.; Fox, C. H.; Bao, Z. N. Skin-like pressure and strain sensors based on transparent elastic films of carbon nanotubes. *Nat. Nanotechnol.* **2011**, *6* (12), 788–792.

109. Cai, L.; Song, L.; Luan, P. S.; Zhang, Q.; Zhang, N.; Gao, Q. Q.; Zhao, D.; Zhang, X.; Tu, M.; Yang, F.; Zhou, W. B.; Fan, Q. X.; Luo, J.; Zhou, W. Y.; Ajayan, P. M.; Xie, S. S. Super-stretchable, transparent carbon nanotube-based capacitive strain sensors for human motion detection. *Sci. Rep.* **2013**, *3*, 3048.

110. Yao, S. S.; Zhu, Y. Wearable multifunctional sensors using printed stretchable conductors made of silver nanowires. *Nanoscale* **2014**, *6* (4), 2345–2352.

111. Cooper, C. B.; Arutselvan, K.; Liu, Y.; Armstrong, D.; Lin, Y.; Khan, M. R.; Genzer, J.; Dickey, M. D. Stretchable capacitive sensors of torsion, strain, and touch using double helix liquid metal fibers. *Adv. Funct. Mater.* **2017**, 1605630.

112. Kaltenbrunner, M.; Adam, G.; Glowacki, E. D.; Drack, M.; Schwodiauer, R.; Leonat, L.; Apaydin, D. H.; Groiss, H.; Scharber, M. C.; White, M. S.; Sariciftci, N. S.; Bauer, S. Flexible high power-per-weight perovskite solar cells with chromium oxide-metal contacts for improved stability in air. *Nat. Mater.* **2015**, *14* (10), 1032–1039.

113. Chen, T.; Peng, H. S.; Durstock, M.; Dai, L. M. High-performance transparent and stretchable all-solid supercapacitors based on highly aligned carbon nanotube sheets. *Sci. Rep.* **2014**, *4*, 3612.

114. Padmajan Sasikala, S.; Lee, K. E.; Lim, J.; Lee, H. J.; Koo, S. H.; Kim, I. H.; Jung, H. J.; Kim, S. O. Interface-confined high crystalline growth of semiconducting polymers at graphene fibers for high-performance wearable supercapacitors. *ACS Nano* **2017**, *11* (9), 9424–9434.

115. Yu, J.; Lu, W.; Smith, J. P.; Booksh, K. S.; Meng, L.; Huang, Y.; Li, Q.; Byun, J. H.; Oh, Y.; Yan, Y.; Chou, T.-W. A high performance stretchable asymmetric fiber-shaped supercapacitor with a core-sheath helical structure. *Adv. Energy Mater.* **2017**, *7* (3), 1600976.

116. Chen, T.; Hao, R.; Peng, H.; Dai, L. High-performance, stretchable, wire-shaped supercapacitors. *Angew. Chem.* **2015**, *54* (2), 618–622.

117. Xu, P.; Gu, T.; Cao, Z.; Wei, B.; Yu, J.; Li, F.; Byun, J. H.; Lu, W.; Li, Q.; Chou, T.-W. Carbon nanotube fiber based stretchable wire-shaped supercapacitors. *Adv. Energy Mater.* **2014**, *4* (3), 1300759.

118. Zhang, Z.; Deng, J.; Li, X.; Yang, Z.; He, S.; Chen, X.; Guan, G.; Ren, J.; Peng, H. Superelastic supercapacitors with high performances during stretching. *Adv. Mater.* **2015**, *27* (2), 356–362.

119. Wang, J. X.; Yan, C. Y.; Chee, K. J.; Lee, P. S. Highly stretchable and self-deformable alternating current electroluminescent devices. *Adv. Mater.* **2015**, *27* (18), 2876–2882.

120. Pyo, J. B.; Kim, B. S.; Park, H.; Kim, T. A.; Koo, C. M.; Lee, J.; Son, J. G.; Lee, S. S.; Park, J. H. Floating compression of Ag nanowire networks for effective strain release of stretchable transparent electrodes. *Nanoscale* **2015**, *7* (39), 16434–16441.

121. Chen, P.; He, S.; Xu, Y.; Sun, X.; Peng, H. Electromechanical actuator ribbons driven by electrically conducting spring-like fibers. *Adv. Mater.* **2015**, *27* (34), 4982–4988.

122. Wang, R.; Xu, Z.; Zhuang, J.; Liu, Z.; Peng, L.; Li, Z.; Liu, Y.; Gao, W.; Gao, C. Highly stretchable graphene fibers with ultrafast electrothermal response for low-voltage wearable heaters. *Adv. Electron. Mater.* **2017**, *3* (2), 1600425.

123. An, B. W.; Gwak, E. J.; Kim, K.; Kim, Y. C.; Jang, J.; Kim, J. Y.; Park, J. U. Stretchable, transparent electrodes as wearable heaters using nanotrough networks of metallic glasses with superior mechanical properties and thermal stability. *Nano Lett.* **2016**, *16* (1), 471–478.

124. Hong, S.; Lee, H.; Lee, J.; Kwon, J.; Han, S.; Suh, Y. D.; Cho, H.; Shin, J.; Yeo, J.; Ko, S. H. Highly stretchable and transparent metal nanowire heater for wearable electronics applications. *Adv. Mater.* **2015**, *27* (32), 4744–4751.

# 3

---

# Components and Devices

Neil Graddage

*National Research Council of Canada*

## CONTENTS

## 3.1 Introduction

It could be stated that conventional electronics fabrication is one of humankind's greatest achievements. The ability to fabricate components with nanometre scale features, and to do so with high yield, is a triumph of modern-day collaborative science and engineering. Routinely fabricating devices that can collect, process and communicate data in a variety of ways has enabled countless leaps forward in almost every conceivable aspect of life. However, conventional electronics fabrication can still be limiting in many ways. Restrictions in substrate material, device size and manufacturing methods result in difficulties applying electronic functionalities to flexible, stretchable, or wearable applications. Developments in materials and manufacturing methods in recent years have begun to enable the fabrication of electronics onto a wide range of substrates, which now allows the integration of electronics into such applications.

The aim of this chapter is to give the reader a flavour of what is possible with flexible, stretchable and wearable electronics and some of the challenges therein. This chapter is designed to be an introduction to some of the myriad of electronic components and devices which can be fabricated in such forms. This is not meant to be an exhaustive list of electronic components and devices, nor a comprehensive discussion of any such component. Many of the components and devices discussed in this chapter will be discussed in more detail later in this book. This chapter shall assume some background knowledge, but every effort will be made to ensure accessibility.

## 3.2 Electrical Components and Circuits

Electrical circuit components are discrete devices, which enable the control of electron flow. There are many such components in conventional electronics, many of which can be fabricated in forms suitable for flexible, wearable and stretchable electronics. This section discusses the principles, characterisation and fabrication of a selection of major circuit components.

### 3.2.1 Conductive Traces

The simplest component in a circuit is the conductive trace. As trivial as it may seem, it is an essential element of every electronic circuit. Conductive traces enable electron flow between components, as well as being an essential part of many other components such as inductors or antennae. Many of the principles for fabricating conductive traces for flexible, stretchable and wearable electronics also apply to electrodes within components.

#### 3.2.1.1 Principles, Materials and Fabrication

In order to enable efficient electron flow between two points, a material with low resistivity should be used to create a path between the two points. The trace should have as low a resistance as possible to minimise losses. The resistance is dependent on the resistivity (or the inverse, conductivity) of the material, as well as the volume of the wire itself. If the resistance is too high, losses occur due to Joule heating, which may lead to a potential drop along the trace and eventual failure. For more complex applications where an alternating current may be applied, such as an antenna, the impedance of the trace becomes important.

Regarding the volume of the trace, this will often be dictated by design and fabrication considerations. The minimum size and separation will be a major factor in dictating the integration density of many circuits; therefore, the aim should generally be to minimise these dimensions while maintaining a suitable resistance and fabrication yield. It must be noted that reducing separation between traces may give rise to further issues in more complex circuits, such as induced currents causing crosstalk.

Resistivity is a fundamental material property; therefore, material selection is a major consideration. In order to minimise the trace resistance, selection of a low-resistivity material is crucial. Some common conductive materials used in flexible electronic applications are carbon, silver, gold, copper, or poly(3,4-ethylenedioxythiophene)-poly(styrenesulfonate) (PEDOT:PSS). Material selection depends on performance, processing and cost requirements. The actual resistivity of the material in a trace may deviate from the ideal due to many factors such as impurities, grain boundaries, or material density. Different fabrication techniques may result in varying conductor resistivity due to differences in material processing. For example, silver deposited by thermal evaporation will have a resistivity close to that of bulk silver, as the film will have high purity and low porosity. A silver film produced by screen-printing of a microsilver-based ink will have a much higher resistivity, as the final film will be a composite of silver flakes and polymer binder, with the total resistivity being affected by inter-particle contact effects. The resistivity of printed materials is often quoted as a multiple of the bulk material resistivity, which is a useful method to rapidly assess the purity and porosity of the printed trace.

The addition of flexibility and stretchability as factors introduces several additional considerations. The act of flexing or stretching a trace will affect its shape and, hence, the resistance. In addition, when applying stress or strain to a material, it is likely that the resistivity will be affected. For example, flexing of a trace made from a microparticle ink may affect the contact area between conductive particles and, hence, the effective resistivity of the composite material. Flexing and stretching of conductive materials may also introduce defects such as cracks, which will increase resistance and may be irreversible. In order to reduce the resistance change due to deformation of a trace, specific materials, such as conductive polymers or nanowires, can be employed. Nanowire networks enable flexibility and stretchability by maintaining inter-particle contact during deformation.

Fabrication of conductive features for flexible, stretchable and wearable electronic components can be achieved by a myriad of techniques. The choice of technique will depend on material, performance and cost requirements. Printing techniques such as inkjet, flexography, screen or gravure are commonly employed for deposition of conductive features due to the wide variety of materials available and the compatibility with any planar substrate. Alternative techniques that are compatible with low glass transition temperature $(T_g)$ substrates include thermal evaporation, atomic layer deposition (ALD) and solution-coating techniques, although some of these may require subsequent patterning steps.

It must be noted that the performance of conductive features can also be affected by any processing subsequent to deposition. For example, printed inks typically require a curing step in order to achieve a functional film. The final performance of the component, in this case the resistivity of the material, will depend on this treatment. In particular, nanoparticle-based inks such as nano-silver conductive inks (a common choice for printed conductive features) require specific sintering steps in order to become conductive. This may include thermal treatment or photonic sintering such as near-infrared or pulsed light sintering.

When connecting a conductive trace to an external circuit there are a number of options, including physical connectors (e.g. flexible flat connectors), conductive adhesive or soldering. Each of these options requires suitable traces to connect to. The key challenge is to ensure that the trace is sufficiently adhered to the substrate. If this is not the case then physical connectors may fail due to scratching or soldered connectors may fail due to poor adhesion. Soldering to printed traces can be challenging due to thermal and chemical issues with both the conductor and substrate.

Finally, the ability to fabricate flexible and stretchable electronic traces also enables new rigid form factors for electronics via vacuum or thermoforming. This field is known as in-mould electronics (IME). In these applications, the traces need to be able to withstand the stresses imparted during the moulding process, wherein the substrate is heated above its glass transition temperature and stretched into a specified shape.

### 3.2.1.2 Examples

Conductive tracks are an essential component in every circuit and device; therefore, the printing of such features for flexible electronics has been the subject of much research interest. Many developments are reported with regards to a specific application; however, the findings are often more widely applicable. For example, Chu et al. [1] reported a novel method to fabricate ultra-narrow conductive lines, which use an ink-substrate interaction to embed the lines into the substrate. Such narrow embedded lines are of interest when fabricating transparent electrodes, where they can be used as current collectors to reduce the sheet resistivity of the film, as shown by Zhou et al. [2], who demonstrated a method to fabricate embedded silver networks for flexible lighting. By embedding the lines into the substrate, the surface roughness is reduced, which reduces the risk of shorting in multi-layer devices.

It is desirable to be able to fabricate conductive networks over large areas for mass production of flexible lighting or photovoltaics; to this end, Deganello et al. [3] demonstrated the use of flexography to fabricate non-embedded conductive lines with an average width of 74.6 µm at a printing speed of 5 m/min, as shown in Figure 3.1. Other fabrication methods have also been explored, such as the use of laser printing to achieve a lift-off process at low cost over large areas developed by Gupta et al. [4]. This was used to demonstrate a number of applications, including large area transparent heaters and photovoltaics.

For wearable electronics, flexible conductive traces are often not sufficient, with stretchability often required for many applications. Yokus et al. [5] explored the fabrication of stretchable interconnects by encapsulating a meandering printed silver track between thermoplastic polyurethane films and laminating it onto fabric. Other techniques to achieve stretchability include use of materials such as nanowires to form macroscopic lines, such as those fabricated by Liang et al. [6]. They demonstrated good printability, conductivity and strain performance. Conductivity can also be incorporated into textiles by use of conductive threads; for example, Atwa et al. [7] demonstrated a method for coating a thread in silver nanowires. They claimed good flexibility compared to a commercial alternative, along with stretchability and wash resistance.

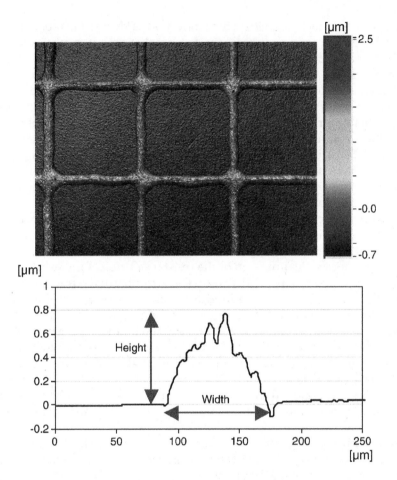

**FIGURE 3.1**
Example of conductive traces printed using flexography. (Reproduced with permission from Deganello et al. [3].)

## 3.2.2 Resistors

Resistors are a ubiquitous circuit component used to control current flow. They are commonly used to limit current, divide voltages or load a circuit. As current flow is impeded, the electrical energy is converted to thermal energy, which can enable resistive heating applications. In addition, they can often be used as a sensing architecture, as the resistance can be affected by a morphology or material change.

### 3.2.2.1 Principles, Materials and Fabrication

The resistance of a component is dependent upon the component size and the material resistivity. Resistors are usually made using a material with high resistivity in order to enable a high resistance to be achieved within a smaller area, increasing the potential integration density of the components. For applications where integration density is not such an issue or a large area is required, resistance can also be increased by increasing the planar aspect ratio of conductive tracks.

Printing of resistor materials, either for fabrication of discrete components or directly onto PCBs, has been a common process for decades; however, the fabrication of resistors onto flexible or stretchable substrates introduces a number of new challenges. Traditionally, resistors are fabricated using screen or stencil printing, using metal oxide-based inks. These inks often need to be cured at temperatures in excess of 800°C, meaning that they are unsuitable for flexible or stretchable substrates. Carbon-based inks with polymer binders are more suitable for such substrates, at the cost of reliability and stability. In addition, other conductive materials such as PEDOT:PSS may be preferable for some applications. Many inks, such as carbon-based screen-printing pastes, will often come in a variety of concentrations which each have a different nominal resistivity. These can then be blended to achieve the desired resistivity. Screen-printing is typically the preferred fabrication method for such components due to the high film thickness possible and ease of integration with PCB manufacturing. Although fabrication of resistive components is possible using a myriad of techniques, the commercial availability of materials formulated for use with printing techniques other than screen-printing is limited at this time.

For many applications, the precision of the resistance value of the resistor is important. This presents a problem for printed resistors, which may have a tolerance of ±20% or more [8]. Post-processing may be required if precise resistor values are necessary. This is commonly achieved using laser trimming, wherein over-sized (with correspondingly lower resistance) devices are fabricated and laser ablation is used to selectively remove material until the resistance is raised to the desired level [9]. The poor tolerance of printed resistors is due to many factors, such as inherent variability in the printing process and ink evolution during processing. Process variation can be minimised in design where possible, such as increasing component dimensions to minimise variation due to line width variation. Component matching to minimise process variation may also be required, depending on the circuit requirements [8].

Resistor precision and stability may be major factors when using the resistors for sensing applications. Resistors can be fabricated to be sensitive to many factors, with common applications including strain and temperature. By applying a strain to a resistor, the morphology of the component will be affected, with a corresponding effect on the resistance value. This is typically characterised in terms of the ratio of resistance change with respect to the mechanical strain, often known as the gauge factor.

The resistivity of a material is also dependent on temperature, which can enable use of resistors as a temperature sensor (known as a thermistor). The dependence of the resistance upon temperature can be positive or negative, though most materials typically used with flexible electronics display a negative temperature coefficient. This dependence is typically linear within a small temperature range. Positive temperature coefficient materials can act as a self-regulating heater, as the temperature rise increases the resistance, which limits the maximum current and hence temperature.

It is also possible to sensitise the resistive material to adsorbed chemicals, enabling a resistance change when exposed to target chemicals. This enables the use of resistor structures for chemical sensing. This tends to be achieved using nanomaterials such as carbon nanotubes or graphene, largely due to the increased surface area resulting in improved sensitivity.

### 3.2.2.2 Examples

As screen-printed resistors have been commonplace in PCBs for some time, much recent research has focused on novel applications. Graddage et al. [8] demonstrated the fabrication

**FIGURE 3.2**
An array of screen-printed resistors, as used in the multiplexor described in [8]. (Image reproduced with permission of the National Research Council Canada, © 2016 Her Majesty the Queen in Right of Canada.)

of screen-printed carbon resistors for use as a multiplexing circuit for digital inputs. By connecting several resistors in parallel with a switch in series with each resistor, the total resistance of the circuit would be dependent on the state of the switches. The limiting factor of this technique was the manufacturing tolerance of the resistor value, as too large a variation would result in overlap between possible total resistance values. They demonstrated a relative standard deviation of 3.7%, which enabled a multiplexing factor of 3. Detailed analysis of resistor printing indicated the major contributor to the variation was resistivity variation caused by solvent evaporation during printing, an issue to which screen-printing is especially prone. Figure 3.2 shows a photograph of an example printed resistor array used in this work.

Common applications for flexible, stretchable and wearable resistors are sensors or heaters. Claramunt et al. [10] deposited carbon nanofibres decorated with metal nanoparticles onto a flexible substrate for use as a resistive sensing architecture. On the underside of the substrate, a resistive heater was printed in order to enable desorption of chemical species from the sensor. It was found that the sensor response time was dependent upon temperature. Strain sensors were fabricated using inkjet printing of a silver ink, with a gauge factor of 21 achieved. Higher gauge factor wearable resistive sensors were reported by Liao et al. [11], who screen printed silver ink onto pre-stretched textiles and demonstrated a gauge factor of ~2000 with a sensing rage of 60% strain. The potential for these sensors in wearable applications was demonstrated by integration of the sensors into a glove, which used finger motion to control various systems. The use of printing to fabricate flexible sensors enables large area devices such as piezoresistive pressure sensor arrays, as demonstrated by Ahmad et al. [12] who developed a system for monitoring pressure distribution in a wheelchair for patient monitoring.

Other application areas have also been developed, including the use of printed resistors as memory. Leppäniemi et al. [13] developed a write-once-read-many memory device, which could be fully printed onto flexible substrates. By depositing nano-silver inks between two electrodes without fully sintering the ink, a high resistance is observed. The ink can then be sintered by means of resistive heating induced by applying a voltage between the two electrodes. This irreversible process reduces the resistance of the structure. The high/low state of the resistance can then be interpreted as a bit of information. This work was initially demonstrated using inkjet printing, but later work demonstrated volume roll-to-roll fabrication using flexography [14].

### 3.2.3 Capacitors

Capacitors are components in which potential energy is stored by an electric field. This is typically achieved by induction of charges on electrodes separated by a dielectric material. They have a myriad of uses in circuits, including energy storage, filtering and sensing.

#### 3.2.3.1 *Principles, Materials and Fabrication*

In the simplest case, capacitors consist of two conductive electrodes separated by an insulating material. When a potential difference is applied across the two electrodes, an electric field develops across the separation region, inducing the two electrodes to become oppositely charged. If the potential difference is removed and no load is applied, an ideal capacitor will store the energy indefinitely. This energy can be discharged by application of a load. This behaviour allows capacitors to block steady potentials while allowing transient/alternating signals to pass, a property known as capacitive coupling. The capacitance of a capacitor is calculated from the overlap area of the electrodes, the gap between the electrodes and the relative permittivity (or dielectric constant) of the material separating the electrodes. In addition to high relative permittivity, the dielectric material should also have a suitable breakdown voltage for the desired application.

Capacitors for thin film flexible, stretchable and wearable applications tend to either have an interdigitated electrode or parallel plate architecture. Interdigitated electrodes consist of two electrodes in the same plane, each having a 'comb'-like shape and interlocking with each other. In this case, the overlap area is restricted to the length and thickness of the electrodes and the electrode separation is limited by the maximum resolution of the patterning process, as illustrated in Figure 3.3a. Thus, capacitors with this architecture tend to have very low capacitance, but are simple to fabricate.

Larger capacitance devices are enabled by the parallel plate architecture. In the simplest case, three layers are required, increasing the complexity of fabrication compared to interdigitated electrodes. These layers consist of two electrodes separated by a dielectric material. The key parameters affecting the capacitance are the overlap area (which may be limited by overall circuit design) and the thickness and relative permittivity of the dielectric material, as illustrated in Figure 3.3b.

Various dielectric materials have been demonstrated in thin film capacitors for flexible devices. For solution-processed devices, the majority of these can be split into two groups, organic polymers or inorganic composites. Organic polymers such as polyvinylpyrrolidone (PVP) [15, 16], polyimide (PI) [17] or poly(methyl methacrylate) (PMMA) [18] have all been used in printed capacitors. It is possible to fabricate thin layers of such materials (as low as 70 nm has been demonstrated [19]); however, the dielectric constant tends to be low (typically, between 2 and 4). Addition of inorganic materials into a polymer composite matrix can enable much higher dielectric constants, but typically results in increased layer thickness and roughness. A popular option is barium titanate ($BaTiO_3$), which has been used to demonstrate dielectric films with a dielectric constant higher than 100 [20].

## (a) Interdigitated Structure

Electrode
Spacing

## (b) Planar Parallel Plate Structure

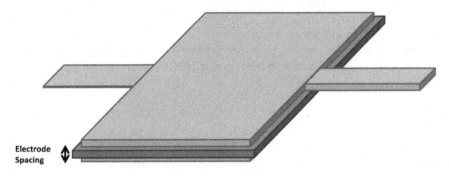

Electrode
Spacing

**FIGURE 3.3**
Illustrations of (a) interdigitated and (b) planar capacitor structures for flexible electronics, with electrode spacing dimension highlighted.

The dielectric properties of these films are dependent upon the particle size, with higher dielectric constants observed when larger particle sizes are used (for particle diameters between 100 and 500 nm [21]). This means that there can be a trade-off between film thickness, roughness and dielectric constant.

Other fabrication processes compatible with flexible substrates may enable alternative materials, such as $Al_2O_3$ or $HfO_2$ [22, 23], which can be grown at low temperatures (between 120 and 300 °C) using ALD.

The choice of dielectric material may also depend on application. For circuit applications, the priority would likely be a material with suitable capacitance, stability and process compatibility. If the capacitor structure is the result of a crossover (crossing of two conductive traces) in a circuit design, then parasitic capacitance should be minimised by use of a thick layer of low dielectric constant material. For force-sensing applications, a material which can enhance the response would be desirable, such as an inorganic composite, which can increase in relative permittivity if the packing fraction increases under applied pressure. Materials which display a high ferroelectric polarizability, such as certain phases of crystalline poly(vinylidene fluoride) (PVDF), enable the use of capacitor structures as data storage [24, 25].

The use of capacitors in flexible or stretchable applications results in further factors to consider. As the capacitance is inversely proportional to the dielectric thickness, the capacitor behaviour will be affected by applied strain. For sensing applications, this may be desirable; for circuit applications, it is likely to be undesirable. Mitigation measures may include tolerant circuit design or mechanical design to minimise component deformation, such as patterning components onto rigid islands separated by stretchable interconnects.

The choice of fabrication technique for production of capacitors for flexible electronics depends on a number of factors, including desired capacitance, size and material selection. Screen-printing has been used for deposition of dielectric materials in traditional capacitor fabrication for decades [26], but fabrication of thin film flexible capacitors introduces new material challenges. Capacitor fabrication has been demonstrated using all major printing techniques, including inkjet [19], screen [27] and gravure [28].

The key challenges in parallel-plate capacitor fabrication are controlling the surface roughness of the bottom electrode and fabricating a thin, uniform and pinhole-free dielectric layer. Regarding the bottom electrode, a smooth film is required to avoid potential shorts between the bottom and top electrodes. Assuming no spike-like features, lower roughness of the bottom electrode enables thinner dielectric layers. Obtaining a thin and uniform dielectric layer without pinholes is challenging for many printing processes and requires careful process control. It may often be necessary to deposit multiple layers of dielectric to minimise the risk of pinholes, at the cost of total layer thickness. Care must also be taken to ensure solvent compatibility between all layers.

### 3.2.3.2 Examples

The key parameters to improve capacitor performance are the relative permittivity and thickness of the dielectric layer. Much research has been performed on dielectric materials for flexible capacitive devices, some of which was discussed previously. There has also been work on device architectures and processes to enable thin and uniform dielectric films. Capacitors fabricated on flexible substrates have been demonstrated in a number of applications. Graddage et al. [19] demonstrated a method to fabricate thin and uniform dielectric layers, which takes advantage of the coffee ring effect, wherein fluid flows to the edge of a feature during drying. This enabled dielectrics as thin as 70 nm to be fabricated, with a capacitance per unit area of 255 pF/mm$^2$. An example structure of such a device is shown in Figure 3.4.

There has also been much research on applications for flexible capacitive devices. Rivadeneyra et al. [29] demonstrated an inkjet-printed humidity sensor using an interdigitated structure on a polyimide film. A similar device was also previously demonstrated using gravure printing by Reddy et al. [30]. The development of pressure-sensing arrays is of significant interest for many wearable electronics applications such as smart skins. Such a pressure-sensing array was demonstrated by Woo et al. [31] who developed highly elastic conductive ink for use as the electrodes in a capacitor array, and a silicone elastomer as the dielectric layer. These were fabricated using micro-contact printing and spin-coating.

The use of ferroelectric materials as the dielectric enables the use of capacitors as data storage devices. Such devices were demonstrated by Bhansali et al. [32], who inkjet printed poly(vinylidenefluoride-trifluoroethylene) P(VDF-TrFE) in capacitor arrays with PEDOT:PSS electrodes to form transparent organic ferroelectric memory devices. These devices showed a remnant polarisation of 6.5 μC/cm$^2$, which after $10^5$ stress cycles retained 45% of its initial value.

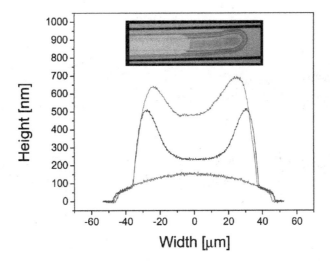

**FIGURE 3.4**
Cross-sectional profile of an inkjet-printed capacitor. The coffee-ring effect enables a thin dielectric layer at the centre. Inset: Optical microscope image of such a device. (Authors' work reproduced from [19], image © 2016 Her Majesty the Queen in Right of Canada.)

### 3.2.4 Transistors

Transistors are active components, which allow for switching or amplifying of electrical signals. A variety of transistor architectures are available; however, with regards to flexible electronics the thin-film transistor (TFT) is ubiquitous, as it can be substrate independent. Other device architectures such as vertical field effect transistors (VFETs) have shown promise [33], but the traditional TFT architecture is the most common.

#### 3.2.4.1 Principles, Materials and Fabrication

In its simplest form, a printed TFT includes three electrodes, commonly referred to as the source, drain and gate. The conductivity between the source and drain electrodes can be modulated by the application of a potential difference between the gate and the source electrodes. In traditional transistors, the behaviour of the device as a result of this electric field is dependent on the majority charge carrier in the semiconductor. If the majority charge carriers are electrons, the semiconductor is considered n-type and, therefore, the application of a positive voltage to the gate electrode is required to induce mobile charges in the semiconductor film. If the majority charge carriers are holes, the material is considered p-type and, therefore, the application of a negative voltage is required to induce mobile charges. Some semiconductor materials display ambipolar characteristics, meaning that mobile charges can be induced by positive or negative voltage. This is typically undesirable for circuit applications, though it is sometimes possible to quench ambipolar characteristics [34]. This terminology remains used for novel semiconductor-based devices, such as organic TFTs, even if the definition is not strictly accurate due to different charge transport mechanisms [35].

The simplest structure of a TFT device consists of four layers, as shown in the diagram in Figure 3.5. This includes an electrode layer comprised of the source and drain electrode which are separated by a gap (known as the channel length), which is bridged with a semiconducting material. The channel length and width dimensions are illustrated further

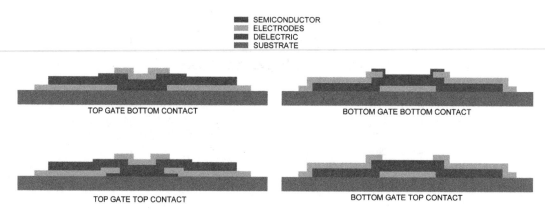

**FIGURE 3.5**
Common TFT structures.

in Figure 3.6. This is then insulated from a gate electrode layer by a dielectric. Aside from the dielectric, the arrangement of these layers can vary depending on the process and application. The gate electrode may be at either the bottom or top of the stack, and the source/drain electrodes may be fabricated above or below the semiconducting layer. Top gate configurations are often preferred for organic semiconductor-based TFTs, as the structure inherently provides some encapsulation for the semiconductor layer. However, layers fabricated subsequently to the semiconductor layer must be compatible with this layer to ensure optimal performance. If the semiconductor layer is sensitive to subsequent fabrication steps then a bottom gate configuration may be preferred. This may also be the case if the device is to be used for sensing applications. Note that the layer order may have a significant effect on device performance due to material interactions and interface effects.

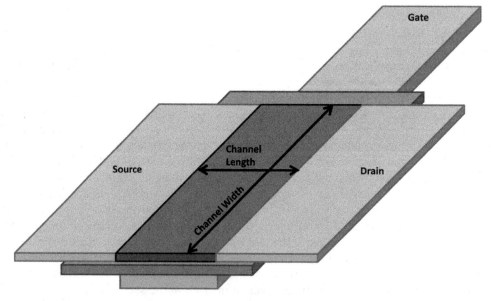

**FIGURE 3.6**
Illustration showing a bottom gate TFT highlighting the channel length and width dimensions.

A large number of parameters are used to characterise TFTs, with mobility and on/off voltage commonly used as key metrics. Mobility is a material property, which indicates the ease at which a charge carrier can move through the material in the presence of an electric field. In the case of TFT characterisation, the mobility quoted is often calculated from the device performance; hence, this should be referred to as the 'device mobility' as the calculation can be affected by a myriad of device factors. Higher device mobility allows for higher transconductance, enabling higher operating frequencies. The on/off ratio is simply the ratio of the current between the source and drain electrodes when the TFT is in an on or off state. This is a major factor in many flexible TFT devices, as the semiconducting materials used often have significant conductivity when the device is in the off state. A low on/off ratio will limit potential circuit complexity and may result in increased power consumption. Because these parameters are determined from device characterisation results, many factors can influence them. Care should be taken during device characterisation and parameters should be reported in accordance with established standards [36].

Each layer has unique challenges for fabrication and materials selection; however, there are many parallels with components previously discussed. Materials selection is inherently a compromise between performance, fabrication process and compatibility with other layers.

The electrode layers are a prime example of this compromise. Fabrication using solution-processing methods restricts available conductive materials to those described in the prior conductive trace section, with silver being the most common option for TFT applications. However, silver may not be an ideal electrode material for many semiconductors if an ohmic contact cannot be formed. This issue can sometimes be reduced by use of interfacial layers to minimise any potential barrier at the junction, at the cost of increased device fabrication complexity [37]. In addition to material issues, the use of solution processing may limit the resolution of the electrode design, as printing narrow gaps between features can be challenging. This results in larger channel lengths, which has a corresponding negative effect on device performance. This may be partially compensated for by selection of semiconductor materials that are more suited to larger channel lengths. Smaller channel lengths and more suitable electrode materials may be possible using other deposition technologies compatible with flexible and stretchable substrates, such as thermal evaporation; however, these have their own drawbacks in terms of manufacturing throughput and cost.

Fabrication of small channel lengths also introduces further manufacturing challenges, chiefly of which is the avoidance of parasitic capacitance. Fabrication of the gate electrode requires that the electrode be aligned with the channel, while avoiding excessive overlap with the source and drain electrodes. Any overlap causes parasitic capacitance, which negatively affects the switching speed. This can be especially problematic in solution-processed electronics; printing methods can be limited in terms of feature size and registration accuracy. A number of alignment methods have been proposed to solve this, including surface modification [38, 39] and imprinting [40, 41]; however, these all introduce further fabrication complexities. Regarding materials, the key challenge with the gate electrode is compatibility with prior layers in a top gate configuration, in particular solvent orthogonality with the dielectric layer is essential to avoid damage or potential shorts.

The dielectric layer faces similar material and fabrication issues as the equivalent layer in capacitors discussed previously, with the addition of interface challenges. Carrier traps and surface dipoles at the interfaces (and in the bulk) may negatively affect device performance. These effects may be reduced by addition of interfacial layers, at the cost of increased complexity [42]. Again, as with any multilayer device fabrication, care must be taken to ensure compatibility with other layers in the stack, both prior and subsequent to the dielectric layer. This includes solvent orthogonality (for solution-processed layers) and ensuring suitable surface properties (e.g. roughness, surface energy) for subsequent layers.

The semiconductor layer provides a number of unique challenges for fabrication of devices on flexible substrates. Semiconductor materials for these applications can broadly be split into three groups: organic semiconductors, carbon nanostructures and inorganic semiconductors. Material selection is based on desired performance, application and fabrication process.

Organic semiconductors can be based on small molecules or conjugated polymers. Vapour phase processing is usually used for small molecule deposition, whereas conjugated polymers can be deposited in vapour or solution phase. The array of available materials is vast, with some common options being pentacene or poly(3-hexylthiophene-2,5-diyl) (P3HT)-based polymers [43]. A full review of organic semiconductor materials is beyond the scope of this chapter; however, the there are several excellent review articles which discuss the subject [44–46]. The major advantage of organic semiconductors is the ease of solution processing, with many semiconducting polymers showing good solubility (although chlorinated solvents may be required). In addition, organic semiconductors have also demonstrated good flexibility [47]. However, there are several limitations, chiefly that organic semiconductors are limited in mobility, with values between 0.001 and 10 cm²/Vs common [43]. Device mobility is heavily dependent upon the film formation, with higher mobility devices typically achieved with crystalline films, at the cost of process sensitivity and device-to-device uniformity [45]. Amorphous films are simpler to process, but demonstrate significantly lower mobility. Also, stability is often an issue with organic semiconductors, especially when exposed to oxygen and water vapour [48]. However, the mobility values may be sufficient for many applications and stability issues can be minimised by material selection and encapsulation.

Carbon nanostructures have shown significant promise for use in TFTs. In particular, devices based on carbon nanotubes (CNTs) or graphene have been demonstrated, which have been able to utilise some of the exceptional properties of these nanomaterials on a more macroscopic scale. Although it is possible to grow these nanostructures and transfer them onto flexible substrates [49], most devices demonstrated for flexible applications have been fabricated by deposition from solution. CNTs are comprised of carbon atoms arranged in a hexagonal lattice, in the form of a cylinder. The lattice can form a cylinder in multiple ways, which is expressed as the chirality. The chirality determines the electrical properties of the nanotube, with a third of nanotubes having metallic properties and the remaining two-thirds being semiconducting. A number of methods have been developed to either selectively produce or to separate these two groups; however, 100% purity is unlikely to be possible [50]. Individual semiconducting CNTs have been reported to have mobility [51] as high as 79,000 cm²/Vs, but manufacturing devices with single nanotubes is a challenging task. All CNT-based TFT devices for flexible electronics have been based upon the formation of a percolating network of CNTs between the sources and drain electrodes.

Although CNTs can be difficult to disperse in solution, there are a number of methods to achieve this which enable solution deposition methods to be employed in device fabrication [52]. The density of the network has a major influence on device performance, with the critical density where a path can form along the CNTs between two points known as the percolation threshold. If the network density drops below the percolation threshold, the device will not be functional. If the network density is low but above the percolation threshold, the on current will be restricted. If the network density is too high, even the small number of metallic CNTs in a purified network could provide a conducting pathway between the electrodes. As a result, CNT-based devices are also sensitive to the channel length, as the shorter the channel length the higher the chance of a metallic pathway through the network. Therefore, it may be preferable to use larger channel lengths for CNT-based devices, depending on material purity. CNT-based TFTs fabricated using solution processing have demonstrated [53] mobilities of 43 $cm^2$/Vs with an on/off ratio of $10^4$. The negative effects of metallic impurities can also be somewhat mitigated by designing circuits with increased tolerance to such effects, as demonstrated by Hill et al. [54] in their (rigid) CNT microprocessor demonstration.

Like CNTs, graphene is also a structure comprised of carbon atoms in a hexagonal lattice; however, unlike CNTs it is in a planar form. Graphene is a zero bandgap semiconductor, often referred to as a semi-metal. Because of this, the on/off ratio of graphene-based devices may be low, unless the graphene is modified to engineer a bandgap (commonly by doping or spatial confinement [55]). Like CNTs, the mobility of graphene is very high (over 200,000 $cm^2$/Vs for suspended sheets [56]), although, again this does not automatically translate to high device mobility. Interest in graphene as a material for use in flexible TFTs rose significantly when it was demonstrated that graphene can be manufactured in solution [57]. This approach was used to develop inkjet printable graphene solutions for use in TFT devices [58]. Graphene has also been deposited on substrates suitable for flexible electronics via a transfer process, demonstrating flexible devices [59].

In addition to graphene, recent research has demonstrated fabrication of a number of other 2D materials in solution. This has resulted in printed TFTs using semiconducting materials such as $MoS_2$ and $WS_2$ [60].

Inorganic-oxide-based semiconductors have shown significant promise in TFT structures for flexible electronics. They offer a number of advantages, with the potential for high-performance and good stability compared to organic devices. Such materials have been rapidly commercialised on rigid substrates and are typically deposited using sputtering before annealing at temperatures in excess of 400°C [61]. However, significant processing and material challenges remain before this performance can be fully exploited at low processing temperatures. Similar to organic semiconductors, these materials are typically employed in either crystalline or amorphous films. Crystalline films can be formed on flexible substrates via nanoparticle or precursor routes. Nanoparticle solutions can be simpler to deposit due to reduced sensitivity to processing parameters and lower temperature processing. However, the performance of such devices is typically lower than equivalent films formed from precursors, largely due to impurities in the film resulting from additives required to form a stable and printable solution. Some common crystalline-oxide-based semiconductor nanoparticle materials include ZnO and $In_2O_3$ for n-type semiconductors. Note that crystallinity (and hence performance) tends to be highly process dependent. Devices processed at temperatures below 200°C tend to show mobility values between 1 and 5 $cm^2$/Vs [62] for both ZnO and $In_2O_3$ films, with values over 100 $cm^2$/Vs reported for $In_2O_3$ at higher processing temperatures.

Amorphous semiconductor films may be employed to reduce processing sensitivity and improve device uniformity, however this is generally at the expense of performance.

Common solution-processed routes to achieve this include aqueous precursors or deep UV photochemical activation of sol-gel films. The aqueous route has demonstrated ZnO [63] and $In_2O_3$ [64] films processed at below 125°C with mobilities between 2 and 3 cm$^2$/Vs. The UV photochemical activation route has demonstrated indium gallium zinc oxide (IGZO) [65], indium zinc oxide (IZO) [66] and $In_2O_3$ [66] films processed at below 150°C with mobilities as high as 34 cm$^2$/Vs.

In addition to the aforementioned processing challenges, another key issue facing the adoption of inorganic-oxide-based semiconductors is the lack of suitable p-type materials of comparable performance to the demonstrated n-type materials. Some solution-processable p-type materials include $Cu_2O$ or NiO; however, few reports of processing at temperatures below 200°C are available. Liu et al. [67] demonstrated solution-processed $NiO_x$ devices fabricated at temperatures as low as 150°C. However, the lowest processing temperature at which functional devices were fabricated was 200°C, resulting in mobility values of 0.004 cm$^2$/Vs, rising to 0.078 at 300°C. The lack of suitable p-type materials and process sensitivity is a hindrance to the development of efficient complementary logic circuits; however, unipolar logic can be employed (which will be discussed further in Section 3.2.5.1.

Deposition of semiconductor films can be challenging for a multitude of reasons. Semiconductor material solutions tend to have a limited window of rheological properties, typically due to solubility limitations. The use of rheological modifiers will add impurities to the film, degrading performance. Therefore, semiconductor material solutions tend to have viscosities that are below the ideal range for many printing processes. As such, techniques such as inkjet have proven popular for semiconductor deposition as the typical ink viscosity of an inkjet ink is in the range of 1 to 10 mPa.s [68]. However, deposition of high aspect ratio nanomaterials such as CNTs can be challenging for inkjet due to nozzle clogging issues. There is much research effort dedicated to improving the morphology of printed films for semiconductor applications [69, 70].

Finally, it must be highlighted that solution processing is not the only route for low-temperature deposition of semiconductor films. For example, ALD has also been shown to be a promising method for oxide-based devices [71] and vapour-phase evaporation is common for small molecule organic semiconductors [45].

### 3.2.4.2 Examples

The development of TFTs for flexible electronics has been a subject of intense research interest for many years. Garnier et al. [72] first demonstrated an all-polymer all-printed TFT on a flexible substrate in 1994. The key enabler for flexible devices was the use of processing techniques that do not expose the substrate to high temperatures. The use of printing techniques has been of particular interest due to the additive nature and potential throughput; with inkjet [19, 72–75], flexography [76, 77], direct gravure [78, 79], reverse off-set gravure [80], screen [27], aerosol [81, 82] and electrohydrodynamic jet [83, 84] all being demonstrated as possible techniques for use in TFT fabrication.

Such a wide variety of materials and processes enables novel transistor characteristics, including flexibility, transparency and stretchability. Highly flexible TFTs are of great interest in the development of wearable electronics and electronic skin. Kaltenbrunner et al. [85] demonstrated what they called 'imperceptible plastic electronics' by fabricating organic TFTs on 1.2 μm-thick PEN substrates. By lamination onto a pre-strained elastomer and ensuring that the devices were at the neutral plane position, functional devices were shown at 5 μm bend radius and up to 223% strain. Fukuda et al. [86] managed to achieve devices on a similar thickness substrate using solution-processing methods and showed stable device characteristics at up to 50% compressive strain.

Fully transparent TFTs are of particular interest for display applications and, therefore, have been a research goal for a number of groups. Many suitable semiconductor materials show good transparency in visible wavelengths, including ZnO and low-density CNT networks. For example, Fortunato et al. [87] demonstrated transparent TFTs fabricated using low temperature processes with a sputtered ZnO active layer. They used indium tin oxide (ITO) and gallium-doped zinc oxide (GZO) for electrodes, with a dielectric consisting of a superlattice of $Al_2O_3$ and $TiO_2$. The downside to metal oxide devices is usually limited flexibility; therefore, for flexible applications CNT networks may be preferred, as demonstrated by Cao et al. [88] who used transfer printing to deposit grown CNT networks onto PET substrates. The CNT networks were used at varying density for both electrodes and semiconductor. Jang et al. [89] have demonstrated fabrication of such devices by inkjet printing.

The plethora of available materials enables other unique functionalities. Use of a ferroelectric material as the dielectric allows for use of TFTs as memory storage devices, not dissimilar in principle to the ferroelectric capacitors described previously. The main advantage of using an FET structure is the memory can be read non-destructively, at the cost of memory retention time compared to a capacitive structure. Park et al. [90] demonstrated solution-processed organic bottom gate TFTs with a polystyrene-brush polymer electret layer on the dielectric and a blended semiconductor. They extrapolated an impressive charge retention time of over 10 years, but this did not consider organic material degradation. Jung et al. [91], who fabricated the TFTs on rigid islands on an elastomer substrate, have also fabricated TFT-based memory devices in stretchable forms.

Functionalisation of the semiconductor layer in a device with a bottom gate structure enables use of TFTs for sensing applications, as chemical adsorption onto the semiconductor can affect charge transport. In particular, nanoparticle semiconductors such as CNTs or graphene have demonstrated promise in this field due to the large surface area available for sensing. Roberts et al. [92] deposited semiconducting CNTs from solution to form TFT devices, which were used to detect 2,4,6-trinitrotoluene (TNT) and dimethyl methylphosphonate (DMMP) in aqueous solutions. One major issue with such devices is poor selectivity, Fu et al. [93] used a combination of CNT-based TFTs and resistors to form an 'electronic nose' to differentiate between carbon monoxide and nitric oxide.

Compared to flexible device fabrication, the field of stretchable electronics is still in its infancy. As discussed previously, Kaltenbrunner et al. [85] demonstrated stretchable devices by engineering the substrate and device structure. Further increase in stretchability requires materials which are inherently stretchable. To that end, Oh et al. [94] developed a stretchable organic semiconducting polymer which enabled fabrication of TFT arrays. These arrays could be stretched by 100% with a corresponding reversible linear decrease in device mobility observed. Devices were fabricated using a combination of spin coating and thermal evaporation.

Development of circuits using flexible and stretchable transistors will improve integration into wearable applications compared to discrete rigid electronics. Further improvements could be achieved by integration of the devices directly into textiles. One way to achieve this is the use of a planarization layer on the textile to enable direct fabrication, which is the approach taken by Carey et al. [95]. Using inkjet printing to deposit a graphene channel and boron nitride dielectric, they formed a 2D heterojunction device, which showed good performance, even after bending and washing tests. Rather than use the textile as a substrate, Kim et al. [96] developed a conductive textile, which was used as the gate electrode in a flexible organic TFT. Scalable realisation of fibre-based TFTs is challenging due to the performance dependence on certain device dimensions (such as dielectric thickness). To overcome this, Hamedi et al. [97] investigated the use of wire electrochemical transistors. They patterned electrodes with regular gaps onto a fibre before coating with an organic semiconductor. Another fibre acts as the gate electrode, with an ionic liquid as the dielectric.

### 3.2.5 Integrated Circuits

An integrated circuit is the combination of electrical components onto a common substrate that would otherwise have been made of discrete components. Common functions include collection, processing or communication of data.

#### 3.2.5.1 Principles, Materials and Fabrication

Integrated circuits are often grouped into two categories, digital or analogue. Analogue circuits process continuously variable signals, whereas digital circuits process discrete values. Analogue circuits are more challenging from a circuit component perspective, due to the component consistency and stability required for reliable operation. Digital circuits can be designed to tolerate some variations; however, it must be noted that this limits circuit complexity. As a result, most demonstrations of circuits fabricated on flexible substrates have been digital circuits.

The simplest implementations of a digital circuit are logic gates. These are the building blocks of all further digital circuits and are an electronic representation of Boolean logic. There are a number of ways to achieve logic gates electronically, the choice of which depends on the available components, in particular the transistor. Common families of logic gates are PMOS, NMOS or CMOS, referring to the use of p-type semiconductor devices, n-type semiconductor devices or combinations of the two (complementary), respectively. Note that the MOS (metal-oxide-semiconductor) acronym is not always technically correct for TFTs; however, the term is still commonly used. PMOS and NMOS logics are simpler to manufacture, at the cost of higher static power dissipation compared to CMOS logic. Due to fabrication and stability challenges, most demonstrated flexible circuits have used either PMOS or NMOS logic. Most logic gates require between one and four transistors. These gates can then be used to construct exponentially more complex circuits, with a corresponding increase in transistor count. Note that designing fault tolerance into a circuit increases the transistor count further. This poses a significant challenge for flexible electronics as it stands today, as obtaining a high yield of consistent devices is difficult.

A popular application area for flexible circuits is displays. The desire for durable, flexible displays requires the development of a flexible active matrix backplane to address individual pixels, as passive matrix-based displays typically have undesirable crosstalk. A number of architectures exist to achieve this, the simplest being an array of pixels, each containing one transistor. The array will be connected to the gate electrode in one axis, while in the other axis the array will connect to the source electrode. Each pixel can therefore be individually addressed. This method is suitable for any pixel, which is stable without applied energy, such as an electrophoretic display. The use of emissive pixels, such as organic light-emitting diodes (OLEDs), requires energy to be stored to

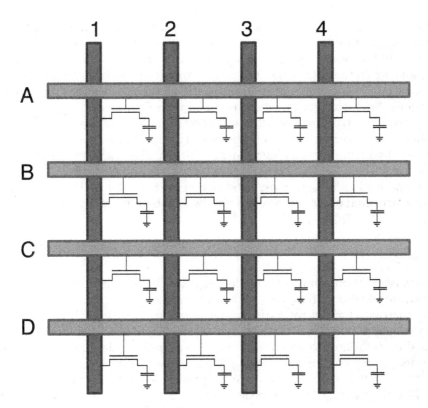

**FIGURE 3.7**
Illustration of a simple active matrix layout.

power the pixel between addressing, in which case a capacitor is commonly included at each pixel, as shown in Figure 3.7. Fabrication of high-resolution active matrix backplanes using methods common in flexible electronics is a challenge. This is largely due to two issues—integration density and transistor performance. The pixel size is directly proportional to the size of the components required; in the simplest case of a bi-stable electrophoretic display, this means the transistor. At the same time, the transistor must be capable of suitable performance to drive the pixel, in terms of speed, power and stability. Transistors fabricated using techniques common in the flexible electronics field tend to have large dimensions due to poor patterning resolution and semiconductor performance.

Fabrication of complex circuits requires a robust and systematic design process. Circuit design in traditional electronics is a mature field, with many tools developed to enable design, simulation and verification of circuits of varying complexity. The success of silicon electronics is due in no small part to the decoupling of the design and manufacturing processes, which enables designers to focus on circuit development. This is not yet the case for flexible electronics, where materials and processes are still too immature. When this decoupling happens in flexible electronics significant and exciting leaps can be expected. There is much research effort devoted to improving this situation, including developing predictive models of device and process behaviour [98, 99]. However, significant progress in this field will not occur without a concerted effort centred on a specific selection of materials and processes.

Note that this discussion has centred on the fabrication of integrated circuits directly onto flexible substrates. It is also common to use rigid Si-based electronics in flexible applications, which can be integrated in a number of ways. The circuit can be connected as a rigid section of the device, as is common in wearable electronics where a detachable circuit is the simplest solution to the challenge of surviving harsh washing conditions. Such circuits can then interface with flexible or stretchable elements. Flexible Si-based electronics is possible, either by thinning a crystalline wafer or growing an amorphous layer on a flexible substrate. Thinning of a crystalline wafer can be accomplished by etching the backside and results in a wafer that is flexible, if fragile [100]. Amorphous Si (a-Si:H) can be grown at temperatures below the $T_g$ of most flexible substrates, and is already commonly used in display backplanes. However, challenges remain for fabrication of flexible devices on low-$T_g$ substrates, primarily related to processing and stability [101].

### 3.2.5.2 Examples

A wide variety of flexible and stretchable integrated circuits have been demonstrated. Common demonstration circuits for such research include logic gates, ring oscillators and active matrices. One of the earlier notable examples is the work of Gelinck et al. [102] who demonstrated flexible active matrix electrophoretic displays and shift registers based on solution-processed organic TFTs on polyimide film. The devices were fabricated using spin coating and thermal evaporation, patterned using photolithography. The active matrix consisted of an array of $64 \times 64$ pixels measuring 5 cm × 5 cm and could be driven at 50 Hz. In order to reduce the number of interconnects, the authors also demonstrated shift registers, including a 32-stage device containing 1888 TFTs, which operated at a clock frequency of 5 kHz. Developing complementary logic circuits proved challenging due to the need to have well-balanced device characteristics, Baeg et al. [103] demonstrated inkjet-printed complementary logic organic TFTs, which enabled inverters with gain values over 30 and ring oscillators with a frequency of approximately 50 kHz.

Significant research effort has centred on developing high-throughput fabrication processes for flexible circuits. The seminal work in this field was published by Sirringhaus et al. [73], who demonstrated fully inkjet-printed inverters, the fabrication of which included vias (using solvent dissolution) and printed resistors. Similar devices have been subsequently produced using gravure and screen-printing [27]. Of particular note has been the work presented by Jung et al. [104] who demonstrated fully printed 13.56 MHz 1-bit RF tags on a flexible substrate. Such a device includes an antenna, diodes, transistors, capacitors and resistors. These were fabricated using a combination of roll-to-roll gravure, inkjet and pad-printing techniques. The same group has also demonstrated fully gravure printed circuits, with Noh et al. [105] demonstrating D flip-flops, albeit with less-than desirable performance. They noted that performance could be improved by reducing parasitic capacitances caused by challenges with layer registration. One solution to this is to use self-alignment techniques to minimise electrode overlap, which has been demonstrated in a roll-to-roll environment [106].

For a circuit to perform more complex data-processing tasks, some form of memory is often required. To this end, Ng et al. [107] demonstrated nonvolatile memory arrays consisting of addressing transistors and ferroelectric memory transistors printed onto flexible PET substrates. Memory retention time was at least 8 hours, with extrapolated data indicating 1 month was possible.

The challenges inherent in developing flexible electronic circuits are considerable. The ability to stretch such circuits is a significant additional complication; however, there is motivation to meet this challenge as it would be a major enabler for technologies such as artificial skin and conformable electronics. As previously discussed, a common technique is to have rigid 'islands' within the stretchable device and locate any complex electronics within this region [108]. Stretchable Si electronics have been demonstrated by transferring thin Si ribbons to a pre-strained elastomer substrate [109–111]; however, this is a complex and costly process. Sekitani et al. [112] demonstrated a 16 × 16 stretchable active matrix display by integrating printed stretchable CNT-based interconnects with transistors, capacitors and OLEDs fabricated using vacuum evaporation, CVD and laser pattering. These devices could be stretched by up to 50% without negative effects. There has been little demonstration of more complex stretchable circuits fabricated by high throughput solution-processing methods.

It must be noted that the field of flexible circuits is relatively mature in age if not technology. Therefore, there has been a number of companies, which are commercialising the fabrication of flexible circuits. At the time of writing, such companies include Samsung, LG, PragmatIC, FlexEnable and ThinFilm, among many others.

## 3.3 Photovoltaics

Any functional electronic circuit or device requires a power source. Power can be sourced from either a storage device, such as a battery, or from a generating device, such as a photovoltaic cell. Flexible energy storage is a highly researched area, with flexible printed batteries being a mature technology. However, energy capacity is limited and secondary (rechargeable) battery cells have proven challenging [113]. Therefore, for many applications a flexible energy generation solution is required, for which photovoltaic devices are commonly used.

### 3.3.1 Background

Photovoltaic devices are those which use the photovoltaic effect to convert energy from light into electricity. In its simplest form, this occurs when photons are absorbed by a semiconductor, which excites an electron from the valence band to the conduction band where the charge can be driven to an electrode. The charge can then be used to perform work. The charge carrier excitation process typically occurs at the interface between a p-type and n-type semiconductor (called a p-n junction) which provides a depletion region where electron-hole pairs can be separated.

Photovoltaic devices can be characterised in a number of ways, with the key figures of merit being power conversion efficiency ($\eta$, %), fill factor (FF), short-circuit current ($I_{SC}$, A) and open-circuit voltage ($V_{OC}$, V). These are extracted from I-V curves of the device measured under illumination. Typically, devices are illuminated by a calibrated source with a spectrum that matches the AM1.5 standard, which is the solar spectrum on the earth's surface at a solar zenith angle of 48.2°, (corresponding to 1.5 atmosphere thickness). From the illuminated I-V measurement, the voltage at which the total current is zero ($V_{OC}$) and the current at zero bias ($I_{SC}$) can be extracted. The FF is the ratio between the maximum power ($P_{MAX}$, W) and the product of $V_{OC}$ and $J_{SC}$. This ratio is an indicator of the quality of the device, as a higher FF implies that the device is closer to its theoretical maximum output. The efficiency of the cell is simply the ratio between the $P_{MAX}$ and the light power applied. One other common measurement when characterising devices is the quantum efficiency, whether external (EQE) or internal (IQE). The EQE is the ratio of charge carriers generated to the number of photons incident to the device for a given wavelength (or energy) while the IQE is the ratio of charge carriers generated to the number of photons absorbed. These are typically reported as a curve across the width of the AM1.5 spectrum.

There are a multitude of different photovoltaic technologies, which are commonly grouped into generations. First-generation devices include cells based on crystalline silicon and are a relatively mature technology at this time, with efficiencies for single junction crystalline cells reaching 26.7% [114]. The second generation encompasses commercialised thin-film solar cells, which are fabricated by growth of the active semiconductor layer onto a supporting substrate. This includes technologies such as amorphous thin-film silicon (a-Si) and copper indium gallium selenide (CIGS). Some second-generation technologies can be made flexible [115]; however, fabrication of flexible devices can be challenging due to processing temperature limitations, with CIGS layers typically fabricated by thermal evaporation and vapour annealing. Advances in low-temperature processing have resulted in devices with efficiencies of over 18% on flexible polymer films [116]; however, flexibility of such devices is limited [117]. Third-generation technologies are typically described by the broad definition of thin-film technologies which are currently in the research phase. These include many technologies which use materials and processes that are less energy intensive than those used in first- and second-generation devices. Such processes include printing and coating from solution, which could enable higher volume fabrication at lower cost, reducing the energy payback time [118]. Third-generation technologies include dye sensitized solar cells, organic photovoltaics (OPV) and perovskite photovoltaics. Of these technologies, OPV has shown the most progress with regards to flexible, stretchable and wearable applications; therefore, the rest of this section will focus on this technology.

### 3.3.2 Organic Photovoltaics

The development of organic semiconductor materials enabled the fabrication of OPV, wherein the active layer is comprised of one or more of such materials. OPV devices have shown significant promise for flexible electronics due to the inherent flexibility of many organic semiconductors. This enables devices which are lightweight, have novel form factors and can be manufactured using high-volume solution-processing techniques. In particular, the potential for low-cost fabrication of devices with low-embedded energy is appealing for solar power generation, as it reduces the energy payback time.

#### 3.3.2.1 Principles, Materials and Fabrication

OPV devices work on similar principles to inorganic devices, as described earlier. The key difference is that the absorbed light generates excitons (the electrostatically bound state of an electron and hole) rather than free electron-hole pairs. This occurs because of the weak interaction between organic molecules compared to a covalently bonded inorganic semiconductor. This weak interaction causes localisation of charge carriers [119]. As excitons are mobile excited states, to generate free charge carriers in an OPV cell the exciton must dissociate. This typically occurs at a heterointerface, as a result interface engineering is a major factor in OPV cell design.

Early demonstrations of OPV used a single semiconductor layer sandwiched between two dissimilar electrodes. In this case, the excitons should be disassociated at the rectifying Schottky contact formed at the semiconductor/electrode interface. However, this field is not strong enough to reliably disassociate the excitons, which tend to recombine resulting in very low IQE. By using two dissimilar semiconductors, which are electron donors and acceptors respectively, the field at the interface can be increased. Such bilayer cells showed significant performance increase; however, performance was still limited as the layer thickness was a trade-off between absorption and diffusion length. Further improvements were made by blending the donor and acceptor semiconductors into a bulk heterojunction (BHJ), a technique pioneered by Yu et al. [120]. This increases the interfacial area, resulting in an increase in exciton generation within the diffusion length of the heterointerface. The exact morphology of the BHJ is crucial to device performance, as after disassociation the charge carriers need still to be extracted [121].

A typical organic BHJ solar cell will consist of five layers. The BHJ layer is sandwiched between two electrodes, at least one of which must be transparent. Interlayers between the photoactive layer and the electrodes are typically included to aid charge carrier collection. The order of the layers depends on the desired structure, with most OPV devices using an inverted structure which provides improved stability (due to the higher stability anode being the top exposed layer), as shown in Figure 3.8. The rest of this section will focus on this inverted BHJ structure. In this case, the electron transport layer (ETL) separates the photoactive layer from the transparent cathode, with a hole transport layer (HTL) between the photoactive layer and anode. These interlayers should enable an ohmic contact with low resistance along with charge selectivity, which reduces undesirable charge recombination.

There are a myriad of available materials for each layer within an OPV cell, with each layer the subject of intense research interest. The choice of materials will define device performance, mechanical properties, processability and stability. Note that, as with any multilayer device, compatibility between layers is a key consideration; therefore, each layer cannot be considered in isolation.

The selection of a transparent conductor material is a compromise between transparency, conductivity and mechanical properties. The most common material is ITO, which offers high transparency and conductivity, at the expense of poor flexibility and high cost. High-performance ITO films are fabricated by sputtering, which is compatible with flexible substrates and can be achieved in a roll-to-roll fashion. Patterning of ITO is typically subtractive in nature, with laser or chemical etching common. Other alternatives include PEDOT:PSS, which is flexible and easier to process than ITO at potentially lower cost; however, conductivity and stability are comparatively poor. Silver nanowire (AgNW) networks

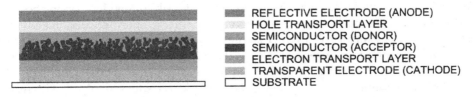

**FIGURE 3.8**
Illustration of a typical inverted bulk heterojunction solar cell structure.

with a low enough percolation density to enable light transmission have shown promise; however, cost is high.

The electron transport layer should enable efficient charge carrier collection by reducing the potential barrier between the photoactive layer and the transparent cathode while providing good electron transport. In addition, the layer should have high transparency and chemical stability. Typical solution processable materials include ZnO [122], AZO [123] and $TiO_2$ [124]. Excellent reviews of electron and hole transporting layers have been written by Lattante [125] and Yin et al. [126].

Selection of materials to form the bulk heterojunction photoactive layer is complex and a full discussion of such materials is beyond the scope of this chapter (the reader is directed to one of the many excellent review articles on the subject [127, 128]). In summary, such layers typically consist of a polymer:fullerene blend, such as poly(3-hexylthiophene-2,5-diyl):phenyl-C61-butyric acid methyl ester (P3HT:PCBM) or poly[2,6-(4,4-bis-(2-ethylhexyl)-4H-cyclopenta[2,1-b;3,4-b']-dithiophene)-alt-4,7-(2,1, 3-benzothiadiazole):phenyl-C61-butyric acid methyl ester ($PCDTBT:PC_{61}BM$). In recent years, non-fullerene acceptor molecules have shown promise [128], enabling higher efficiencies and more tuneable absorption characteristics (useful for efficiency and aesthetic purposes). A desirable photoactive layer will demonstrate a good EQE resulting from broad and high absorption characteristics and efficient charge generation. The charge generation is directly related to the bulk heterojunction morphology, as well as the energy levels of the donor and acceptor materials. Another factor to consider is material solubility and solvent selection. This has a direct effect on heterojunction morphology as well as processability.

The hole transport layer should also enable efficient and selective charge carrier collection. A common choice is $MoO_x$, which is usually deposited by thermal evaporation; however, it can be deposited from solution [129]. Other solution processable alternatives include $V_2O_x$ [130] and PEDOT:PSS [131]. However, the acidity and hygroscopic nature of PEDOT:PSS reduces device stability.

The anode is typically a reflective metal, which enables efficient charge transport due to a low sheet resistance. Reflectivity is important for increased light absorption in the device, as part of the light which was not absorbed on the first path through the photoactive layer, will be reflected back through the device. In solution-processed devices, this layer is typically silver, although the use of vacuum processing methods enables the use of aluminium. Note that this electrode need not be reflective. Use of a transparent electrode enables semi-transparent devices, which may be desirable in certain applications such as building integration. Such devices suffer from less optimal light paths through the photoactive layer, as well as limited material options for the cathode.

The choice of materials and fabrication techniques for OPV devices are not independent decisions. Although each layer has unique challenges, all layers require thin, uniform and pinhole free films with accurately defined thickness. As well as the aforementioned carrier diffusion limitations, device performance is highly dependent upon the optical properties of the layer stack. It has been shown that interference between incoming and reflected light plays a major role in device efficiency [132], which affects the optimal layer thicknesses. For example, devices based on $PCDTBT:PC_{61}BM$ have an optimal layer thickness of 75 nm [133]. Obtaining such films uniformly over large areas is challenging for solution-based fabrication processes due to the multiple dynamics involved, with research ongoing into materials which can enable thicker films [134]. Fabrication of layers in OPV devices has been demonstrated using all major printing and coating techniques, including inkjet [135], screen [136], flexography [137], gravure [138], spray [139] and slot die coating [140]. These techniques are all capable of use in roll-to-roll fabrication, a goal which many research groups have targeted. Among these, the leading research group has been that of Krebs et al., who have published many demonstrations of roll-to-roll OPV fabrication [140]. These and others are summarised in several review articles on solution processing of OPV devices [141, 142].

For large area applications, it is not practical to fabricate one single cell. OPV panels tend to consist of multiple smaller cells connected in series, in order to minimise resistive losses and maximise reliability. It is important to minimise the gap between cells in panels, as panel efficiency is proportional to the active area of the panel. This is a common challenge in roll-to-roll processed panels, as compounding inaccuracies in layer registration require suitable margin for error in the design.

Note that photovoltaic devices can also be adapted for use as photodetectors. Organic polymers exhibit a broad spectral range, making them useful for certain photodetector applications [143]. Although in principle such devices are similar to photovoltaic cells, photodetectors have more stringent requirements for low dark current and rapid response. With the addition of scintillator particles which absorb X-ray radiation and emit visible photons, such organic photodetectors (OPDs) can be indirectly sensitised to X-ray radiation [144].

### 3.3.2.2 Examples

As previously discussed, the demonstration of roll-to-roll manufacture of OPV modules was a major milestone in the field of OPV. Andersen et al. [145] took this a step further by demonstrating roll-to-roll fabrication and encapsulation of tandem cells in an ambient atmosphere. Tandem cells enable higher efficiencies by stacking multiple cells, each of which is optimised for different sections of the solar spectrum. In this way, the EQE can be significantly increased. However, fabrication of such devices is challenging, especially using solution processing where layer compatibility is a major constraint. The demonstrated devices consisted of 14 solution-processed layers and modules with eight cells had an active area of over 50 cm$^2$. The PCE of the modules was 1.8%, which is significantly lower than the state of the art; however, the value is impressive given the fabrication processes employed.

Beyond the scalability of roll-to-roll processing, the use of printing techniques in the fabrication of OPV cells and panels also enables patterning of devices onto various substrates, resulting in unique form factors. This is a major advantage of OPV compared to more established technologies, as such patterning can be achieved in an additive fashion, enabling design customisation while minimising additional process complexity. One such example is the leaf structure fabricated by Välimäki et al. [138] (shown in Figure 3.9a), which was fabricated by a combination of screen printing, gravure printing and thermal evaporation. They discuss a number of the compromises in the final design, which are predominantly a result of alignment limitations, a major factor in roll-to-roll fabrication. Another example is the tree shape fabricated by Eggenhuisen et al. [135] (shown in Figure 3.9b) who used inkjet to fabricate semi-transparent devices. They observed a decrease in device performance compared to reference cells fabricated with a traditional rectangular shape, which they attributed to non-optimal design, especially with regards to the silver busbars used for charge collection.

Further design customisation is possible by tuning the absorption properties of the device in order to define the colour. Lee et al. [146] achieved this in hybrid a-Si/organic

(a)

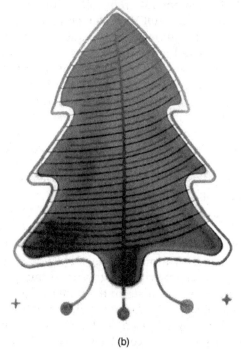

(b)

**FIGURE 3.9**

Functional devices which demonstrate design customisation possibilities enabled by organic photovoltaics. (a) Leaf design (image reproduced with permission from Välimäki et al. [138] under the terms of the Creative Commons Attribution 4.0 International License (*http://creativecommons.org/licenses/by/4.0/*)). (b) Tree design (reproduced with permission from Eggenhuisen et al. [135]).

devices by varying the thickness of the a-Si layer, which can enable resonant optical transmission. Because the a-Si film is very thin (between 6 and 31 nm), the optical effect shows good angle insensitivity. Cui et al. [147] achieved semi-transparent devices of various colours by varying the donor molecule in the bulk heterojunction. They developed a non-fullerene acceptor which had weak absorption in the visible, resulting in the colour being dominated by the absorption of the donor molecules.

Beyond device colour tuning, selection of suitable materials can enable OPV devices with high transmittance in visible wavelengths, therefore appearing transparent. By using materials which primarily absorb in the ultraviolet and near-infrared, Lunt et al. [148] achieved devices with an efficiency of 1.3% at a visible transmission of 65%. Selection of electrode material is challenging for transparent devices, as sheet resistivity can be high in many transparent conductors. Chen et al. [149] demonstrated devices with a solution-processed AgNW top electrode, demonstrating a power conversion efficiency of 4% at an average light transmission of 61%. Such transparent devices are of particular interest for building integrated photovoltaic (BIPV) applications, especially in office structures where there commonly exists a large area of tinted windows. In this case, the sealed nature of windowpanes and transmission properties of glass are major factors in device performance and stability. Due to the device dependence on UV and near-infrared wavelengths for acceptable performance, polymer substrates can be challenging as they tend to show high absorption in the UV; therefore, encapsulation options are limited. Flexible glass provides an alternative option for such devices.

Integration of photovoltaic energy harvesting into wearable applications will require innovations in form factor. To this end, OPV devices have a number of advantages over alternative photovoltaic energy harvesting technologies. Materials developments have enabled the fabrication of stretchable photovoltaics, which could be compatible with wearable and conformable electronics. Fabrication of such devices is challenging, with each layer in the device requiring suitable mechanical properties. Most crystalline materials will develop cracks and fail when stretched; therefore, alternative morphologies are required such as nanowire networks, pre-strained substrates or inherently stretchable materials [150]. Lipomi et al. [151] demonstrated stretchable OPV devices fabricated by spin coating active layers and electrodes onto a pre-strained substrate. When the strain was released the layers buckled, enabling repeated stretching without exceeding the mechanical limits of the layers. The authors expected that the buckled structure would affect the photovoltaic properties due to the resulting changes in optical path; however, minimal effects were observed. A power conversion efficiency of 1.2% was observed in strained devices (18.5% strain) after repeated cycles. Straining devices beyond the point at which the substrate was pre-strained resulted in cracking and device failure. Fullerene-based bulk heterojunction materials tend to show limited stretchability due to cracking [152]; therefore, all-polymer-based devices are preferable for stretchable applications. Kim et al. [153] demonstrated devices which showed elastic limits which were over 60 times that of fullerene-based control samples, although performance at such strains was not discussed.

Another major form factor development towards wearable energy harvesting is the development of photovoltaic fibres. One early example of such devices is the work of O'Connor et al. [154], who thermally evaporated devices onto a polymer-coated silica fibre. Efficiency losses due to the geometry of the fibre were expected to be 17%; however, fabricated devices were seen to be 34% less efficient than the planar controls, with a peak PCE of 0.5% achieved.

The integration of OPV devices into wearable applications is especially challenging due to the mechanical strain and water exposure inherent in such a scenario. Jinno et al. [155] demonstrated OPV panels, which could withstand repeated mechanical compression and water exposure. By fabricating the devices at the neutral strain position between 1 μm thick parylene films and encapsulating between 0.5-mm thick pre-stretched acrylic elastomer, they were able to achieve impressive device performance after immersion, with a PCE of 7.9%. When immersed in water for 120 min, a 5.4% decrease in efficiency was observed. The devices still showed some degradation after simulated washing (100 min of water exposure and mechanical compression of 52% for 20 cycles), with efficiency values reduced by 20%; however, this is still an impressive result.

### 3.3.3 Summary

The use of electronics in flexible, stretchable and wearable applications requires a suitable power source. Harvesting energy from light has long been an effective method to generate electricity and several photovoltaic technologies offer the potential of mechanical flexibility. Of these, OPV has shown the most promise for flexible, stretchable and wearable applications due to its unique combination of inherently flexible materials and compatibility with low-temperature roll-to-roll fabrication processes. These advantages result in the potential for novel form factors, reduced embedded energy and scalable fabrication of large area devices.

However, many challenges need to be overcome before successful commercialisation of OPV will be realised. The power conversion efficiency of OPV devices is still poor when compared to other photovoltaic technologies and the long-term stability is still considered insufficient for many applications.

Note that photovoltaics are not the only energy-harvesting methods that show promise for flexible, stretchable and wearable electronics. Alternative energy generating mechanisms include piezoelectric (applied stress), thermoelectric (temperature gradient) and triboelectric (separating surfaces), among others. Each technology has its own advantages and challenges and all have potential applications.

### 3.4 Luminescent Devices

Illumination is used in a variety of applications, from area lighting to displays. Traditional general lighting solutions are limited in form factor, with most incandescent bulbs and fluorescent tubes requiring reflectors and shades to distribute diffuse light. The ability to make flexible and stretchable diffuse light sources of arbitrary shape would enable a myriad of form factors and subsequent designs. The same advantages can be utilised in display applications, reducing the limitations of rigidity and planar form.

### 3.4.1 Background

There are many mechanisms to produce light, one of which is where a material emits photons due to radiative recombination of an excited charge carrier, known as electroluminescence. This mechanism is used in most solid-state lighting, such as LEDs. There are multiple methods to achieve electroluminescence in flexible and stretchable forms, with the most common being either alternating-current thin-film electroluminescence (ACEL) or organic light emitting diodes (OLEDs).

Although the mechanism for achieving electroluminescence may differ, these devices share many metrics. The output of light sources is typically characterised in terms of luminance and spectral output (or colour). The luminance of the device is the luminous intensity per unit area ($cd/m^2$) and ideally should be measured using an integrating sphere. The colour of the lamp can be measured using a spectrophotometer and is usually reported as CIE coordinates or a spectrum. The operating lifetime is also a key metric for light sources and can be defined in many ways, but typically is reported as the number of hours for the luminance of the device to reach 50% of its initial value ($t_{50}$).

### 3.4.2 ACEL

ACEL devices are based on the emission from films of phosphorescent particles when exposed to an alternating electric field. This is a mature technology, having been used in segmented displays since the mid-twentieth century. Significant research was performed towards the goal of ACEL displays, but they were not competitive compared to LCD technologies. However, their use in segmented displays has continued, and the technology has shown promise in the field of flexible, stretchable and wearable luminescence due to the robustness of the materials and potential for deposition from solution.

#### 3.4.2.1 Principles, Materials and Fabrication

ACEL can be achieved in a number of ways, the most relevant being the capacitive structure wherein a phosphor layer is sandwiched between two dielectric layers and conductive electrodes, as shown in Figure 3.10. Printed ACEL often omits one or both of the discrete dielectric layers, with the polymer binder in the phosphor ink providing the same role. Light emission is produced when the electric field across the device reaches a point wherein charges are injected into the phosphor layer from traps and defects in the phosphor/insulator interface. At this point, electrons collide with the phosphor particles, which are typically doped inorganic semiconductors. This results in excitation and subsequent decay, in which a photon may be emitted. When the alternating field reverses, the process repeats. In practice, the emission mechanism is significantly more complex than this, and the reader is directed to a number of excellent articles if more detail is required [156, 157].

The capacitive structure shown in Figure 3.10 has a number of advantages. As no charges are injected from the electrodes, materials and morphology requirements are simplified.

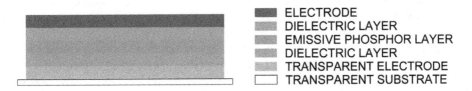

ELECTRODE
DIELECTRIC LAYER
EMISSIVE PHOSPHOR LAYER
DIELECTRIC LAYER
TRANSPARENT ELECTRODE
TRANSPARENT SUBSTRATE

**FIGURE 3.10**
Illustration showing the structure of a printed capacitive ACEL device. Many devices merge the dielectric and phosphor layers by incorporation of the phosphor powder into a dielectric material. Note that if a non-transparent substrate is used, the electrodes should be swapped.

Devices are tolerant of thick films, which improves manufacturability and yield as the chances of pinholes are reduced. Finally, the structure is inherently encapsulating of the emissive material, improving device stability.

The development of inorganic phosphors over the previous century has resulted in wide availability of phosphor powders, which can be incorporated into a polymer matrix to form a printable ink. Many such materials are available commercially from major ink suppliers. Typically, the selection of phosphor material depends on the colour required, with doped ZnS being the most common material [158]. Doping with Cu enables emission in the red, while doping with Cu and Cl provides emission in the blue.

Key metrics for ACEL lamps are the aforementioned luminance, spectral output and lifetime as well as the corresponding driving voltage, frequency and waveform. Note that as the lamps age these metrics will change, it is common for ACEL devices to increase the voltage and/or frequency to correct for a decrease in luminance. Performance should be reported in line with established standards [159].

The key aspects of dielectric material selection are film uniformity and dielectric constant. As the device luminance is dependent upon the field generated within the structure, a higher dielectric constant enables a higher field at the same voltage. The reflective properties of the dielectric also affect the emission, as the phosphor emission is non-directional. As such, a transparent dielectric and reflective electrode are preferred. Inorganic $BaTiO_3$-based dielectrics are commonly employed for ACEL devices due to the high dielectric constant that can be achieved. As discussed in Section 3.2.3, $BaTiO_3$ has been used to demonstrate dielectric films with a dielectric constant higher than 100 [20].

Transparent electrodes are primarily ITO or PEDOT:PSS based. ITO offers higher conductivity and transmittance at the expense of flexibility and stretchability. AgNWs have been demonstrated for use in stretchable applications, as the morphology of the nanowire network enables continued conduction while under mechanical strain. If non-uniform emission is observed due to the resistance of the transparent electrode, silver bus bars may be used to minimise electric field variation. The non-transparent electrode materials in flexible devices are typically carbon or silver, as these are the most common printable conductors. Silver may be preferred for larger devices due to the higher conductivity reducing field variation across the device.

Screen-printing is the most common method for device fabrication, as thicker films are often preferred due to reliability and particle size considerations. The robustness of the screen printing process, coupled with its existing usage in electronic and textile fabrication, means that it is an ideal choice for fabrication of devices for wearable electronics.

Although ACEL devices offer a robust and low-cost route to flexible, stretchable and wearable luminescence, they have a number of disadvantages, primarily due to the voltage-dependent behaviour. Operating voltage and frequency depend on device size, luminance and required lifetime, with typical values in the range of 50 to 100 V and 100 Hz to 1 kHz. The output can be increased by increasing the voltage or frequency; however, this negatively affects the device lifetime. The device behaviour can also be impacted by the AC characteristics, with a sawtooth wave being the most efficient. As the luminance of ACEL devices is low, they are typically unsuitable for use in bright environments. Additionally, for flexible devices, care must be taken to avoid undesirable acoustic effects caused by vibration related to the operating frequency. That being said, the robustness, manufacturability and mechanical characteristics make ACEL a popular choice for flexible, stretchable and wearable lighting applications.

### 3.4.2.2 Examples

The maturity of the technology has not limited research groups from exploring new materials and fabrication techniques to expand the use of ACEL to flexible, stretchable and wearable electronics. In particular, the goal of stretchable devices has been an area of significant research interest. Many groups achieve stretchable devices by using polydimethylsiloxane (PDMS)-based composites for each layer. This material can be optically transparent and stretchable, making it an ideal candidate for a binder material for functional composites. Wang et al. [160] used AgNWs spray-coated onto PDMS film, followed by spin-coating of a ZnS:Cu/PDMS composite. Another PDMS/AgNW layer then encapsulated this. Device performance was demonstrated at strains of up to 100%. Stauffer et al. [161] were able to increase luminosity by incorporation of high dielectric constant $BaTiO_3$ particles into the luminescent layer to increase the electric field. This increased luminosity by up to 700%, while maintaining stretchability up to 50% strain.

Hu et al. [162] also fabricated PDMS-based devices in fibre form, as shown in Figure 3.11. They used AgNWs coated on a PDMS core, with a phosphor/PDMS composite as the emissive layer. Devices showed uniform radial emission and stable performance up to 4000 cycles up to 50% strain.

However, PDMS has some drawbacks. It has a hydrophobic surface and may absorb many organic solvents, which complicates multilayer device fabrication. You et al. [163] demonstrated a method to encapsulate AgNW/PDMS composites with a polyurethane urea layer, resulting in stretchable electrodes, which provide a good moisture barrier.

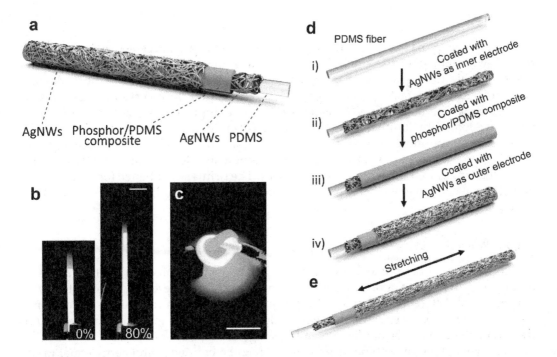

**FIGURE 3.11**
Stretchable fibre EL devices. (a) Device structure. (b, c) Device functionality demonstrated under various strain conditions. (d) Device fabrication process flow. (e) Illustration of device when stretched. (Reproduced without modification and with permission of the author from Hu et al. [162], under the Creative Commons Attribution License.)

Their devices were still functional after 5000 stretching cycles (up to 100%) and water immersion for 30 min, which are promising results for washable wearable applications.

Thermoplastic elastomers offer an alternative to PDMS for stretchable applications. Zhou et al. [164] used thermoplastic polyurethane (TPU) to fabricate ACEL devices, which could be strained up to 100%. They claimed that the use of TPU enables lower operating voltages, as the dielectric constant of the TPU is higher than that of PDMS.

An alternative strategy to achieve very high stretchability was pursued by Yang et al. [165], who developed devices based on ionic conducting hydrogels. These were fabricated by sandwiching phosphor particles between two layers of pre-stretched dielectric and hydrogels. These were then connected to non-stretchable electronic conductors outside the stretchable area of the device. Very high device stretchability of up to 1500% was demonstrated. However, care must be taken to avoid electrolysis at the conductor interfaces, and hydrogels must be encapsulated to maintain their moisture content. Device performance was said to be maintained for over 8 months.

A similar hydrogel approach was used by Larson et al. [166], who demonstrated the use of ACEL in a soft robotic application. The stretchable ACEL devices were integrated into pneumatic chambers, where the capacitance was used to determine the chamber pressure based on the change caused by the ACEL device stretching. A series of these chambers were connected, enabling worm-like movement if they were actuated in sequence. The use of the capacitive sensor was combined with the luminescence to emit light during movement.

### 3.4.3 OLED

OLED technology has become one of the major commercial successes of organic electronics, with its use in the front plane of displays now commonplace. This success is due to the high contrast, low power requirements and the structure of such devices enabling better form factors when compared to competitor liquid crystal display (LCD) technologies. The use of OLED displays in flexible applications has been repeatedly demonstrated, with commercialisation efforts ongoing at the time of writing. OLEDs are also being commercialised for applications beside displays, such as indoor lighting and signage. However, despite this intense commercial interest, there remains a number of challenges for OLED-based technologies. The primary challenge for such devices is lifetime, a challenge which is only exacerbated when flexibility and stretchability is introduced.

#### 3.4.3.1 Principles, Materials and Fabrication

In a conventional LED, photons are emitted when recombination of an electron and hole occurs in a p-n junction, a phenomenon known as electroluminescence. When a forward-bias potential difference is applied to a p-n junction, electrons from the n-doped semiconductor are repelled towards the semiconductor junction, likewise for the holes from the n-type material. If the applied potential is high enough, charge carriers can cross the junction where they recombine. A photon can then be emitted with energy equal to the band gap between the conduction and valence band of the material, resulting in the device being a monochromatic light source.

One of the key differences between an LED and an OLED is the carrier transport mechanism. Due to the localisation of charge carriers caused by the weak interaction

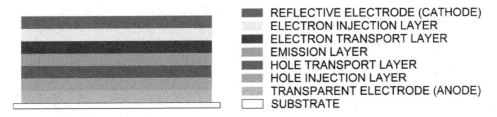

REFLECTIVE ELECTRODE (CATHODE)
ELECTRON INJECTION LAYER
ELECTRON TRANSPORT LAYER
EMISSION LAYER
HOLE TRANSPORT LAYER
HOLE INJECTION LAYER
TRANSPARENT ELECTRODE (ANODE)
SUBSTRATE

**FIGURE 3.12**
Layer structure of a typical bottom-emission SMOLED.

between organic molecules, the concept of conduction and valance bands is not directly applicable. In OLEDs, the attraction between carriers forms an exciton (as discussed previously with regards to OPV), the recombination of which in the emission layer can produce a photon. The energy of the photon will depend on the energy difference between the highest occupied molecular orbital (HOMO) and lowest unoccupied molecular orbital (LUMO) of the emitter material. It is clear from this process that the operation of an OLED is highly dependent upon careful matching of the electrical, optical and physical properties of each individual layer. For further details on the device physics of OLEDs, the reader is directed towards several excellent books and articles [167, 168].

There are two common types of OLED, small molecule OLED (SMOLED) and polymer LED (PLED); both named for the class of materials used in the emission layers of the device. SMOLED are typically fabricated using vacuum processes, which results in limited throughput and significant cost, especially with regards to material wastage during deposition. PLED devices can be fabricated using solution processing; however, luminous efficiency and lifetime are both poor compared to SMOLED. The typical structure of a bottom-emission SMOLED is shown in Figure 3.12. Note that this structure has evolved over years of research in order to optimise the efficiency and lifetime of the devices. Simpler structures can be used at the expense of device performance. Depending on the desired colour output, the emission layer may consist of several layers, the output of which can combine to form a broader output spectrum than possible with a single emissive layer. This one of the methods used to produce white OLED emission.

Key OLED metrics are the aforementioned luminance, spectral output and operating lifetime, as well as luminous efficiency and EQE. The luminous efficiency is a measure of the power required to drive the OLED (cd/A). By combining the luminance and luminous efficiency, the device power requirement can be obtained. The EQE is the ratio of photons emitted from the device to the number of injected electrons.

Due to the sensitivity of SMOLED materials to oxygen and moisture, as well as the sensitivity of device performance upon layer thickness, vacuum deposition tools are typically used for fabrication. Such tools are capable of uniform layer deposition with accuracy in the nm range, which is required for optimal device performance. Many PLED materials can be deposited by solution processes, and even in ambient conditions [169]. However, obtaining uniform films of precise thickness is challenging and solvent compatibility between layers may limit material options. For applications where additive patterning is not required, coating methods such as slot die [170–172] and blade [169] coating are commonly used. If additive fabrication with patterning is desired, OLED fabrication has been demonstrated using gravure [173, 174] and inkjet [175] printing.

Materials selection for OLED devices is a complex topic, which is summarised well in many review articles [176–181]. Due to the acute dependence of performance upon matching of energy levels between layers, no material selection can be made in isolation. This is further complicated by the mechanical and stability requirements required for flexible, stretchable or wearable electronics. For example, ITO is not ideal for use as a flexible transparent conductor as it will crack during the mechanical stresses of such applications, which increases the sheet resistance. PEDOT:PSS has been demonstrated as a replacement [182]; however, its poor stability and limited conductivity impede potential usage. Nanowire networks (CNTs [183] or AgNWs [184]) have shown promise, but roughness issues are a concern. In addition, the low work function of such materials necessitates the use of an interlayer to lower the injection barrier, as such hybrids with PEDOT:PSS are common. Cathodes are similarly problematic, especially if solution processing is desired. Aluminium is commonly used for the cathode layer, and if thin enough can be flexible [185]; however, it requires vacuum processing. PEDOT:PSS has also been demonstrated as a cathode material; however, interlayers of ZnO and polyethylenimine (PEI) were used in order to enable electron injection [186].

The emission layers tend to be either fluorescent or phosphorescent, with phosphorescent materials preferred due to higher efficiencies. For solution-processed devices, these have commonly taken the form of organometallic complex and polymer blends [180]. Solution processing of small molecule emitter materials based on tetraphenylsilane with pyridine moieties has also been demonstrated [187].

A major challenge for OLED devices is degradation, and this challenge increases significantly when targeting flexible, stretchable or wearable applications. Most OLED materials show high sensitivity to oxygen and water due to oxidation and trap formation [188]. Extending device lifetime requires improved material and interfacial stability alongside effective encapsulation. ALD deposition of alternating inorganic films has been shown to be effective [189], but the throughput of the ALD process is low and such films have limited flexibility.

OLED may not be as mechanically robust or as easy to manufacture as ACEL; however, the efficiency and luminance are significantly higher. These advantages incentivise ongoing research into materials, structures and processes to achieve stable, flexible and potentially stretchable OLED devices.

### 3.4.3.2 Examples

The challenge of introducing flexibility, stretchability and wearable robustness into OLED devices is clearly a significant one. However, many research groups have been focussed on solving these challenges. Novel material solutions are required to enable flexible devices, as highlighted by the work of Han et al. [187], who used graphene for the anode material in flexible OLEDs to replace the inflexible ITO. This was challenging as it required modification of the graphene to achieve a suitable work function for efficient hole injection. This was achieved by designing a gradiated hole injection layer using PEDOT:PSS and a tetrafluoroethylene-perfluoro-3,6-dioxa-4-methyl-7-octenesulphonic acid copolymer. In addition, the graphene was doped using $AuCl_3$ to reduce the sheet resistance. This electrode was demonstrated in fluorescent and phosphorescent green OLEDs, with performance comparable to ITO-based devices. Devices showed stable performance after 1000 bending cycles at a radius of 7.5 mm, whereas ITO-based devices failed after 800 cycles. Fluorescent white OLEDs up to 25 $cm^2$ were also fabricated, demonstrating potential for large area lighting applications.

Focussing on such large area lighting applications, the use of solution processing for manufacture is desirable due to the scalability of the techniques. Perumal et al. [190] focussed on the hole transport and emission layers, highlighting the requirement for orthogonal solvents to ensure process compatibility. They used copper thiocyanate (CuSCN, dissolved in diethyl sulphide) as the hole injection layer instead of the more common PEDOT:PSS. The CuSCN devices showed good performance compared to the PEDOT:PSS-based devices at lower luminance values; however, the PEDOT:PSS devices performed better at higher luminance. They attributed this difference to the lower mobility of the CuSCN. Zhang et al. [186] made all-solution-processed devices using PEDOT:PSS for the anode and cathode. Electron injection from the cathode was achieved by use of a PEI and ZnO bilayer, while $WO_3$ was used as a hole injection layer for the anode. AgNWs were also used to enhance the conductivity of the upper anode electrode. All layers apart from the ITO layer were fabricated using spin coating or drop casting, with masking used for rudimentary patterning. This structure is interesting as it is transparent in nature, which may enable novel applications. Devices were compared to a rigid and opaque (Al electrode) structure on glass, with similar performance observed.

Efforts towards use of more scalable solution deposition techniques were made by Raupp et al. [191], who investigated the use of slot die coating and flexography for deposition of various OLED layers. Slot die coating was found to result in more uniform PEDOT:PSS-based hole injection layers than flexography. Slot die coating was also used for the deposition of a small molecule green emitter layer. Devices with active areas of 27 $cm^2$ were fabricated to demonstrate the scalability of the technique. Abbel et al. [170] demonstrated roll-to-roll fabrication of large area OLEDs using a sophisticated roll-to-roll slot die coating line with two coating stations. The deposition section of the line was in cleanroom conditions and a nitrogen atmosphere, and the substrate was only contacted on one side during coating. This line was used to deposit a hole injection layer and the emissive layer with orthogonal solvents. Samples were compared to devices fabricated using sheet-to-sheet fabrication in a controlled atmosphere, with lower performance observed in the roll-to-roll samples. This was attributed to higher oxygen and moisture content in the atmosphere during roll-to-roll deposition.

If flexible OLED devices are to be reliable, a flexible encapsulation solution is required to protect against oxygen and moisture. Typical barrier films comprise of multiple oxide layers in a stack to minimise permittivity. Such films typically have limited flexibility, as the mechanical stresses cause cracks, which enable permeation of oxygen and moisture. Park et al. [192] demonstrated a scalable method of lamination wherein a thin metal film (either 0.6 mm stainless steel or 0.04 mm Fe-Ni alloy) was used as the primary barrier layer, which was laminated onto the sample using PDMS as an intermediary, which enabled good encapsulation of the device. Encapsulation performance was observed to be equivalent to glass, with good bending stability observed due to the locating of the ITO layer at the neutral plane.

Enabling stretchable OLED devices is a major challenge due to the dependence of the device performance upon layer thickness. One example of stretchable PLEDs was achieved by White et al. [185], who fabricated ultra-thin devices on PET, which were then laminated onto a pre-stretched elastomer. This technique is similar to that used for stretchable OPV discussed previously. When the strain is released from the pre-stretched substrate, the device conforms to the random folds which form as it compresses. Device functionality was demonstrated at up to 100% tensile strain, the same strain at which the pre-stretched substrate was subjected during lamination. The key to this functionality is the ultra-thin devices (2 μm including PET substrate), which allowed sufficient flexibility to conform to the folds when laminated to the elastomer.

Towards wearable OLED devices, a number of groups have investigated methods for incorporating OLED functionality onto textiles. Kim et al. [193] reported OLED devices fabricated directly onto fabrics. They used PU and PVA layers to planarise the polyester fabric substrate, as shown in Figure 3.13. The PU film was laminated onto the fabric, followed by spin coating of two layers of PVA in order to reduce the surface roughness to a level at which functional devices could be fabricated. This process increased the stiffness of the fabric; however, devices remained functional after 1000 bending cycles at a radius of 5 mm. Choi et al. [194], who also encapsulated devices with an aluminium oxide and silane-based barrier film, took a similar planarization approach. The encapsulation technique also served to reduce the substrate roughness, leading to a reduction in leakage

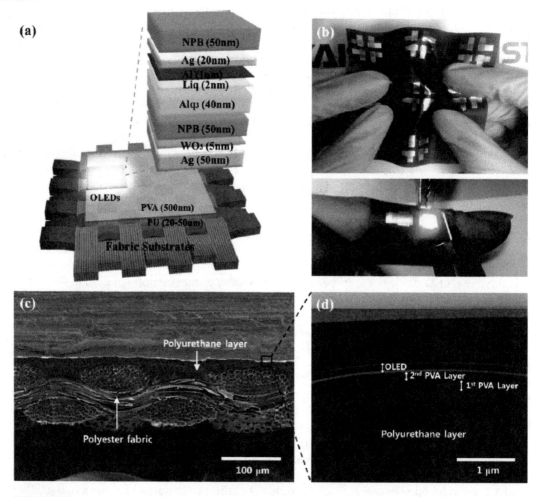

**FIGURE 3.13**
OLED devices fabricated onto a fabric substrate using a planarization layer. (a) Diagram showing the device structure and materials. (b) Demonstration of flexibility and functionality. (c) SEM image showing a cross-section of the device stack. (d) SEM image showing a cross-section of the planarization layer and OLED device. (Image reproduced from Kim et al. [193]).

current and corresponding increase in current efficiency. High humidity (90%) testing results were encouraging but inconclusive due to defects.

### 3.4.4 Summary

Both ACEL and OLED offer promise in the drive towards flexible, stretchable and wearable luminescence; however, neither technology is ideal at present. ACEL offers a mechanically robust technology, which is easy to manufacture; however, the low luminance and requirement for high voltage AC limits applications. OLED is capable of high luminance and efficiency; however, relatively speaking manufacturing can be complex and mechanical durability is poor. Both technologies are the subject of continued research. ACEL, although considered a mature technology, has recently seen a reinvigoration of research interest as its potential for use in stretchable electronic applications is better understood. OLED has been the subject of intense research interest for many years, with no signs of this slowing.

Other technologies, such as light-emitting electrochemical cells [195], have been also been demonstrated in flexible forms. In addition, traditional lighting solutions such as LEDs can also be adapted to flexible, stretchable and wearable applications using the rigid-islands technique discussed previously. This can even be achieved in a roll-to-roll fabrication process [196]; however, such solutions result in discrete sources of light and not the large area diffuse lighting provided by ACEL and OLED devices.

## 3.5 Conclusions

The aim of this chapter was to provide the reader with a flavour of what is imaginable in the field of flexible, stretchable and wearable electronics. It is not feasible in the scope of a single chapter to achieve an exhaustive list of possibilities, if such an exhaustive list could be compiled in such an active area of research interest. However, every effort has been made to provide a selection of examples, which highlight both the possibilities and challenges in the field.

It must be emphasised that the components and devices discussed are merely the building blocks of more complex systems for a variety of applications. One excellent example of this is the work of Lee et al. [197], who demonstrated a pulse-oximetry-sensing patch which utilised flexible OLED and OPD components to enable monitoring of blood oxygenation. Not only did the use of flexible devices enable improvements in form factor compared to traditional oximetry sensors, but the power consumption was also significantly reduced. Such devices have been enabled by decades of research into materials systems, fabrication processes and device physics. This multidisciplinary research continues at pace, with new developments such as flexible organic-inorganic halide perovskite solar cells demonstrating the potential for a significant leap in PCE compared to flexible OPV [198]. Looking beyond the research lab, the complexity of commercial flexible electronics is also rapidly advancing, with OLED displays already routinely being fabricated onto flexible substrates for rigid applications, and the first flexible applications becoming available.

Flexible, stretchable and wearable electronics enables designers to imagine a world where electronic functionality can be integrated into any object. This revolution in how we interact with the world around us will be the next big step in the remarkable achievement that is modern electronics.

# Bibliography

[1] T. Y. Chu, Z. Zhang and Y. Tao, "Printing silver conductive inks with high resolution and high aspect ratio," *Advanced Materials Technologies*, vol. 3, p. 1700321, 2018.

[2] L. Zhou, H. Y. Xiang, S. Shen, Y. Q. Li, J. D. Chen, H. J. Xie, I. A. Goldthorpe, L. S. Chen, S. T. Lee and J. X. Tang, "High-performance flexible organic light-emitting diodes using embedded silver network transparent electrodes," *ACS Nano*, vol. 8, pp. 12796–12805, 12 2014.

[3] D. Deganello, J. A. Cherry, D. T. Gethin and T. C. Claypole, "Patterning of micro-scale conductive networks using reel-to-reel flexographic printing," *Thin Solid Films*, vol. 518, pp. 6113–6116, 2010.

[4] R. Gupta, S. Walia, M. Hösel, J. Jensen, D. Angmo, F. C. Krebs and G. U. Kulkarni, "Solution processed large area fabrication of Ag patterns as electrodes for flexible heaters, electrochromics and organic solar cells," *Journal of Materials Chemistry A*, vol. 2, pp. 10930–10937, 28 2014.

[5] M. A. Yokus, R. Foote and J. S. Jur, "Printed stretchable interconnects for smart garments: Design, fabrication, and characterization," *IEEE Sensors Journal*, vol. 16, pp. 7967–7976, 11 2016.

[6] J. Liang, K. Tong and Q. Pei, "A water-based silver-nanowire screen-print ink for the fabrication of stretchable conductors and wearable thin-film transistors," *Advanced Materials*, vol. 28, pp. 5986–5996, 5 2016.

[7] Y. Atwa, N. Maheshwari and I. A. Goldthorpe, "Silver nanowire coated threads for electrically conductive textiles," *Journal of Materials Chemistry C*, vol. 3, pp. 3908–3912, 2015.

[8] N. Graddage, H. Ding, C. Py, J. Lee and Y. Tao, "Manufacturability of a printed resistance-based multiplexing scheme for smart drug packaging," *IEEE Transactions on Components, Packaging and Manufacturing Technology*, vol. 6, pp. 335–345, 3 2016.

[9] P. Sandborn and P. A. Sandborn, "A random trimming approach for obtaining high-precision embedded resistors," *IEEE Transactions on Advanced Packaging*, vol. 31, pp. 76–81, 2 2008.

[10] S. Claramunt, O. Monereo, M. Boix, R. Leghrib, J. D. Prades, A. Cornet, P. Merino, C. Merino and A. Cirera, "Flexible gas sensor array with an embedded heater based on metal decorated carbon nanofibres," *Sensors and Actuators B: Chemical*, vol. 187, pp. 401–406, 2013.

[11] X. Liao, W. Song, X. Zhang, H. Huang, Y. Wang and Y. Zheng, "Directly printed wearable electronic sensing textiles towards human–machine interfaces," *Journal of Materials Chemistry C*, vol. 6, pp. 12841–12848, 2018.

[12] J. Ahmad, H. Andersson and J. Siden, "Screen printed piezoresistive sensors for monitoring pressure distribution in wheelchair," *IEEE Sensors Journal*, pp. 2055–2063, 2018.

[13] J. Leppäniemi, M. Aronniemi, T. Mattila, A. Alastalo, M. Allen and H. Seppa, "Printed WORM memory on a flexible substrate based on rapid electrical sintering of nanoparticles," *IEEE Transactions on Electron Devices*, vol. 58, pp. 151–159, 1 2011.

[14] J. Leppäniemi, T. Mattila, T. Kololuoma, M. Suhonen and A. Alastalo, "Roll-to-roll printed resistive WORM memory on a flexible substrate," *Nanotechnology*, vol. 23, p. 305204, 7 2012.

[15] A. Brodeala, A. Bonea, A. Alcade, B. Mihailescu, A. Vasile and P. Svasta, "Electrical characterization of ink-jet printed organic capacitors on flexible substrate," in *Advanced Topics in Optoelectronics, Microelectronics, and Nanotechnologies VI*, 2012.

[16] B. J. Kang, C. K. Lee and J. H. Oh, "All-inkjet-printed electrical components and circuit fabrication on a plastic substrate," *Microelectronic Engineering*, vol. 97, pp. 251–254, 9 2012.

[17] Y. Liu, T. Cui and K. Varahramyan, "All-polymer capacitor fabricated with inkjet printing technique," *Solid-State Electronics*, vol. 47, pp. 1543–1548, 9 2003.

[18] Y. Y. Noh and H. Sirringhaus, "Ultra-thin polymer gate dielectrics for top-gate polymer field-effect transistors," *Organic Electronics*, vol. 10, pp. 174–180, 2 2009.

[19] N. Graddage, T. Y. Chu, H. Ding, C. Py, A. Dadvand and Y. Tao, "Inkjet printed thin and uniform dielectrics for capacitors and organic thin film transistors enabled by the coffee ring effect," *Organic Electronics*, vol. 29, pp. 114–119, 2 2016.

[20] T. Lehnert, P. Herbeck-Engel, J. Adam, G. Klein, T. Kololuoma and M. Veith, "Dielectric properties of a printed sol–gel matrix composite," *Advanced Engineering Materials*, vol. 12, pp. 379–384, 2010.

[21] J. Lim, J. Kim, Y. J. Yoon, H. Kim, H. G. Yoon, S. N. Lee and J. H. Kim, "Improvement on relative permittivity of inkjet-printed $BaTiO_3$–resin hybrid film by manipulating ceramic powder size-distribution," *International Journal of Applied Ceramic Technology*, vol. 9, pp. 199–205, 2011.

[22] C. H. Woo, C. H. Ahn, Y. H. Kwon, J. H. Han and H. K. Cho, "Transparent and flexible oxide thin-film-transistors using an aluminum oxide gate insulator grown at low temperature by atomic layer deposition," *Metals and Materials International*, vol. 18, pp. 1055–1060, 12 2012.

[23] J. Yang, J. K. Park, S. Kim, W. Choi, S. Lee and H. Kim, "Atomic-layer-deposited ZnO thin-film transistors with various gate dielectrics," *Physica Status Solidi (A)*, vol. 209, pp. 2087–2090, 7 2012.

[24] S. J. Kang, Y. J. Park, I. Bae, K. J. Kim, H. C. Kim, S. Bauer, E. L. Thomas and C. Park, "Printable ferroelectric PVDF/PMMA blend films with ultralow roughness for low voltage non-volatile polymer memory," *Advanced Functional Materials*, vol. 19, pp. 2812–2818, 9 2009.

[25] P. Heremans, G. H. Gelinck, R. Müller, K. J. Baeg, D. Y. Kim and Y. Y. Noh, "Polymer and organic nonvolatile memory devices†," *Chemistry of Materials*, vol. 23, pp. 341–358, 2 2011.

[26] L. C. Hoffman and T. Nakayama, "Screen printed capacitor dielectrics," *Microelectronics Reliability*, vol. 7, pp. 131–135, 5 1968.

[27] J. Chang, X. Zhang, T. Ge and J. Zhou, "Fully printed electronics on flexible substrates: High gain amplifiers and DAC," *Organic Electronics*, vol. 15, pp. 701–710, 3 2014.

[28] N. L. Vaklev, R. Müller, B. V. O. Muir, D. T. James, R. Pretot, P. Schaaf, J. Genoe, J.-S. Kim, J. H. G. Steinke and A. J. Campbell, "High-performance flexible bottom-gate organic field-effect transistors with gravure printed thin organic dielectric," *Advanced Materials Interfaces*, vol. 1, p. 1300123, 2 2014.

[29] A. Rivadeneyra, J. Fernández-Salmerón, M. Agudo, J. A. López-Villanueva, L. F. Capitan-Vallvey and A. J. Palma, "Design and characterization of a low thermal drift capacitive humidity sensor by inkjet-printing," *Sensors and Actuators B: Chemical*, vol. 195, pp. 123–131, 5 2014.

[30] A. S. G. Reddy, B. B. Narakathu, M. Z. Atashbar, M. Rebros, E. Rebrosova and M. K. Joyce, "Fully printed flexible humidity sensor," *Procedia Engineering*, vol. 25, pp. 120–123, 2011.

[31] S. J. Woo, J. H. Kong, D. G. Kim and J. M. Kim, "A thin all-elastomeric capacitive pressure sensor array based on micro-contact printed elastic conductors," *Journal of Materials Chemistry C*, vol. 2, pp. 4415–4422, 2014.

[32] U. S. Bhansali, M. A. Khan and H. N. Alshareef, "Organic ferroelectric memory devices with inkjet-printed polymer electrodes on flexible substrates," *Microelectronic Engineering*, vol. 105, pp. 68–73, 5 2013.

[33] B. Lüssem, A. Günther, A. Fischer, D. Kasemann and K. Leo, "Vertical organic transistors," *Journal of Physics: Condensed Matter*, vol. 27, p. 443003, 10 2015.

[34] J. W. Chang, P. W. Liang, M. W. Lin, T. F. Guo, T. C. Wen and Y. J. Hsu, "An ambipolar to n-type transformation in pentacene-based organic field-effect transistors," *Organic Electronics*, vol. 12, pp. 509–515, 3 2011.

[35] S. Ahmad, "Organic semiconductors for device applications: Current trends and future prospects," *Journal of Polymer Engineering*, vol. 34, 1 2014.

[36] IEEE 1620-2008 - IEEE Standard for Test Methods for the Characterization of Organic Transistors and Materials, 2008.

[37] C. Liu, Y. Xu and Y. Y. Noh, "Contact engineering in organic field-effect transistors," *Materials Today*, vol. 18, pp. 79–96, 3 2015.

[38] C. W. Sele, T. Werne, R. H. Friend and H. Sirringhaus, "Lithography-free, self-aligned inkjet printing with sub-hundred-nanometer resolution," *Advanced Materials*, vol. 17, pp. 997–1001, 4 2005.

[39] W. T. Park and Y. Y. Noh, "A self-aligned high resolution patterning process for large area printed electronics," *Journal of Materials Chemistry C*, vol. 5, pp. 6467–6470, 2017.

[40] S. Li, W. Chen, D. Chu and S. Roy, "One step self-aligned multilayer patterning process for the fabrication of organic complementary circuits in combination with inkjet printing," *Organic Electronics*, vol. 13, pp. 737–743, 5 2012.

[41] W. J. Hyun, E. B. Secor, F. Z. Bidoky, S. B. Walker, J. A. Lewis, M. C. Hersam, L. F. Francis and C. D. Frisbie, "Self-aligned capillarity-assisted printing of top-gate thin-film transistors on plastic," *Flexible and Printed Electronics*, vol. 3, p. 035004, 8 2018.

[42] D. Liu and Q. Miao, "Recent progress in interface engineering of organic thin film transistors with self-assembled monolayers," *Materials Chemistry Frontiers*, vol. 2, pp. 11–21, 2018.

[43] H. Sirringhaus, "25th Anniversary article: Organic field-effect transistors: The path beyond amorphous silicon," *Advanced Materials*, vol. 26, pp. 1319–1335, 1 2014.

[44] C. Wang, H. Dong, W. Hu, Y. Liu and D. Zhu, "Semiconducting-conjugated systems in field-effect transistors: A material odyssey of organic electronics," *Chemical Reviews*, vol. 112, pp. 2208–2267, 11 2011.

[45] J. Mei, Y. Diao, A. L. Appleton, L. Fang and Z. Bao, "Integrated materials design of organic semiconductors for field-effect transistors," *Journal of the American Chemical Society*, vol. 135, pp. 6724–6746, 4 2013.

[46] H. Dong, X. Fu, J. Liu, Z. Wang and W. Hu, "25th Anniversary article: Key points for high-mobility organic field-effect transistors," *Advanced Materials*, vol. 25, pp. 6158–6183, 9 2013.

[47] T. Sekitani, U. Zschieschang, H. Klauk and T. Someya, "Flexible organic transistors and circuits with extreme bending stability," *Nature Materials*, vol. 9, pp. 1015–1022, 11 2010.

[48] H. Sirringhaus, "Device physics of solution-processed organic field-effect transistors," *Advanced Materials*, vol. 17, pp. 2411–2425, 10 2005.

[49] S. H. Hur, O. O. Park and J. A. Rogers, "Extreme bendability of single-walled carbon nanotube networks transferred from high-temperature growth substrates to plastic and their use in thin-film transistors," *Applied Physics Letters*, vol. 86, p. 243502, 6 2005.

[50] A. S. R. Bati, L. Yu, M. Batmunkh and J. G. Shapter, "Synthesis, purification, properties and characterization of sorted single-walled carbon nanotubes," *Nanoscale*, vol. 10, pp. 22087–22139, 2018.

[51] T. Dürkop, S. A. Getty, E. Cobas and M. S. Fuhrer, "Extraordinary mobility in semiconducting carbon nanotubes," *Nano Letters*, vol. 4, pp. 35–39, 1 2004.

[52] S. M. Fatemi and M. Foroutan, "Recent developments concerning the dispersion of carbon nanotubes in surfactant/polymer systems by MD simulation," *Journal of Nanostructure in Chemistry*, vol. 6, pp. 29–40, 11 2015.

[53] C. W. Lee, S. K. R. Pillai, X. Luan, Y. Wang, C. M. Li and M. B. Chan-Park, "High-performance inkjet printed carbon nanotube thin film transistors with high-k HfO2 dielectric on plastic substrate," *Small*, vol. 8, pp. 2941–2947, 7 2012.

[54] G. Hills, C. Lau, A. Wright, S. Fuller, M. D. Bishop, T. Srimani, P. Kanhaiya, R. Ho, A. Amer, Y. Stein, D. Murphy, Arvind, A. Chandrakasan and M. M. Shulaker, "Modern microprocessor built from complementary carbon nanotube transistors," *Nature*, vol. 572, pp. 595–602, 8 2019.

[55] D. M. Sun, C. Liu, W.-C. Ren and H. M. Cheng, "A review of carbon nanotube- and graphene-based flexible thin-film transistors," *Small*, vol. 9, pp. 1188–1205, 2013.

[56] K. I. Bolotin, K. J. Sikes, Z. Jiang, M. Klima, G. Fudenberg, J. Hone, P. Kim and H. L. Stormer, "Ultrahigh electron mobility in suspended graphene," *Solid State Communications*, vol. 146, pp. 351–355, 6 2008.

[57] Y. Hernandez, V. Nicolosi, M. Lotya, F. M. Blighe, Z. Sun, S. De, I. T. McGovern, B. Holland, M. Byrne, Y. K. GunKo, J. J. Boland, P. Niraj, G. Duesberg, S. Krishnamurthy, R. Goodhue, J. Hutchison, V. Scardaci, A. C. Ferrari and J. N. Coleman, "High-yield production of graphene by liquid-phase exfoliation of graphite," *Nature Nanotechnology*, vol. 3, pp. 563–568, 8 2008.

[58] F. Torrisi, T. Hasan, W. Wu, Z. Sun, A. Lombardo, T. S. Kulmala, G. W. Hsieh, S. Jung, F. Bonaccorso, P. J. Paul, D. Chu and A. C. Ferrari, "Inkjet-printed graphene electronics," *ACS Nano*, vol. 6, pp. 2992–3006, 2012.

[59] J. H. Chen, M. Ishigami, C. Jang, D. Hines, M. Fuhrer and E. Williams, "Printed graphene circuits," *Advanced Materials*, vol. 19, pp. 3623–3627, 11 2007.

[60] A. G. Kelly, T. Hallam, C. Backes, A. Harvey, A. S. Esmaeily, I. Godwin, J. Coelho, V. Nicolosi, J. Lauth, A. Kulkarni, S. Kinge, L. D. A. Siebbeles, G. S. Duesberg and J. N. Coleman, "All-printed thin-film transistors from networks of liquid-exfoliated nanosheets," *Science*, vol. 356, pp. 69–73, 4 2017.

[61] W. Xu, H. Li, J. B. Xu and L. Wang, "Recent advances of solution-processed metal oxide thin-film transistors," *ACS Applied Materials & Interfaces*, vol. 10, pp. 25878–25901, 3 2018.

[62] S. K. Garlapati, M. Divya, B. Breitung, R. Kruk, H. Hahn and S. Dasgupta, "Printed electronics based on inorganic semiconductors: From processes and materials to devices," *Advanced Materials*, vol. 30, p. 1707600, 6 2018.

[63] Y. H. Lin, H. Faber, K. Zhao, Q. Wang, A. Amassian, M. McLachlan and T. D. Anthopoulos, "High-performance ZnO transistors processed via an aqueous carbon-free metal oxide precursor route at temperatures between 80–180°C," *Advanced Materials*, vol. 25, pp. 4340–4346, 6 2013.

[64] Y. H. Hwang, J. S. Seo, J. M. Yun, H. Park, S. Yang, S. H. K. Park and B. S. Bae, "An 'aqueous route' for the fabrication of low-temperature-processable oxide flexible transparent thin-film transistors on plastic substrates," *NPG Asia Materials*, vol. 5, pp. e45–e45, 4 2013.

[65] Y. H. Kim, J. S. Heo, T. H. Kim, S. Park, M. H. Yoon, J. Kim, M. S. Oh, G. R. Yi, Y. Y. Noh and S. K. Park, "Flexible metal-oxide devices made by room-temperature photochemical activation of sol–gel films," *Nature*, vol. 489, pp. 128–132, 9 2012.

[66] R. A. John, N. A. Chien, S. Shukla, N. Tiwari, C. Shi, N. G. Ing and N. Mathews, "Low-temperature chemical transformations for high-performance solution-processed oxide transistors," *Chemistry of Materials*, vol. 28, pp. 8305–8313, 11 2016.

[67] A. Liu, G. Liu, H. Zhu, B. Shin, E. Fortunato, R. Martins and F. Shan, "Hole mobility modulation of solution-processed nickel oxide thin-film transistor based on high-k dielectric," *Applied Physics Letters*, vol. 108, p. 233506, 6 2016.

[68] D. Jang, D. Kim and J. Moon, "Influence of fluid physical properties on ink-jet printability," *Langmuir*, vol. 25, pp. 2629–2635, 3 2009.

[69] Y. Diao, L. Shaw, Z. Bao and S. C. B. Mannsfeld, "Morphology control strategies for solution-processed organic semiconductor thin films," *Energy & Environmental Science*, vol. 7, pp. 2145–2159, 2014.

[70] H. Lian, L. Qi, J. Luo and K. Hu, "Experimental study and mechanism analysis on the effect of substrate wettability on graphene sheets distribution morphology within uniform printing droplets," *Journal of Physics: Condensed Matter*, vol. 30, p. 335001, 7 2018.

[71] J. Sheng, J. H. Lee, W. H. Choi, T. Hong, M. Kim and J. S. Park, "Review article: Atomic layer deposition for oxide semiconductor thin film transistors: Advances in research and development," *Journal of Vacuum Science & Technology A*, vol. 36, p. 060801, 11 2018.

[72] F. Garnier, R. Hajlaoui, A. Yassar and P. Srivastava, "All-polymer field-effect transistor realized by printing techniques," *Science*, vol. 265, pp. 1684–1686, 9 1994.

[73] H. Sirringhaus, T. Kawase, R. H. Friend, T. Shimoda, M. Inbasekaran, W. Wu and E. P. Woo, "High-resolution inkjet printing of all-polymer transistor circuits," *Science*, vol. 290, pp. 2123–2126, 12 2000.

[74] H. Minemawari, T. Yamada, H. Matsui, J. Tsutsumi, S. Haas, R. Chiba, R. Kumai and T. Hasegawa, "Inkjet printing of single-crystal films," *Nature*, vol. 475, pp. 364–367, 7 2011.

[75] M. Singh, H. M. Haverinen, P. Dhagat and G. E. Jabbour, "Inkjet printing-process and its applications," *Advanced Materials*, vol. 22, pp. 673–685, 2 2010.

[76] J. Leppäniemi, O. H. Huttunen, H. Majumdar and A. Alastalo, "Flexography-printed In2O3Semiconductor layers for high-mobility thin-film transistors on flexible plastic substrate," *Advanced Materials*, vol. 27, pp. 7168–7175, 10 2015.

[77] T. Cosnahan, A. A. R. Watt and H. E. Assender, "Flexography printing for organic thin film transistors," *Materials Today: Proceedings*, vol. 5, pp. 16051–16057, 2018.

[78] G. Grau and V. Subramanian, "Fully high-speed gravure printed, low-variability, high-performance organic polymer transistors with sub-5 V operation," *Advanced Electronic Materials*, vol. 2, p. 1500328, 1 2016.

[79] C. M. Homenick, R. James, G. P. Lopinski, J. Dunford, J. Sun, H. Park, Y. Jung, G. Cho and P. R. L. Malenfant, "Fully printed and encapsulated SWCNT-based thin film transistors via a combination of R2R gravure and inkjet printing," *ACS Applied Materials & Interfaces*, vol. 8, pp. 27900–27910, 10 2016.

[80] Y. Takeda, Y. Yoshimura, R. Shiwaku, K. Hayasaka, T. Sekine, T. Okamoto, H. Matsui, D. Kumaki, Y. Katayama and S. Tokito, "Organic complementary inverter circuits fabricated with reverse offset printing," *Advanced Electronic Materials*, vol. 4, p. 1700313, 11 2017.

[81] K. Hong, S. H. Kim, A. Mahajan and C. D. Frisbie, "Aerosol jet printed p- and n-type electrolyte-gated transistors with a variety of electrode materials: Exploring practical routes to printed electronics," *ACS Applied Materials & Interfaces*, vol. 6, pp. 18704–18711, 11 2014.

[82] C. S. Jones, X. Lu, M. Renn, M. Stroder and W.-S. Shih, "Aerosol-jet-printed, high-speed, flexible thin-film transistor made using single-walled carbon nanotube solution," *Microelectronic Engineering*, vol. 87, pp. 434–437, 3 2010.

[83] S. Lee, J. Kim, J. Choi, H. Park, J. Ha, Y. Kim, J. A. Rogers and U. Paik, "Patterned oxide semiconductor by electrohydrodynamic jet printing for transparent thin film transistors," *Applied Physics Letters*, vol. 100, p. 102108, 3 2012.

[84] Y. G. Lee and W. S. Choi, "Electrohydrodynamic jet-printed zinc–tin oxide TFTs and their bias stability," *ACS Applied Materials & Interfaces*, vol. 6, pp. 11167–11172, 7 2014.

[85] M. Kaltenbrunner, T. Sekitani, J. Reeder, T. Yokota, K. Kuribara, T. Tokuhara, M. Drack, R. Schwödiauer, I. Graz, S. Bauer-Gogonea, S. Bauer and T. Someya, "An ultra-lightweight design for imperceptible plastic electronics," *Nature*, vol. 499, pp. 458–463, 7 2013.

[86] K. Fukuda, Y. Takeda, Y. Yoshimura, R. Shiwaku, L. T. Tran, T. Sekine, M. Mizukami, D. Kumaki and S. Tokito, "Fully-printed high-performance organic thin-film transistors and circuitry on one-micron-thick polymer films," *Nature Communications*, vol. 5, 6 2014.

[87] E. Fortunato, P. Barquinha, A. Pimentel, A. Gonçalves, A. Marques, L. Pereira and R. Martins, "Fully transparent ZnO thin-film transistor produced at room temperature," *Advanced Materials*, vol. 17, pp. 590–594, 3 2005.

[88] Q. Cao, S.-H. Hur, Z. T. Zhu, Y. Sun, C. J. Wang, M. Meitl, M. Shim and J. Rogers, "Highly bendable, transparent thin-film transistors that use carbon-nanotube-based conductors and semiconductors with elastomeric dielectrics," *Advanced Materials*, vol. 18, pp. 304–309, 2 2006.

[89] J. Jang, H. Kang, H. C. N. Chakravarthula and V. Subramanian, "Fully inkjet-printed transparent oxide thin film transistors using a fugitive wettability switch," *Advanced Electronic Materials*, vol. 1, p. 1500086, 5 2015.

[90] Y. Park, K. J. Baeg and C. Kim, "Solution-processed nonvolatile organic transistor memory based on semiconductor blends," *ACS Applied Materials & Interfaces*, 2 2019.

[91] S. W. Jung, J. B. Koo, C. W. Park, B. S. Na, N. M. Park, J. Y. Oh, Y. G. Moon, S. S. Lee and K. W. Koo, "Non-volatile organic ferroelectric memory transistors fabricated using rigid polyimide islands on an elastomer substrate," *Journal of Materials Chemistry C*, vol. 4, pp. 4485–4490, 2016.

[92] M. E. Roberts, M. C. LeMieux and Z. Bao, "Sorted and aligned single-walled carbon nanotube networks for transistor-based aqueous chemical sensors," *ACS Nano*, vol. 3, pp. 3287–3293, 9 2009.

[93] D. Fu, H. Lim, Y. Shi, X. Dong, S. G. Mhaisalkar, Y. Chen, S. Moochhala and L. J. Li, "Differentiation of gas molecules using flexible and all-carbon nanotube devices," *The Journal of Physical Chemistry C*, vol. 112, pp. 650–653, 1 2008.

[94] J. Y. Oh, S. Rondeau-Gagné, Y. C. Chiu, A. Chortos, F. Lissel, G. J. N. Wang, B. C. Schroeder, T. Kurosawa, J. Lopez, T. Katsumata, J. Xu, C. Zhu, X. Gu, W. G. Bae, Y. Kim, L. Jin, J. W. Chung, J. B. H. Tok and Z. Bao, "Intrinsically stretchable and healable semiconducting polymer for organic transistors," *Nature*, vol. 539, pp. 411–415, 11 2016.

[95] T. Carey, S. Cacovich, G. Divitini, J. Ren, A. Mansouri, J. M. Kim, C. Wang, C. Ducati, R. Sordan and F. Torrisi, "Fully inkjet-printed two-dimensional material field-effect heterojunctions for wearable and textile electronics," *Nature Communications*, vol. 8, 10 2017.

[96] Y. Kim, Y. Kwon, K. Lee, Y. Oh, M. K. Um, D. Seong and J. Lee, "Flexible textile-based organic transistors using graphene/Ag nanoparticle electrode," *Nanomaterials*, vol. 6, p. 147, 8 2016.

[97] M. Hamedi, L. Herlogsson, X. Crispin, R. Marcilla, M. Berggren and O. Inganäs, "Fiber-embedded electrolyte-gated field-effect transistors for e-textiles," *Advanced Materials*, vol. 21, pp. 573–577, 12 2008.

[98] D. Soltman, B. Smith, S. J. S. Morris and V. Subramanian, "Inkjet printing of precisely defined features using contact-angle hysteresis," *Journal of Colloid and Interface Science*, vol. 400, pp. 135–139, 6 2013.

[99] G. Grau, W. J. Scheideler and V. Subramanian, "High-resolution gravure printed lines: Proximity effects and design rules," in *Printed Memory and Circuits*, 2015.

[100] A. M. Hussain and M. M. Hussain, "CMOS-technology-enabled flexible and stretchable electronics for internet of everything applications," *Advanced Materials*, vol. 28, pp. 4219–4249, 11 2015.

[101] G. B. Raupp, "Flexible thin film transistor arrays as an enabling platform technology: Opportunities and challenges," in *3rd International Conference on Semiconductor Technology for Ultra Large Integrated Circuits and Thin Film Transistors*, 2011.

[102] G. H. Gelinck, H. E. A. Huitema, E. Veenendaal, E. Cantatore, L. Schrijnemakers, J. B. P. H. Putten, T. C. T. Geuns, M. Beenhakkers, J. B. Giesbers, B.-H. Huisman, E. J. Meijer, E. M. Benito, F. J. Touwslager, A. W. Marsman, B. J. E. Rens and D. M. Leeuw, "Flexible active-matrix displays and shift registers based on solution-processed organic transistors," *Nature Materials*, vol. 3, pp. 106–110, 1 2004.

[103] K. J. Baeg, D. Khim, D. Y. Kim, S. W. Jung, J. B. Koo, I. K. You, H. Yan, A. Facchetti and Y. Y. Noh, "High speeds complementary integrated circuits fabricated with all-printed polymeric semiconductors," *Journal of Polymer Science Part B: Polymer Physics*, vol. 49, pp. 62–67, 9 2010.

[104] M. Jung, J. Kim, J. Noh, N. Lim, C. Lim, G. Lee, J. Kim, H. Kang, K. Jung, A. D. Leonard, J. M. Tour and G. Cho, "All-printed and roll-to-roll-printable 13.56-MHz-operated 1-bit RF tag on plastic foils," *IEEE Transactions on Electron Devices*, vol. 57, pp. 571–580, 3 2010.

[105] J. Noh, M. Jung, K. Jung, G. Lee, J. Kim, S. Lim, D. Kim, Y. Choi, Y. Kim, V. Subramanian and G. Cho, "Fully gravure-printed D flip-flop on plastic foils using single-walled carbon-nanotube-based TFTs," *IEEE Electron Device Letters*, vol. 32, pp. 638–640, 5 2011.

[106] M. Vilkman, T. Ruotsalainen, K. Solehmainen, E. Jansson and J. Hiitola-Keinänen, "Self-aligned metal electrodes in fully roll-to-roll processed organic transistors," *Electronics*, vol. 5, p. 2, 1 2016.

[107] T. N. Ng, B. Russo, B. Krusor, R. Kist and A. C. Arias, "Organic inkjet-patterned memory array based on ferroelectric field-effect transistors," *Organic Electronics*, vol. 12, pp. 2012–2018, 12 2011.

[108] C. W. Park, J. B. Koo, C. S. Hwang, H. Park, S. G. Im and S. Y. Lee, "Stretchable active matrix of oxide thin-film transistors with monolithic liquid metal interconnects," *Applied Physics Express*, vol. 11, p. 126501, 10 2018.

[109] D. Y. Khang, "A stretchable form of single-crystal silicon for high-performance electronics on rubber substrates," *Science*, vol. 311, pp. 208–212, 1 2006.

[110] D. H. Kim, J. H. Ahn, W. M. Choi, H. S. Kim, T. H. Kim, J. Song, Y. Y. Huang, Z. Liu, C. Lu and J. A. Rogers, "Stretchable and foldable silicon integrated circuits," *Science*, vol. 320, pp. 507–511, 4 2008.

[111] S. Wang, J. Xu, W. Wang, G.-J. N. Wang, R. Rastak, F. Molina-Lopez, J. W. Chung, S. Niu, V. R. Feig, J. Lopez, T. Lei, S. K. Kwon, Y. Kim, A. M. Foudeh, A. Ehrlich, A. Gasperini, Y. Yun, B. Murmann, J. B.-H. Tok and Z. Bao, "Skin electronics from scalable fabrication of an intrinsically stretchable transistor array," *Nature*, vol. 555, pp. 83–88, 2 2018.

[112] T. Sekitani, H. Nakajima, H. Maeda, T. Fukushima, T. Aida, K. Hata and T. Someya, "Stretchable active-matrix organic light-emitting diode display using printable elastic conductors," *Nature Materials*, vol. 8, pp. 494–499, 5 2009.

[113] W. Liu, M. S. Song, B. Kong and Y. Cui, "Flexible and stretchable energy storage: Recent advances and future perspectives," *Advanced Materials*, vol. 29, p. 1603436, 11 2016.

[114] M. A. Green, Y. Hishikawa, E. D. Dunlop, D. H. Levi, J. Hohl-Ebinger, M. Yoshita and A. W. Y. Ho-Baillie, "Solar cell efficiency tables (version 53)," *Progress in Photovoltaics: Research and Applications*, vol. 27, pp. 3–12, 12 2018.

[115] F. Kessler and D. Rudmann, "Technological aspects of flexible CIGS solar cells and modules," *Solar Energy*, vol. 77, pp. 685–695, 12 2004.

[116] A. Chirilă, S. Buecheler, F. Pianezzi, P. Bloesch, C. Gretener, A. R. Uhl, C. Fella, L. Kranz, J. Perrenoud, S. Seyrling, R. Verma, S. Nishiwaki, Y. E. Romanyuk, G. Bilger and A. N. Tiwari, "Highly efficient Cu(In,Ga)Se$_2$ solar cells grown on flexible polymer films," *Nature Materials*, vol. 10, pp. 857–861, 9 2011.

[117] A. Gerthoffer, F. Roux, F. Emieux, P. Faucherand, H. Fournier, L. Grenet and S. Perraud, "CIGS solar cells on flexible ultra-thin glass substrates: Characterization and bending test," *Thin Solid Films*, vol. 592, pp. 99–104, 10 2015.

[118] S. Lizin, S. V. Passel, E. D. Schepper, W. Maes, L. Lutsen, J. Manca and D. Vanderzande, "Life cycle analyses of organic photovoltaics: A review," *Energy & Environmental Science*, vol. 6, p. 3136, 2013.

[119] B. A. Gregg and M. C. Hanna, "Comparing organic to inorganic photovoltaic cells: Theory, experiment, and simulation," *Journal of Applied Physics*, vol. 93, pp. 3605–3614, 3 2003.

[120] G. Yu, J. Gao, J. C. Hummelen, F. Wudl and A. J. Heeger, "Polymer photovoltaic cells: Enhanced efficiencies via a network of internal donor-acceptor heterojunctions," *Science*, vol. 270, pp. 1789–1791, 12 1995.

[121] P. K. Watkins, A. B. Walker and G. L. B. Verschoor, "Dynamical Monte Carlo modelling of organic solar cells: The dependence of internal quantum efficiency on morphology," *Nano Letters*, vol. 5, pp. 1814–1818, 9 2005.

[122] S. Alem, J. Lu, R. Movileanu, T. Kololuoma, A. Dadvand and Y. Tao, "Solution-processed annealing-free ZnO nanoparticles for stable inverted organic solar cells," *Organic Electronics*, vol. 15, pp. 1035–1042, 5 2014.

[123] M. Prosa, M. Tessarolo, M. Bolognesi, O. Margeat, D. Gedefaw, M. Gaceur, C. Videlot-Ackermann, M. R. Andersson, M. Muccini, M. Seri and J. Ackermann, "Enhanced ultraviolet stability of air-processed polymer solar cells by Al doping of the ZnO interlayer," *ACS Applied Materials & Interfaces*, vol. 8, pp. 1635–1643, 1 2016.

[124] C. Waldauf, M. Morana, P. Denk, P. Schilinsky, K. Coakley, S. A. Choulis and C. J. Brabec, "Highly efficient inverted organic photovoltaics using solution based titanium oxide as electron selective contact," *Applied Physics Letters*, vol. 89, p. 233517, 12 2006.

[125] S. Lattante, "Electron and hole transport layers: Their use in inverted bulk heterojunction polymer solar cells," *Electronics*, vol. 3, pp. 132–164, 3 2014.

[126] Z. Yin, J. Wei and Q. Zheng, "Interfacial materials for organic solar cells: Recent advances and perspectives," *Advanced Science*, vol. 3, p. 1500362, 2 2016.

[127] C. J. Brabec, S. Gowrisanker, J. J. M. Halls, D. Laird, S. Jia and S. P. Williams, "Polymer-fullerene bulk-heterojunction solar cells," *Advanced Materials*, vol. 22, pp. 3839–3856, 8 2010.

[128] C. Yan, S. Barlow, Z. Wang, H. Yan, A. K. Y. Jen, S. R. Marder and X. Zhan, "Non-fullerene acceptors for organic solar cells," *Nature Reviews Materials*, vol. 3, p. 18003, 2 2018.

[129] K. Zilberberg, H. Gharbi, A. Behrendt, S. Trost and T. Riedl, "Low-temperature, solution-processed MoOx for efficient and stable organic solar cells," *ACS Applied Materials & Interfaces*, vol. 4, pp. 1164–1168, 2 2012.

[130] T. Kololuoma, J. Lu, S. Alem, N. Graddage, R. Movileanu, S. Moisa and Y. Tao, "Flexo printed sol-gel derived vanadium oxide films as an interfacial hole-transporting layer for organic solar cells," in *Oxide-based Materials and Devices VI*, 2015.

[131] J. G. Tait, B. J. Worfolk, S. A. Maloney, T. C. Hauger, A. L. Elias, J. M. Buriak and K. D. Harris, "Spray coated high-conductivity PEDOT:PSS transparent electrodes for stretchable and mechanically-robust organic solar cells," *Solar Energy Materials and Solar Cells*, vol. 110, pp. 98–106, 3 2013.

[132] L. S. Roman, W. Mammo, L. A. A. Pettersson, M. R. Andersson and O. Inganäs, "High quantum efficiency polythiophene," *Advanced Materials*, vol. 10, pp. 774–777, 7 1998.

[133] T. -Y. Chu, S. Alem, P. G. Verly, S. Wakim, J. Lu, Y. Tao, S. Beaupré, M. Leclerc, F. Bélanger, D. Désilets, S. Rodman, D. Waller and R. Gaudiana, "Highly efficient polycarbazole-based organic photovoltaic devices," *Applied Physics Letters*, vol. 95, p. 063304, 8 2009.

[134] A. Armin, A. Yazmaciyan, M. Hambsch, J. Li, P. L. Burn and P. Meredith, "Electro-optics of conventional and inverted thick junction organic solar cells," *ACS Photonics*, vol. 2, pp. 1745–1754, 11 2015.

[135] T. M. Eggenhuisen, Y. Galagan, A. F. K. V. Biezemans, T. M. W. L. Slaats, W. P. Voorthuijzen, S. Kommeren, S. Shanmugam, J. P. Teunissen, A. Hadipour, W. J. H. Verhees, S. C. Veenstra, M. J. J. Coenen, J. Gilot, R. Andriessen and W. A. Groen, "High efficiency, fully inkjet printed organic solar cells with freedom of design," *Journal of Materials Chemistry A*, vol. 3, pp. 7255–7262, 2015.

[136] F. C. Krebs, M. Jørgensen, K. Norrman, O. Hagemann, J. Alstrup, T. D. Nielsen, J. Fyenbo, K. Larsen and J. Kristensen, "A complete process for production of flexible large area polymer solar cells entirely using screen printing—first public demonstration," *Solar Energy Materials and Solar Cells*, vol. 93, pp. 422–441, 4 2009.

[137] S. Alem, N. Graddage, J. Lu, T. Kololuoma, R. Movileanu and Y. Tao, "Flexographic printing of polycarbazole-based inverted solar cells," *Organic Electronics*, vol. 52, pp. 146–152, 1 2018.

[138] M. Välimäki, E. Jansson, P. Korhonen, A. Peltoniemi and S. Rousu, "Custom-shaped organic photovoltaic modules—freedom of design by printing," *Nanoscale Research Letters*, vol. 12, 2 2017.

[139] T. Wang, N. W. Scarratt, H. Yi, A. D. F. Dunbar, A. J. Pearson, D. C. Watters, T. S. Glen, A. C. Brook, J. Kingsley, A. R. Buckley, M. W. A. Skoda, A. M. Donald, R. A. L. Jones, A. Iraqi and D. G. Lidzey, "Fabricating high performance, donor-acceptor copolymer solar cells by spray-coating in air," *Advanced Energy Materials*, vol. 3, pp. 505–512, 2 2013.

[140] F. C. Krebs, S. A. Gevorgyan and J. Alstrup, "A roll-to-roll process to flexible polymer solar cells: Model studies, manufacture and operational stability studies," *Journal of Materials Chemistry*, vol. 19, p. 5442, 2009.

[141] F. C. Krebs, "Fabrication and processing of polymer solar cells: A review of printing and coating techniques," *Solar Energy Materials and Solar Cells*, vol. 93, pp. 394–412, 4 2009.

[142] R. R. Søndergaard, M. Hösel and F. C. Krebs, "Roll-to-roll fabrication of large area functional organic materials," *Journal of Polymer Science Part B: Polymer Physics*, vol. 51, pp. 16–34, 10 2012.

[143] X. Gong, M. Tong, Y. Xia, W. Cai, J. S. Moon, Y. Cao, G. Yu, C. L. Shieh, B. Nilsson and A. J. Heeger, "High-detectivity polymer photodetectors with spectral response from 300 nm to 1450 nm," *Science*, vol. 325, pp. 1665–1667, 8 2009.

[144] P. Büchele, M. Richter, S. F. Tedde, G. J. Matt, G. N. Ankah, R. Fischer, M. Biele, W. Metzger, S. Lilliu, O. Bikondoa, J. E. Macdonald, C. J. Brabec, T. Kraus, U. Lemmer and O. Schmidt, "X-ray imaging with scintillator-sensitized hybrid organic photodetectors," *Nature Photonics*, vol. 9, pp. 843–848, 11 2015.

[145] T. R. Andersen, H. F. Dam, M. Hösel, M. Helgesen, J. E. Carlé, T. T. Larsen-Olsen, S. A. Gevorgyan, J. W. Andreasen, J. Adams, N. Li, F. Machui, G. D. Spyropoulos, T. Ameri, N. Lemaître, M. Legros, A. Scheel, D. Gaiser, K. Kreul, S. Berny, O. R. Lozman, S. Nordman, M. Välimäki, M. Vilkman, R. R. Søndergaard, M. Jørgensen, C. J. Brabec and F. C. Krebs, "Scalable, ambient atmosphere roll-to-roll manufacture of encapsulated large area, flexible organic tandem solar cell modules," *Energy & Environmental Science*, vol. 7, p. 2925, 6 2014.

[146] J. Y. Lee, K. T. Lee, S. Seo and L. J. Guo, "Decorative power generating panels creating angle insensitive transmissive colors," *Scientific Reports*, vol. 4, 2 2014.

[147] Y. Cui, C. Yang, H. Yao, J. Zhu, Y. Wang, G. Jia, F. Gao and J. Hou, "Efficient semitransparent organic solar cells with tunable color enabled by an ultralow-bandgap nonfullerene acceptor," *Advanced Materials*, vol. 29, p. 1703080, 10 2017.

[148] R. R. Lunt and V. Bulovic, "Transparent, near-infrared organic photovoltaic solar cells for window and energy-scavenging applications," *Applied Physics Letters*, vol. 98, p. 113305, 3 2011.

[149] C. C. Chen, L. Dou, R. Zhu, C. H. Chung, T. B. Song, Y. B. Zheng, S. Hawks, G. Li, P. S. Weiss and Y. Yang, "Visibly transparent polymer solar cells produced by solution processing," *ACS Nano*, vol. 6, pp. 7185–7190, 7 2012.

[150] D. J. Lipomi and Z. Bao, "Stretchable, elastic materials and devices for solar energy conversion," *Energy & Environmental Science*, vol. 4, p. 3314, 2011.

[151] D. J. Lipomi, B. C. K. Tee, M. Vosgueritchian and Z. Bao, "Stretchable organic solar cells," *Advanced Materials*, vol. 23, pp. 1771–1775, 2 2011.

[152] N. R. Tummala, C. Bruner, C. Risko, J.-L. Brédas and R. H. Dauskardt, "Molecular-scale understanding of cohesion and fracture in P3HT:Fullerene blends," *ACS Applied Materials & Interfaces*, vol. 7, pp. 9957–9964, 4 2015.

[153] T. Kim, J. H. Kim, T. E. Kang, C. Lee, H. Kang, M. Shin, C. Wang, B. Ma, U. Jeong, T.-S. Kim and B. J. Kim, "Flexible, highly efficient all-polymer solar cells," *Nature Communications*, vol. 6, 10 2015.

[154] B. O'Connor, K. P. Pipe and M. Shtein, "Fiber based organic photovoltaic devices," *Applied Physics Letters*, vol. 92, p. 193306, 5 2008.

[155] H. Jinno, K. Fukuda, X. Xu, S. Park, Y. Suzuki, M. Koizumi, T. Yokota, I. Osaka, K. Takimiya and T. Someya, "Stretchable and waterproof elastomer-coated organic photovoltaics for washable electronic textile applications," *Nature Energy*, vol. 2, pp. 780–785, 9 2017.

[156] P. D. Rack and P. H. Holloway, "The structure, device physics, and material properties of thin film electroluminescent displays," *Materials Science and Engineering: R: Reports*, vol. 21, pp. 171–219, 1 1998.

[157] L. Wang, L. Xiao, H. Gu and H. Sun, "Advances in alternating current electroluminescent devices," *Advanced Optical Materials*, vol. 7, p. 1801154, 1 2019.

[158] M. Bredol and H. S. Dieckhoff, "Materials for powder-based AC-electroluminescence," *Materials*, vol. 3, pp. 1353–1374, 2 2010.

[159] *ASTM F2964-12, Standard Test Method for Determining the Uniformity of the Luminance of an Electroluminescent Lamp or Other Diffuse Lighting Device*, West, Conshohocken, 2012.

[160] J. Wang, C. Yan, K. J. Chee and P. S. Lee, "Highly stretchable and self-deformable alternating current electroluminescent devices," *Advanced Materials*, vol. 27, pp. 2876–2882, 3 2015.

[161] F. Stauffer and K. Tybrandt, "Bright stretchable alternating current electroluminescent displays based on high permittivity composites," *Advanced Materials*, vol. 28, pp. 7200–7203, 6 2016.

[162] D. Hu, X. Xu, J. Miao, O. Gidron and H. Meng, "A stretchable alternating current electroluminescent fiber," *Materials*, vol. 11, p. 184, 1 2018.

[163] B. You, Y. Kim, B. K. Ju and J. W. Kim, "Highly stretchable and waterproof electroluminescence device based on superstable stretchable transparent electrode," *ACS Applied Materials & Interfaces*, vol. 9, pp. 5486–5494, 2 2017.

[164] Y. Zhou, S. Cao, J. Wang, H. Zhu, J. Wang, S. Yang, X. Wang and D. Kong, "Bright stretchable electroluminescent devices based on silver nanowire electrodes and high-k thermoplastic elastomers," *ACS Applied Materials & Interfaces*, vol. 10, pp. 44760–44767, 11 2018.

[165] C. H. Yang, B. Chen, J. Zhou, Y. M. Chen and Z. Suo, "Electroluminescence of giant stretchability," *Advanced Materials*, vol. 28, pp. 4480–4484, 11 2015.

[166] C. Larson, B. Peele, S. Li, S. Robinson, M. Totaro, L. Beccai, B. Mazzolai and R. Shepherd, "Highly stretchable electroluminescent skin for optical signaling and tactile sensing," *Science*, vol. 351, pp. 1071–1074, 3 2016.

[167] P. Kordt, J. J. M. Holst, M. A. Helwi, W. Kowalsky, F. May, A. Badinski, C. Lennartz and D. Andrienko, "Modeling of organic light emitting diodes: From molecular to device properties," *Advanced Functional Materials*, vol. 25, pp. 1955–1971, 1 2015.

[168] A. Buckley, Ed., *Organic Light-Emitting Diodes (OLEDs): Materials, Devices and Applications*, Woodhead Publishing, 2013.

[169] F. Guo, A. Karl, Q. F. Xue, K. C. Tam, K. Forberich and C. J. Brabec, "The fabrication of color-tunable organic light-emitting diode displays via solution processing," *Light: Science & Applications*, vol. 6, p. e17094, 6 2017.

[170] R. Abbel, I. Vries, A. Langen, G. Kirchner, H. H. Gorter, J. Wilson and P. Groen, "Toward high volume solution based roll-to-roll processing of OLEDs," *Journal of Materials Research*, vol. 32, pp. 2219–2229, 6 2017.

[171] S. M. Raupp, L. Merklein, M. Pathak, P. Scharfer and W. Schabel, "An experimental study on the reproducibility of different multilayer OLED materials processed by slot die coating," *Chemical Engineering Science*, vol. 160, pp. 113–120, 3 2017.

[172] K. J. Choi, J. Y. Lee, J. Park and Y.-S. Seo, "Multilayer slot-die coating of large-area organic light-emitting diodes," *Organic Electronics*, vol. 26, pp. 66–74, 11 2015.

[173] G. Hernandez-Sosa, N. Bornemann, I. Ringle, M. Agari, E. Dörsam, N. Mechau and U. Lemmer, "Rheological and drying considerations for uniformly gravure-printed layers: Towards large-area flexible organic light-emitting diodes," *Advanced Functional Materials*, vol. 23, pp. 3164–3171, 2 2013.

[174] P. Kopola, M. Tuomikoski, R. Suhonen and A. Maaninen, "Gravure printed organic light emitting diodes for lighting applications," *Thin Solid Films*, vol. 517, pp. 5757–5762, 8 2009.

[175] H. Gorter, M. J. J. Coenen, M. W. L. Slaats, M. Ren, W. Lu, C. J. Kuijpers and W. A. Groen, "Toward inkjet printing of small molecule organic light emitting diodes," *Thin Solid Films*, vol. 532, pp. 11–15, 4 2013.

[176] M. Y. Wong, "Recent advances in polymer organic light-emitting diodes (PLED) using non-conjugated polymers as the emitting layer and contrasting them with conjugated counterparts," *Journal of Electronic Materials*, vol. 46, pp. 6246–6281, 7 2017.

[177] Q. Wei, N. Fei, A. Islam, T. Lei, L. Hong, R. Peng, X. Fan, L. Chen, P. Gao and Z. Ge, "Small-molecule emitters with high quantum efficiency: Mechanisms, structures, and applications in OLED devices," *Advanced Optical Materials*, vol. 6, p. 1800512, 7 2018.

[178] J. H. Jou, S. Kumar, A. Agrawal, T. H. Li and S. Sahoo, "Approaches for fabricating high efficiency organic light emitting diodes," *Journal of Materials Chemistry C*, vol. 3, pp. 2974–3002, 2015.

[179] D. Zhao, Z. Qin, J. Huang and J. Yu, "Progress on material, structure and function for tandem organic light-emitting diodes," *Organic Electronics*, vol. 51, pp. 220–242, 12 2017.

[180] A. Liang, L. Ying and F. Huang, "Recent progresses of iridium complex-containing macromolecules for solution-processed organic light-emitting diodes," *Journal of Inorganic and Organometallic Polymers and Materials*, vol. 24, pp. 905–926, 10 2014.

[181] C. Zhong, C. Duan, F. Huang, H. Wu and Y. Cao, "Materials and devices toward fully solution processable organic light-emitting diodes," *Chemistry of Materials*, vol. 23, pp. 326–340, 2 2011.

[182] K. Fehse, K. Walzer, K. Leo, W. Lövenich and A. Elschner, "Highly conductive polymer anodes as replacements for inorganic materials in high-efficiency organic light-emitting diodes," *Advanced Materials*, vol. 19, pp. 441–444, 2 2007.

[183] E. C. W. Ou, L. Hu, G. C. R. Raymond, O. K. Soo, J. Pan, Z. Zheng, Y. Park, D. Hecht, G. Irvin, P. Drzaic and G. Gruner, "Surface-modified nanotube anodes for high performance organic light-emitting diode," *ACS Nano*, vol. 3, pp. 2258–2264, 7 2009.

[184] H. Lee, D. Lee, Y. Ahn, E. W. Lee, L. S. Park and Y. Lee, "Highly efficient and low voltage silver nanowire-based OLEDs employing a n-type hole injection layer," *Nanoscale*, vol. 6, pp. 8565–8570, 2014.

[185] M. S. White, M. Kaltenbrunner, E. D. Głowacki, K. Gutnichenko, G. Kettlgruber, I. Graz, S. Aazou, C. Ulbricht, D. A. M. Egbe, M. C. Miron, Z. Major, M. C. Scharber, T. Sekitani, T. Someya, S. Bauer and N. S. Sariciftci, "Ultrathin, highly flexible and stretchable PLEDs," *Nature Photonics*, vol. 7, pp. 811–816, 7 2013.

[186] M. Zhang, S. Höfle, J. Czolk, A. Mertens and A. Colsmann, "All-solution processed transparent organic light emitting diodes," *Nanoscale*, vol. 7, pp. 20009–20014, 2015.

[187] T. H. Han, M. R. Choi, C. W. Jeon, Y. H. Kim, S. K. Kwon and T. W. Lee, "Ultrahigh-efficiency solution-processed simplified small-molecule organic light-emitting diodes using universal host materials," *Science Advances*, vol. 2, p. e1601428, 10 2016.

[188] S. Scholz, D. Kondakov, B. Lüssem and K. Leo, "Degradation mechanisms and reactions in organic light-emitting devices," *Chemical Reviews*, vol. 115, pp. 8449–8503, 7 2015.

[189] Y. Y. Lin, Y. N. Chang, M. H. Tseng, C. C. Wang and F.-Y. Tsai, "Air-stable flexible organic light-emitting diodes enabled by atomic layer deposition," *Nanotechnology*, vol. 26, p. 024005, 12 2014.

[190] A. Perumal, H. Faber, N. Yaacobi-Gross, P. Pattanasattayavong, C. Burgess, S. Jha, M. A. McLachlan, P. N. Stavrinou, T. D. Anthopoulos and D. D. C. Bradley, "High-efficiency, solution-processed, multilayer phosphorescent organic light-emitting diodes with a copper thiocyanate hole-injection/hole-transport layer," *Advanced Materials*, vol. 27, pp. 93–100, 11 2014.

[191] S. Raupp, D. Daume, S. Tekoglu, L. Merklein, U. Lemmer, G. Hernandez-Sosa, H. M. Sauer, E. Dörsam, P. Scharfer and W. Schabel, "Slot die coated and flexo printed highly efficient SMOLEDs," *Advanced Materials Technologies*, vol. 2, p. 1600230, 12 2016.

[192] M. H. Park, J. Y. Kim, T. H. Han, T. S. Kim, H. Kim and T. W. Lee, "Flexible lamination encapsulation," *Advanced Materials*, vol. 27, pp. 4308–4314, 6 2015.

[193] W. Kim, S. Kwon, S. M. Lee, J. Y. Kim, Y. Han, E. Kim, K. C. Choi, S. Park and B.-C. Park, "Soft fabric-based flexible organic light-emitting diodes," *Organic Electronics*, vol. 14, pp. 3007–3013, 11 2013.

[194] S. Choi, S. Kwon, H. Kim, W. Kim, J. H. Kwon, M. S. Lim, H. S. Lee and K. C. Choi, "Highly flexible and efficient fabric-based organic light-emitting devices for clothing-shaped wearable displays," *Scientific Reports*, vol. 7, 7 2017.

[195] A. Sandström, H. F. Dam, F. C. Krebs and L. Edman, "Ambient fabrication of flexible and large-area organic light-emitting devices using slot-die coating," *Nature Communications*, vol. 3, 1 2012.

[196] E. Juntunen, S. Ihme, A. Huttunen and J.T. Makinen, "R2R process for integrating LEDs on flexible substrate," in *2017 IMAPS Nordic Conference on Microelectronics Packaging (NordPac)*, 2017.

[197] H. Lee, E. Kim, Y. Lee, H. Kim, J. Lee, M. Kim, H. J. Yoo and S. Yoo, "Toward all-day wearable health monitoring: An ultralow-power, reflective organic pulse oximetry sensing patch," *Science Advances*, vol. 4, p. eaas9530, 11 2018.

[198] D. Yang, R. Yang, S. Priya and S. F. Liu, "Recent advances in flexible perovskite solar cells: Fabrication and applications," *Angewandte Chemie International Edition*, vol. 58, pp. 4466–4483, 2 2019.

# 4

---

## *Printing Techniques*

---

**Chloé Bois, Marie-Ève Huppé, Michael Rozel, and Ngoc Duc Trinh**
*Printability and Graphic Communications Institute*

## CONTENTS

## 4.1 Introduction

This chapter will provide the reader with a general understanding of how graphic printing processes have evolved over the last two decades to serve the fabrication of printed electronics (PE). As the graphics industry is an established industrial domain with its own resource ecosystem, supply chain and standardized techniques (since the seventh century!), graphic chain optimization has been focused to serve the traditional printing industry first. Consequently, PE manufacturing is frequently constrained by the limitations of a standardized graphic industry; but, its potential for flexible, low-cost mass-production techniques, as demonstrated by many recent academic and industrial developments, is driving a shift in the manufacturing environment.

The following statement, and its illustration in Figure 4.1, summarizes the vision of the authors:

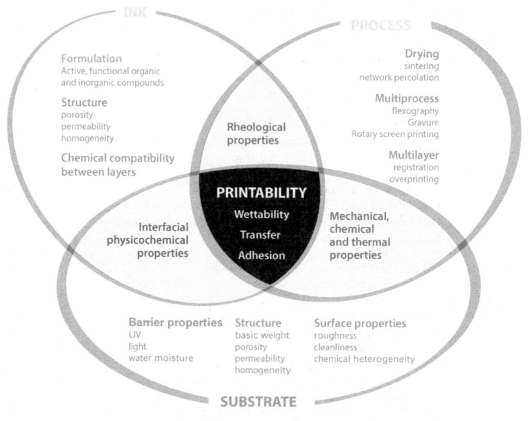

**FIGURE 4.1**
Ink/substrate/process matrix.

For PE or graphic production, the industrial printer's objective is to implement a coherent ink/substrate/process system by controlling rheological, mechanical, interfacial properties, and drying in order to reach the required effect for a targeted application at the highest possible speed and cost effectiveness.

## 4.2 Evolution of Printing

Printing is a disruptive invention which has supported human development since its birth in seventh or eighth century China. With papermaking, also invented in China in the seventh century AD, printing techniques have evolved in their applications, technical aspects, and social value. Therefore, printing has had, at least, three main consecutives, yet coexisting, definitions depending on its usage: (1) to share information, (2) as a tool for visual mass communication, and (3) as an additive manufacturing technique. The history of printing, although undoubtedly deformed by Eurocentrism biases, provides a deeper understanding of the specificities, limitations, and potentials of the current techniques available for either conventional graphics or PE.

### 4.2.1 Development of Printing Processes for Information Sharing

Printing is a set of techniques which allow the reproduction of signs and symbols on a substrate in large quantities, enabling the mass distribution of specific information.

At its invention in China and Korea, ink was transferred from the relief image on an engraved wood plate onto the paper through the application of manual pressure. Typography and moveable type were invented in the eleventh-century China using terra cotta; then, in thirteenth century, type made from metallic alloys was introduced in Korea.

Printing techniques expanded to the west with the Mongol invasions of Europe and the help of Marco Polo (thirteenth century), who brought back from Asia many things previously unseen by Europeans of that time, among them paper that gradually replaced expensive and elitist vellum and parchment made from animal skins. With cheaper material available to convey ideas, books and pamphlets became more accessible to literate people. In this context, Gutenberg, a fifteenth century German metalsmith, worked on the improvement of moveable type by inventing molds to allow the production of multiple batches of moveable type with standardized characteristics such as typefaces, point sizes, line lengths, line-spacing (leading), letter-spacing (tracking), and the adjustment of the space between pairs of letters (kerning). Each set of moveable type was made from the same *fonte*, a French word referring to the process of casting metal type at a foundry, using a cheap metal alloy containing antimony, lead, and tin. While this technique has been replaced by more modern processes, the terminology remains and typefaces are nowadays known as fonts.

His skill as a metalsmith also allowed Gutenberg to create a printing press that allowed the application of constant pressure of the moveable type over the entire surface of paper during the entire run.

The cyclical motion of the ink being pressed into the paper, which is subsequently removed from the press, may scatter droplets of liquid ink onto non-printed areas, a phenomenon well known in modern ink-jet printing as satellite droplets. For this reason, oily, pasty inks were developed to provide good adhesion on the metal type and to limit spreading and leaking during the run.

One of the most famous applications of these principles was the Gutenberg Bible B42, of which 180 copies were printed in the 1450s. Despite the lack of success of this book, the techniques developed for its printing reduced production costs and manufacturing time, starting an era of mass communication of new ideas in various languages, breaking the Latin-only monopoly of the clergy on culture, sparking the Renaissance (fourteenth to seventeenth centuries) and Age of Enlightenment (seventeenth to nineteenth centuries) in Europe.

### 4.2.2 Development of Printing Processes for Visual Mass Communication

In Europe, beginning with the work of physicists during the Age of Enlightenment, printing came to be explained through the law of colorimetry as a technique for the reproduction of shapes and colors.

While various models have been proposed to explain color perception, a simple model, first proposed by Munsell at the turn of the twentieth century and developed by Hunter and the Commission International de l'Éclairage (CIE) some 60 years later, is the illuminant-object-observer model illustrated in Figure 4.2, where:

- The illuminant is the mathematical representation of the spectral power distribution of a light source. In industry, standard illuminants provide comparable light sources, depending on the printing application.

- The object is composed of matter which may absorb, emit, reflect, and/or transform part of the light emitted by the light source as it interacts with its surface and first atomic layers.

- The observer is the representation of the capacity of a receptor and processor to either physically measure the light emissions from the object or physiologically perceive light emissions from the object in the case of a human eye and brain.

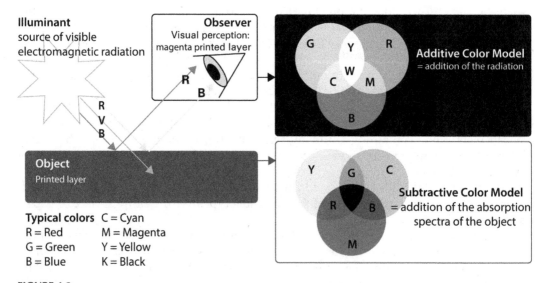

**FIGURE 4.2**

Color perception models: Illuminant/object/observer model (left). Additive and subtractive models (right).

Changing the characteristics of any of these elements will affect the final color perception or measurement. Thus, printing is a group of techniques that modifies the way a substrate (object) interacts with the emission spectrum of a light source, as explained by the subtractive color model in Figure 4.2. On the other hand, the perception or measurement of the emission spectrum from the printed substrate is described by the additive color model in Figure 4.2. This emission spectrum is represented by the brain or calculated by devices as colors and shapes.

Incident light may be either absorbed, reflected, or transmitted by the substrate. The interaction of the substrate, paper or plastic, film for example, with a light source is modified by the ink transferred onto its surface. The functional elements of inks (pigments or dyes) will modify the incident light in a number of ways, depending on their nature:

- Absorb a part of the illuminant spectrum in the case of colorants.
- Absorb energy from the ultraviolet (invisible) region of the spectrum of the illuminant and emit as visible light, as in the case of fluorescent and phosphorescent pigments and dyes.
- Reduce the transmission of incident light through the substrate, in the case of opaque pigments.
- Modify the reflective properties of the substrate as in the case of matting agents, reflective pigments, or metallic flakes.

Two main strategies are used in the graphics industries to reproduce the perception of a given color with a chosen shape. One strategy consists in the use of solid spot colors. The substrate is printed with one ink containing a pigment or mix of pigments which absorb the complementary part of the spectrum for the targeted color. The second strategy is based on the principle of color separation, where the chosen shape is subdivided into very small areas called pixels or rastering. The number of pixels per surface area is called printing resolution. Each pixel has (x, y) coordinates in the plain and a specified color. That color is decomposed into a set of predefined (process) colors. Every pixel is composed of multiple dots of the different process colors, where the number of dots per surface area is called dot density. Empirically, the dot density is limited by the equipment and pressroom supplies available and is four times higher than the highest the printing resolution achievable.

For example, in quadrichromy or four color (CMYK) process, inks composed of cyan, magenta, yellow, and black pigments are printed as dots of various sizes ranging from 0 to 100% of the area of the pixel. As each pixel can be a particular color, various hues and shapes may be reproduced using a different combination of a limited number of process inks.

In process printing, not only the thickness of the ink layer forming the dot will impact visual perception but also the type of pigment and the quality of the ink formulation. For this reason, as various printing processes were developed (detailed in Section 4.7), equipment and supplies were standardized for each of them to both control the size and shape of the printed dots as well as the thickness of the printed layer, optimizing the cost of a given application.

### 4.2.3 Development of Printing Processes as an Additive Manufacturing Technique

In the course of its evolution from its original application as a tool of mass communication, printing has become essentially the combination of additive manufacturing techniques which allow a film of material to be deposited on a substrate with a chosen shape and thickness.

At the fundamental level, printing aims to provide an optimal ink/substrate/printing process system through the control of rheological, mechanical, and interfacial properties in order to produce the targeted application at the highest speed and best cost effectiveness possible.

This definition is the most relevant for PE.

## 4.3 The Graphic Chain versus the Functional Chain

The typical Shannon-Weaver model of communication includes the concepts of information source, message, transmitter, signal, channel, noise, receiver, information destination, probability of error, encoding, decoding, information rate, and channel capacity. The graphic chain is composed of various steps which allow cost-effective communication with limited noise of complex information to the principal receivers (the customers) in conjunction with secondary objectives such as protection (packaging) or marketing (advertising) using a common channel (the print media).

The graphic chain was principally optimized and all the steps are heavily standardized for printed media and packaging, therefore, the use of printing processes as an additive manufacturing techniques for electronics, medical, or other applications presents new challenges. While the input data, supplies, and the primary and secondary functions of the output have imposed drastic changes on the printing industry, many of the distinctive skills and capabilities of the different steps of the graphics chain remain to the benefits of the new applications but sometimes for their detriment. An example of adaptation of the graphics chain for functional applications, called functional chain, is illustrated in Figure 4.3.

### 4.3.1 Prepress: File Preparation and Supplies Selection

For both graphic and functional applications, the prepress step at the beginning of the chain consists of preparing the printing file by:

1. Converting tridimensional devices, such as capacitors or resistors, into a collection of (1) layers with given shape and thickness and (2) layers superimposition information.

2. Translating these targets into digital data in a comprehensive format for the graphic supplies' manufacturer. Traditionally, graphic printing for the most part uses halftones and polka dots to reproduce pictures and images, whereas lines and solids elements are mostly used for text and symbol reproduction. The latter are often created by vector graphics, the format preferred in electronics. However, whatever the type of format used to create the features, printing information is binary and requires a rastering conversion. This step introduces a degradation of the targeted original patterns. Moreover, the file formats favored in the graphics industry and in electronics can differ, which may result in errors of format translation, .ai vs .dvx for example. Therefore, protocols of control for exchanging files should be created to limit information degradation.

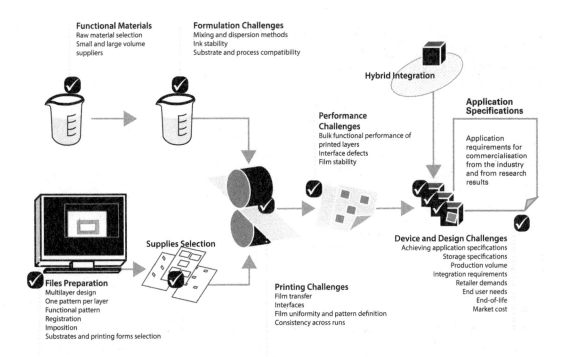

**FIGURE 4.3**
Graphic chain applied to functional printed products.

3. Selecting a printing process for each layer to be manufactured. Contrary to graphics printing, printing process selection for PE is primarily driven by technical limitations such as ink viscosity, drying requirements, and lateral and vertical resolutions, before considering economic aspects. The lateral resolution (x,y) defines the size limitations of the features and is closely related to printing definition, where the vertical resolution represents the size limitations of the thickness (z). Figure 4.4 offers examples of different resolution requirements: batteries and fuel cells require low lateral resolution and very high vertical resolution, while transistors and sensors are at the opposite end of the spectrum.

4. Integrating corrective models to adjust the shapes as a consequence of defects which are inherent in the process (deformation of an image transferred from a cylinder onto a substrate supported by another cylinder, for instance).

5. Adding elements unrelated to the application but required by the production chain, such as cutting bars, automatic registration marks, printing and drying quality control features, etc.

## 4.3.2 Press Configuration and Terminology

Appropriation of technical and empirical terms related to printing manufacturing systems by other industries demonstrates the potential of these additive techniques; but, unfortunately, this has led to some variation in their primary definitions. Therefore, this section attempts to disambiguate the terminology.

The term printer is preferably used for office devices, whereas the term printing press refers to industrial devices, whatever their size. A printing press consists of several printing

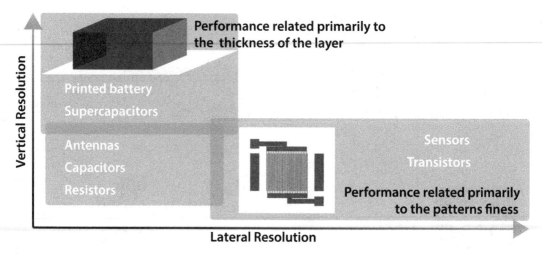

**FIGURE 4.4**
Qualitative examples of lateral and vertical resolution requirements for various applications.

units, each of which may or may not use a different printing process, one or several drying system, and substrate feeding and collecting systems that may be manual or automatic. Lab-scale devices usually consist of only the printing unit.

Each printing unit is composed of:

- An inking system that contains, agitates, and transports the ink from the reservoir to the printing form.
- A printing form, printing plate, or master that details the pattern to be printed with the ink.
- A transfer system that takes the ink from the inking system (for masterless processes) or printing form to the substrate. Diverse methods exist for transferring the ink onto the substrate.
- Non-impact processes which are also masterless, where there is no physical contact between the printing unit and the substrate. The ink-jet process is the most famous digital and nonimpact printing process.
- Processes which contact the substrate require a printing form and a counter pressure system in the printing group. The printing plate holder and the counter-pressure holder can either be flat (flat-bed processes) or cylindrical (rotary processes).

The printing press can be fed either by rolls (roll-to-roll) or sheets (sheet-fed processes). It is also possible to feed rolls and add a die-cutting unit for roll-to-sheet processing. Moreover, a process can be continuous, where there is no stoppage to transfer the ink or discontinuous where the substrate is stopped to allow ink transfer. In the printing industry, the terms roll-to-roll, rotary, and continuous are often considered to be synonymous. Indeed, most of the presses in the graphics industry are actually continuous, roll-to-roll, and rotary with one type of ink and one type of drying system. This expectation from conventional printings industry creates ambiguity and misunderstanding as this is not the case in PE. For example, some higher-scale pieces of equipment for printed battery manufacturing are roll-to-roll discontinuous non-rotary presses using flat-bed screen printing units that require to stop the web from transferring the ink.

Whatever the press configuration and the scale of production, various printing processes were created in order to provide dried ink films with a variety of characteristics adapted to commercial requirements. As the dried film characteristics (and, therefore, performance) are the most important aspects for additive manufacturing of functional layers, a deeper focus in Section 4.7 will follow on the types of deposition systems available.

### 4.3.3 Manufacturing Processes

As discussed previously in Section 4.3, the definition in this chapter used to distinguish printing processes from other deposition techniques describes printing as the combination of additive manufacturing techniques which allow a film of material to be deposited on a substrate in a chosen shape and thickness.

Figure 4.5 highlights the technical differences between some of the processes that allow the transfer of functional dispersions or solutions onto flexible or rigid substrates, based on (1) whether additive or subtractive manufacturing is being used, (2) the need for a printing form, or (3) if the substrate is part of the manufactured product. Additive and subtractive processes can also be differentiated if the ink transfer is achieved with or without applying a contact on the printing substrate. Generally, processes with a master or printing form require a transfer with contact.

Using the definition described in this chapter, manufacturing processes where the substrate is not part of the final product or is nonexistent, as in 3D printing and other additive manufacturing, should not be considered to be printing processes. Proponents of 3D printing highlight the past similarity between the 3D printing ejection heads and the ink-jet print heads, but this manufacturing process has evolved, while holding on to the now anachronous term.

Subtractive processes require a mask to discriminate between the functional area and the non-printed area include blade coating, where the dispersion is scraped onto the

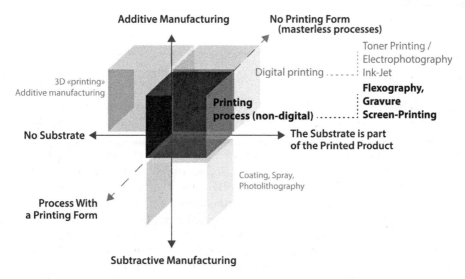

**FIGURE 4.5**
Manufacturing processes commonly used in printed electronics.

PRINTING PROCESSES

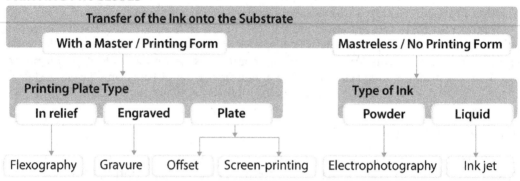

**FIGURE 4.6**
Various printing processes and their printing forms.

substrate with a squeegee (or a doctor blade), electro-spray deposition, where the dispersion is ejected in a conical fine aerosol spray, or photolithography for conventional or flexible electronics. These processes are sometimes mistakenly referred as printing processes in publication or are presented as a step toward printing at larger scale. The following Section 4.5 on formulation will offer points weakening this assumption.

For PE, functional dispersions are generally the most expensive elements of the fabrication process, which promotes the use of mask-less, additive processes, which limit the loss of expensive materials. There are several printing processes, which are suitable for different scales of production and fulfil this requirement with good precision and repeatability.

The main processes used in the printing industry are summarized in Figure 4.6 and the printing processes used in PE will be detailed in Section 4.7.

All of the above-mentioned processes, which allow the transfer of functional dispersions or solutions onto flexible or rigid substrates, require an optimization of the mechanisms involved in printability (to be described in Section 4.4), formulation (Section 4.5), and drying (Section 4.6).

## 4.4 Printability

Printability characterizes the quality of (1) the transfer of an ink onto a printing substrate using a printing process, (2) the ability of the ink to wet the surface, and (3) to demonstrate proper adhesion in liquid and in dry states. It depends on the compatibility of the rheological and energetic properties of the ink with the process, as well as the mechanical, chemical, and thermal properties of the substrate. In the ink/substrate/process system, either for graphics or PE applications, printability is the core objective of printing, while drying (detailed in Section 4.6) is more a process specific, yet still critical challenge.

The first step in enhancing the printability of an ink is to stabilize and optimize its rheological compatibility with the system used to transfer it through the inking system, from a printing plate or nozzles, so that the required and uniform quantity of ink is transferred onto the surface of the substrate. The transfer depends on the optimization of the affinity of the liquid ink for the substrate surface, to enhance planar shape resolution and the transversal layer's ability to form a continuous film, thereby promoting the adhesion of the solid ink layer onto the surface. The mechanisms used by the printer to control the ink/process and ink/substrate interactions, which emphasize the critical role of ink behavior and formulation, will be discussed in Sections 4.4.1 and 4.4.2.

### 4.4.1 Control of Rheological Properties

The behavior of liquid ink as it passes through the inking system during a printing run depends on its rheological properties and stability when exposed to air. Moreover, fluid ink experiences film splitting through the nip in the roll-to-roll process, scraping and screening in flatbed and rotary screen printing, and ejection and flying in ink-jet printing. These strongly different phenomena make it difficult to transfer an ink formula from one printing process to another without making significant adjustments to that formula. In addition to meeting the requirements of the resin and pigment used in the ink to achieve functional performance, solvents must be compatible with the different requirements of the printing processes as detailed in Table 4.1.

The rheological properties strongly depend on the temperature conditions and shear rates applied in the ink system. Practically, relevant ink rheological behavior optimization is performed at least within applicable ranges of these two environmental and manufacturing conditions as detailed in Section 4.5 on ink formulation.

### 4.4.2 Control of Interfacial Properties

Substrate core and surface characteristics are critical for printability. Closed substrates such as ceramics, glass, or plastic films usually allow limited penetration by ink, and their surface is normally finished to promote adhesion by coating with a primer or by surface hydrophilization using plasma or corona treatments. These types of substrates are typically favored for PE applications manufacturing and are selected with consideration for the performance target of the completed device.

**TABLE 4.1**

Common Rheological Properties and Solvent Limitations for Inks for Roll-to-Roll and Ink-Jet Processes

| Printing Process | | Solvent Limitation | Rheological Properties at Printing Conditions |
|---|---|---|---|
| **Flexography** | | Alcohols with max 20 wt% ester, no aromatics or alkanes. There are specific plates or screens that might be compatible with esters and ketones. | 0.01–0.1 Pa·s |
| **Screen- Printing** | Flatbed | | 0.1–10 Pa·s |
| | Rotary | | 0.1–5 Pa·s |
| **Gravure** | | Compatible with all chemical classes including aromatics, alkanes, chlorinated such as trichloroethylene, ketones like MEK or MIBK | 0.01–0.4 Pa·s |
| **Ink-Jet** | | Solvents limited by boiling point, which should be >100°C to avoid cavitation for ink-jet head with thermal chamber. | < 0.001 Pa.s to ~ 0.04 Pa.s |

While paper is the standard for graphics printing, it also demonstrates potential for PE manufacturing. Paper is composed of 90–95%$_{mass}$ of hydrophilic cellulosic fibers and other plant residues, such as lignin, tannin, and hemicellulose that are maintained in a network by hydrogen bonding promoted by some of the 5–10%$_{mass}$ of water remaining after manufacture. Typical cellulosic networks contain high volume of air, which can be calculated and expressed as porosity. That air might form channels expressed as tortuosity and allow fluids (water, air, and oil typically) to cross from one side to the other under gravity and certain other forces, a phenomenon known as permeability. The diameter of these channels is critical for capillarity as well as for the penetration (and possibly loss) of functional elements. When printing on paper using water or hydrophilic solvent inks, it is important that absorption and adsorption be taken into consideration. Though the penetration of nonfunctional elements into the paper might help the manufacturing process by quickening the layer solidification, the penetration of functional elements might cause a loss in the functionality per unit of transferred active material. Substrate roughness, created by fibers or threads at the surface of the paper, is also known to limit the functionality of the dried ink layer.

These interactions are not trivial, and, first and foremost, the application requirements must frame the technical solutions. For instance, substrate roughness is known to increase sheet resistance of a carbon layer in the plain while manufacturing a resistor in a printed circuit. On the other hand, for fuel cell manufacturing, increasing roughness and porosity of the gas diffusion layer network by allowing some carbon particles to enter the network might increase the exchange surface with the electrodes.

Most porous substrates are sold with one or both surfaces modified by calendaring, physically flattening between cylinders, or coating with a dispersion, to optimize pore openings, roughness, and adhesions.

While a substrate's core properties primarily influence its life cycle and processability, its surface characteristics directly impact the wetting of the substrate by an ink and how that ink spreads over the surface. These characteristics can be divided into physical characteristics, such as roughness and cleanliness and physico-chemical properties, led by surface energy.

### 4.4.2.1 Models

The deposition of a liquid droplet ($L$) in air ($V$), onto a flat, chemically homogeneous surface ($S$) forms a typical system described in Figure 4.7. The system is composed of two interfaces, $L/S$ and $L/V$. At the edge of the drop on the surface is an arc of triple phases $L/S/V$. At this triple point, the angle between the interfaces $L/S$ and $L/V$ is measured.

The theoretical contact angle ($\theta$ or $\theta_{theo}$) is the contact angle given by the Young-Dupré equation (4.1). A hypothetical system at equilibrium, which includes an absolutely smooth and flat surface and a pure, chemically homogeneous liquid drop of a given volume, and

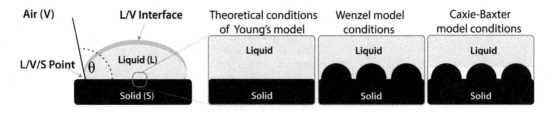

**FIGURE 4.7**
Cross view of drops on a surface. Liquid/solid interface models. From left to right: theoretical conditions of the Young, Wenzel, and Cassie-Baxter models of liquid/solid interfaces.

where the environmental conditions (temperature, relative humidity (RH%), and vapor pressure) are fixed.

In this equation, $\gamma$ is the interfacial tension between two different phases (S, L, or V). The interfacial tension (N m$^{-1}$ or J m$^{-2}$) quantifies the energy required to create an interface between two phases over one square meter.

$$\cos\theta = \frac{\gamma_{SV} - \gamma_{SL}}{\gamma_{LV}} \tag{4.1}$$

Young-Dupré equation (4.1) where L stands for liquid, S, solid, and V, vapor.

In the following text, it is assumed that $\gamma_{SV} = \gamma_S$ and $\gamma_{LV} = \gamma_L$.

Conveniently, in the printing industry, solid surface energy $\gamma_S$ (4.2), and liquid surface tension $\gamma_L$ (4.3) are expressed as their polar and dispersive contributions, which describe the van der Waals interactions. In these equations, $y^p$ includes the hydrogen bonding, acid-base, the Keesom, and Debye interactions, and $y^d$ represents the London interaction.

$$\gamma_S = \gamma_S^p + \gamma_S^d \tag{4.2}$$

$$\gamma_L = \gamma_L^p + \gamma_L^d \tag{4.3}$$

Different models link the contact angle to the dispersive and polar contributions of the interfacial tension. The Owens-Wendt two-parameter equation (4.4) proposes a relation between the contributions $y_S^p$ and $y_S^d$ of the interfacial tension of a solid and the Young's contact angles $\theta$

$$\gamma_L(\cos\theta + 1) = 2\left(\sqrt{\gamma_S^p \cdot \gamma_L^d} + \sqrt{\gamma_S^d \cdot \gamma_L^d}\right) \tag{4.4}$$

However, the solid surfaces are generally rough and covered by chemical heterogeneities. For example, Johnson and Dettre have shown the influence of the roughness on the contact angle of water drops on a wax surface. Consequently, if the Young's contact angle $\theta$ is intrinsic to the surface chemistry, the measure of a contact angle of a liquid on a real surface provides the apparent contact angle $\theta^*$, which considers the other surface characteristics. Two models illustrated in Figure 4.7, Wenzel, in equations (4.5) and (4.6), and Cassie-Baxter help to express the relation between the theoretical contact angle and the apparent contact angle $\theta^*$, introducing possible corrections for rough or chemically heterogeneous surfaces.

In 1936, Wenzel calculated a roughness parameter $r$, where $r$ is the ratio of the area of the surface in contact with the liquid $A_{apparent}$ and the area projected for a perfectly smooth and flat surface is represented by $A_{projected}$.

$$r = \frac{A_{apparent}}{A_{projected}} \tag{4.5}$$

$$\cos\theta^* = r \cdot (\cos\theta) \tag{4.6}$$

The Wenzel state postulates that the droplet of liquid is wetting the entire surface of the irregularities. In the case of flat but chemically heterogeneous surfaces, with n being the number of chemically different surfaces, the Cassie-Baxter equation can be applied to correct $\theta^*$.

$$\cos\theta^* = \varnothing_{S,1}(\cos\theta_1) + \varnothing_{S,2}(\cos\theta_2) + .. + \varnothing_{S,n}(\cos\theta_n) \tag{4.7}$$

In the Cassie-Baxter equation (4.7), $\varnothing_{S,n}$ is the fraction of the solid n in contact with the liquid and $\theta_n$ is the Young's contact angle of the liquid on the chemical surface $n$.

When it is postulated that the liquid is supported by the peaks of the irregularities of a rough surface, in a fakir-droplet state as shown in Figure 4.7, the Cassie-Baxter equation is extended as in equation (4.8), where $\varnothing_{S,1}$ is the fraction of solid/liquid.

$$\cos\theta^* = \varnothing_{S,1}(\cos\theta_1 + 1) - 1 \tag{4.8}$$

Equations (4.7) and (4.8) represent two extremities of a range of behaviors of a liquid over a rough surface, from complete wetting to fakir droplets, respectively.

### 4.4.2.2 Applications

The measure of the interactions between a liquid and a surface could be performed by the deposition of a liquid droplet onto the surface. The deposition of a drop falling vertically onto a horizontal surface has been studied in order to investigate the wetting properties of the liquid for that surface.

Using drops of different liquids with known interfacial tension contributions $y_l^p$ and $y_l^d$, the contact angle measurement allows $y_S^p$ and $y_S^d$ to be estimated, according to the Owens-Wendt model. $y_l^p$ and $y_l$ can also be measured by sessile drops or Denouilh ring techniques.

On site, pressmen rely on dyne pens to estimate the surface energy of substrates. This rapid technique is suitable when printing water-based or solvent-based graphic inks, which have well understood behavior on paper or plastic films. For high-performance functional manufacturing, the polar and dispersive contribution of the surface energy and surface tension must be taken into consideration in order to achieve the targeted wetting and film formation. In this case, calculations based on Owens-Wendt are acceptable. As corrective parameters must be adapted for every combination of liquid and substrate, this method is limited when measuring special substrates, which may be rough, chemically instable, or contain many surface additives. For this reason, the dynamic apparent contact angle $\theta^*_{dyn}$ of the ink on the substrate is used as a straightforward technique for the estimation of wetting. Not only does it provide an indication of the affinity between the ink and the substrate, considering all physical and chemical heterogeneities (not including process conditions), but it also illustrates, at small scale, dynamic mechanisms such as ink absorption by the substrate, porosity, substrate swelling, and deformation or even chemical degradation of the surface.

Figure 4.8 demonstrates different behaviors of a liquid on a theoretical perfect surface, if the liquid is pure water, when $\theta_1 \approx 180°$, the surface is super hydrophobic and the liquid dewets the surface, $\theta_2 > 90°$, the surface is described as hydrophobic, $\theta_3 < 90°$, the surface is considered to be hydrophilic, and $\theta_4 \approx 0°$, the surface is totally hydrophilic.

In printing conditions, the type of wetting desired depends on the ink film thickness and print definition required, while obtaining the best adhesion possible. Typically, the highest thickness and the best printing definition are usually obtained with $\theta^*_{eq} \approx 90°$ at equilibrium. Therefore, using the Owens-Wendt equation leads to the optimization of the energies of the surface and the ink to attain similar values for both the polar and dispersive contributions. The optimization of the polar and dispersive contributions to the surface tension by selecting adequate ink vehicle mix is one of the challenges of the formulation process described in Section 4.5. On the other hand, surface modification by various treatment methods allows the optimization of the polar and dispersive contributions of the surface energy. It should be noted that industrial surface treatment methods are typically less well controlled than what may be available in the laboratory.

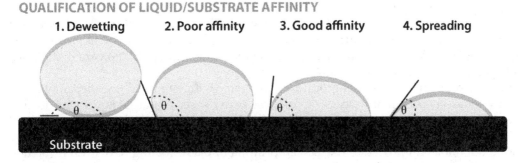

QUALIFICATION OF LIQUID/SUBSTRATE AFFINITY

**1. Dewetting**     **2. Poor affinity**     **3. Good affinity**     **4. Spreading**

Substrate

**FIGURE 4.8**
Examples of different wetting behaviors, from left to right, dewetting, poor affinity, good affinity, and spreading.

Chemical modification treatments include corona treatment, an inline treatment installed on the press immediately before the printing stations, and plasma treatment, a treatment applied to commercial substrates during their manufacture. These treatments create surface oxidation and micro-roughness, each of which increase the hydrophilic behavior of the treated surface. Empirically, the corona treatment is valid only during the press run, whereas plasma treatment is valid for 2 to 3 months or as specified by the substrate manufacturer. Consequently, plasma-treated substrates must be used shortly after manufacture due to their susceptibility to surface degradation and should be tested before every print run (either by dyne pen or preferably by ink contact angle). Physical treatments can be applied during substrate manufacture, allowing the control of roughness. Examples include the creation superhydrophobic surfaces or the calendaring of paper and board, which flattens the fibers present on the surface through the application of heat and pressure between soft and hard cylinders.

Finally, treatment by the addition of material to the substrate, such as the printing an adhesion promoter (primer) under the functional printed pattern is a technique very often used in the printing and paper industries. Such treatments have been optimized to meet the needs of the printing and paper industries and it is usually applied to increase the hydrophilic behavior of the substrate and optimize ink penetration into the paper. Primer layer, when adequately selected and transferred, allows to improve definition and reduce energy consumption during drying in conventional printing.

However, these treatments can be detrimental when printing functional inks, which may not be hydrophilic and might not penetrate or even wet the substrate. For example, old habits of using full corona power as a solution for all printability problems, helping to burn dust and make the ink spread might have bad side effects for PE applications. It is then critical to assess the substrate surface characteristics in order to fully benefit from those treatments.

## 4.5 Ink Formulation

The compatibility of the ink with the process and substrate is the main factor influencing the success of functional layer manufacturing. Transfer, wetting, adhesion, drying, and performance are all affected by this relationship.

The importance of having a specific formulation for each trio of ink/process/substrate cannot be overemphasized. Unfortunately, the multidisciplinary competencies in chemistry, printability, and process required to address the challenges of ink formulation are rarely found in the collegial or university worlds, due almost certainly to a lack of interest in the conventional printing domain. This trend is also adversely affecting other "traditional" manufacturing techniques such as coating for paper manufacturing. Surprisingly, the available courses most closely related to ink formulation are in the fields of formulation of rubber for tires and cosmetics.

Most formulators have a background in chemistry with empirical knowledge of printability amassed over many years of priceless experience. Again, it is difficult to find scientific models for formulation, which often leads to ink formulas being developed with very little potential to fully serve the requirements of the application. As a result, formulation-tolerant processes like screen-printing or blade-coating have been favored to achieve the required level of liquid ink transfer, requiring that the relatively high amount of solvent deposited in these processes be evaporated by drying in an oven at high temperatures for long periods of time. Such a strategy, focused on functional performance, is really optimized for lab-scale technology demonstration. Unfortunately, industrialization requires inks adapted to modern processes, which dry to form a film with repeatable high-performance functionality. These inks must be stable on press and able to print at high speeds and low drying temperatures and have a long shelf life to allow the reuse of leftovers from previous production runs.

### 4.5.1 Inks for PE, Graphics, and Slurries

Inks for PE exist at the border between slurries and inks for graphic applications; the dried ink film must have the same high performance as a film made from the deposition of a slurry but with processability and stability of ink designed for graphics. These contradictory requirements increase the formulation challenge and compromises are inevitable on both aspects. For this reason, the formulator must understand the three main classes of components: solvents (liquid); printability promotors (solid), and functional components (solid). Printability promoters are the resins and additives required to ensure good transfer, wettability, and adhesion as well as shelf life of both the liquid ink and printed, dried ink (functional) film. The ratio of these three components are determined by the type: graphic ink, PE ink, or functional slurry. For instance, as illustrated in Table 4.2, a graphic ink (here, for flexography) has a high liquid to solid ratio, compared to PE inks and slurries, making graphics inks the champions of printability and stability. Slurries generally have a very high ratio of functional components to printability-promoter in comparison with graphics inks, the result of which is a product with poor printability using processes with complex transfer system.

The formulation strategy for PE inks, whether for prototyping or industrialization includes incremental steps and correction loops. While a proper discussion of ink formulation would require a dedicated book of its own, the dominant theme is that, despite what is commonly

**TABLE 4.2**

Typical Ink Composition by Mass% for Solvent-Based Dispersions for Different Applications

| Qualitative Ink Composition by Mass% | Typical Flexographic Graphic Ink | Typical Flexographic Conductive Ink | Typical Conductive Slurry |
|---|---|---|---|
| Solvent(s) | 50 | 48 | Adjusted on demand for viscosity |
| Resin(s) | 25 | 10 | 5 |
| Pigment(s) | 20 | 40 | 95 |
| Additive(s) | 5 | 2 | 0 |
| Total | 100 | 100 | 100 |

referred to as ink in academia, it is very different from the stable dispersion or solution required to print acceptable results on an industrial scale press. Moreover, the compatibility and synergy of all components of the ink is important and their influence on the liquid and dried ink film must be investigated. Finally, the equipment used to manufacture the ink will have an influence over the behavior of the liquid ink during and after printing.

### 4.5.2  Raw Material Selection

Every formulation of a colloidal dispersion or solution for the production of layers starts with the selection of the active/functional material. The active material contributes to the function of the dry printed film only—it should not be considered to be contributing to printability; in fact, it may inhibit printability.

The main criteria for the selection of active material(s) are the target function of the dried film, the energy required for drying and obtaining the desired performance value, and its physico-chemical characteristics. The latter include considerations for the geometry of the material: 3D dimensions, aspect ratio (AR), the size distribution, and modifications to facilitate dispersion.

#### 4.5.2.1  Percolation Threshold and Aspect Ratio

The Aspect Ratio (AR) is the ratio between the length and the diameter of a, spherical and pseudospherical particles typically have an AR close to 1, whereas rod-like particles (think about nanotubes) have a very high AR, while flakes are defined by a low AR.

In the context of PE inks, AR is very useful for the understanding of particle geometry. However, it is limited in some cases and some empirical corrections should be considered. First, certain polymers may stretch, or pack themselves into random coils, depending on their chemical affinity for their surrounding environment. Therefore, in this case, AR is a dynamic value and may be considered to be an indicator of dispersion quality. Another way to picture such colloidal dispersions is by using the maximal volume occupied by the particles without being in contact with other.

For inks where one of the functions is electrical conductivity, the critical parameter is the value of Percolation Threshold (PT) after drying which directly depends on the AR of the selected particle. When applied to dry ink film formation, the percolation threshold can be simplifying as the minimum quantity of functional particles in the dry phase (usually expressed for ink formulation as mass percent, $\%_{mass}$) required for the formation of a connected network of these particles.

The higher the AR, the lower the PT. For instance, all other factors being equal, liquid inks may require $5\%_{mass}$ of silver nanowires or more than $50\%_{mass}$ of silver flakes to achieve comparable electrical performances when dried. It is clear that the cost per unit mass as well as the physico-chemical characteristics of the dried ink film might drastically differ, depending on the material used. It should be noted that high AR particles are generally more expensive due to their difficulty of production and purification, and their use in conventional printing is limited by their compatibility with printing equipment such as mesh opening in screen printing, and the cell volume and opening diameter of the flexographic anilox or gravure cylinder (more details are provided in Section 4.7.3).

#### 4.5.2.2  Active Material Selection

For prototyping purposes, particles are often selected, first, for their intrinsic performance and, second, for their mass cost and last for their geometry. This approach generally forces the

**TABLE 4.3**

General Impacts of Particle Geometry and Dispersibility Formulation Strategy

| | Limitations on Deposition Process | Impact on Formulation Strategy | Impact on Dry Film Performance |
|---|---|---|---|
| Aspect ratio (AR) | Might impact rheology | Adapted dispersion and mixing techniques as a function of AR | Usually high AR particles are more expensive, but less is required to achieve network percolation |
| Average size | Largest particles should be four to ten times smaller than the process dimension limit (refer to Section 4.7.3) | Adapted mixing and dispersing equipment for targeted size and particle size distribution | Larger particles might increase core porosity and surface roughness |
| Size distribution for dispersion | | | Wider size distribution might diminish film porosity, smaller elements acting as filler between bigger elements. |
| Nano/micro materials | Nano materials generally require adapted equipment and substrates with roughness parameters in the same range | | Might impact drying strategy (refer to Section 4.6) |
| Modification to help dispersion | Might impact rheology | Several physical dispersion techniques and chemical dispersion promoters might be required in combination | If chemical dispersion promoters are used, higher energy might be required for sintering, performance may be diminished |

use of additional and costly mixing and dispersion techniques, as well as more additives, as described in Table 4.3. Unfortunately, this approach might not allow the latitude for adapting the rheological properties to the deposition process and scale-up potential may be limited.

Possible constraints due to geometric characteristics and dispersibility of the selected materials are detailed in the Table 4.3.

Another recommended and efficient, but more complex, approach to the selection of active materials for industrialization is to first consider the cost per unit of the main targeted performance value for a given volume or mass unit of the dried printed and to formulate the ink with materials which conform to this limitation. It should be remembered that in this case, the cost of all components of the ink required to produce that mass of dried functional material needs to be considered, including volatiles. Moreover, cost optimization should include manufacturing resources such as energy, human resources, and depreciation of the pieces of equipment to formulate, print, dry, etc.

### 4.5.2.3 Vehicle Composition and Material Selection

The ink vehicle or varnish is composed of:

1. Resin or a mixture of resins, which have the largest influence over the ink properties. Resin determines rheology, adhesion, flexibility, gloss, resistance to abrasion,

and other physical characteristics of the dried film, and influences drying through its release of solvent.

2. Solvent or a mixture of solvents to solubilize the resins. Solvent is the ingredient with the most influence over the drying behavior of the ink.

The vehicle is produced through the dissolution of the resin in the solvent using optimized techniques and protocols, including speed and temperature variations, as well as processing duration. It plays three roles in an ink: the wetting out and stabilization of the pigment; the transfer of the ink to the substrate; the formation of a dried film on the substrate with the properties required for end use. The vehicle may consist of a single resin and solvent or a combination, depending on the required properties. It also depends on the type and conditions of drying (solvent-based, water-based, UV-curable inks and electron beam (EB) curable inks), the type of deposition process and compatibility of the functional compounds/resin/solvents.

Solvent-based ink varnishes consist of resins dissolved in solvent. It is common for solvent-based inks to consist of one resin only or a primary resin with another which modifies the properties of the ink to provide additional characteristics. For example, the addition of nitrocellulose to a polyamide system to provide additional heat resistance. Solvent-based inks are resoluble and their rheological properties can be modified on press by adding the blend of solvents used to produce the varnish.

Presses running solvent-based inks must be explosion-proof, and, unfortunately, drying techniques are limited due to safety concerns: for instance, IR drying is forbidden directly on the press in the printing industry.

The most common water-based ink technology is based on two classes of acrylic resin. Lower molecular weight resins, often referred to as solution resins, are usually composed of acidic resins which are neutralized with ammonia or amines, will stabilize rheology and impart re-solubility to the ink so that it will stay stable on the printing plate. Liquid emulsions composed of higher molecular weight resins are used to impart mechanical properties such as adhesion and resistance, but they are generally not resoluble and must be used in conjunction with a resin solution. Press management of water-based ink is performed by pH control. In the case of PE inks, the pH control may be very sensitive and if not stabilized, the ink is generally lost.

UV or EB-curable inks consist of specific monomers and oligomers that will polymerize when exposed to radiation. In the case of UV-curable inks, the presence of a photoinitiator, which will activate in the presence of photons at a specific wavelength in the near UV region, is required to trigger the reaction. EB-curable inks do not include a photoinitiator but require a considerably higher energy input to achieve curing. In the presence of UV energy, the photoinitiator will form a free radical, which reacts with the monomer to create a larger free radical in the initiation phase. In the propagation phase, that larger radical will then react with other monomers and oligomers to form even larger radicals, eventually forming a continuous, polymerized film. In EB curing, the monomer is stimulated by contact with energized electrons, eliminating the need for a photoinitiator. The technical challenge of energy curable inks is that they, in most cases, do not contain solvent, the implication of which is that all nonfunctional material in the ink is converted to a solid (resin), which may inhibit contact between the functional particles, a very relevant solution for dielectric inks but a drawback for conductive inks, for example.

### 4.5.3 From a Stable Dispersion/Solution to an Ink

Once a dispersion or a solution of active particles in a selected vehicle that corresponds to the requirements of the process and drying is attained, several steps remain to produce a printable, functional, and stable ink. At this stage, various compatible additives might be integrated and the ratio of the components is fine-tuned in response to testing under actual printing conditions at various scales.

1. Stabilization of the formulation

   Stabilizing additives are often incorporated into the functional ink formula to limit oxidation, agglomeration, and sedimentation, and to improve compatibility with the vehicle. It is achieved either by steric stabilization, where layers of polymers and/or surfactants surround the functional element, or by electrostatic stabilization, where layers of charged elements create a double layer of charges around the functional element, inducing a coulombic repulsion force.

   The criteria that are used to evaluate stability in storage or during the use of manual film deposition or proofing systems are:

   - Dispersion/solubilization of compounds in the varnish;
   - Conservation of the functional properties of the compounds;
   - Drying and rheology compatibility with production;
   - Agglomeration with time; and
   - Stabilization of the dispersion to minimize sedimentation/dephasing with time.

2. Evaluation of ink transfer on proofing devices and targeted deposition equipment to evaluate:

   - Film formation on selected substrates;
   - Drying energy required to reach the desired level of functionality;
   - Selected printing method; and
   - Distribution of compounds in the film (x,y,z).

3. Qualification of the ink for the targeted deposition equipment using the following criteria:

   - Viscosity adjustment before printing and on press;
   - Behavior in the inking system;
   - Behavior on the printing plates;
   - Transfer capacity at different speeds onto the substrate;
   - Behavior of formulation over time, during print run and stops;
   - Drying capacity at different speeds; and
   - Definition of printing conditions.

## 4.6 Drying

The transformation of a liquid ink/solution/dispersion into a solid, functional thin film is far from trivial and depends heavily on several process and formulation parameters. Drying consists of two phenomena: solidification (detailed in Section 4.6.1) and functionalization

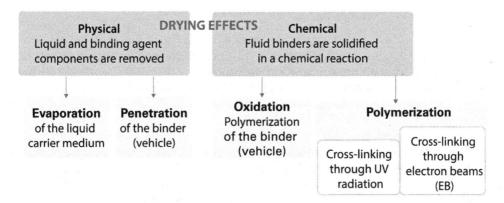

**FIGURE 4.9**
Conventional types of ink drying effects, note that combinations are possible. (Adapted from Kipphan, 2001.)

(Section 4.6.2), which takes place in dedicated drying systems which may be a combination of inline or offline elements. Each dryer is composed of three or four parts:

1. Conduction, convection, or radiation systems that provide the desired energy to the printed layer, such as UV, LED, EB, IR, hot air, and pulse light. The different types of drying effects used in the printing industries are illustrated in Figure 4.9.

2. Gas-evacuation systems to limit vapor saturation and force solvent evaporation.

3. Vapor-treatment systems that correspond to applicable pollution regulations regarding the release of volatile organic compounds (VOCs) into the atmosphere.

4. Special requirements systems are detailed in the Table 4.4 for safety concerns. For example, in the case of printing solvent-based inks, every piece of equipment should work under negative pressure to draw solvent out of the printing environment in order to avoid solvent contamination in the environment and to provide safe and explosion proof conditions for printing.

**TABLE 4.4**

Radiation Ranges and Limitations for Dryers by Application

| | Wave Length | Activity | Drying Applications | Limitations |
|---|---|---|---|---|
| Microwaves | $10^{-1}$ to $10^{-3}$ m | Excitation of dipole vibrations of polar molecules | Applications for water-based formulations | |
| IR | $10^{-3}$ to $0.4 \times 10^{-6}$ m | Vibration excitation of organic molecular bonds | | Forbidden on explosion-proof press for solvent-based inks |
| UV | $0.75 \times 10^{-6}$ to $10^{-8}$ | Electron excitation in molecules, photochemical reaction | UV-curable inks | Release of free radicals Ozone formation |
| X-Ray | $10^{-8}$ to $10^{-11}$ | Ionization of molecule Molecular fission | EB-curing systems | Release of free radicals Requires high-purity inert atmosphere |

### 4.6.1 Solidification Mechanisms

Solidification provides adhesion, stability, durability, and a certain surface state (roughness, for instance) that can give a glossy or matte appearance, and is optimized depending on the visual and lifespan requirements of the products.

First, the removal/evaporation of the majority of the solvents results in a very high viscosity liquid. This occurs through a combination of absorption into the substrate and evaporation. The liquid has to be solidified enough to allow the film to go through the next cylinder nip (in case of roll-to-roll process) or to the next manufacturing step of prototyping.

In the press, the web or sheet is placed into contact with, respectively, a roll or a supporting plate that can be warmed or cooled. When warmed, it helps the system by providing heat by conduction, transferring heat by molecular collision from the verso of the substrate to the top of the printed layer. However, in roll-to-roll industrial scale presses, it is conventionally used to cool the backside of the substrate to limit its deformation under heat, which is applied to the top side by radiation and/or convection equipment.

Solvent evaporation is promoted by the evacuation of the solvents in the gaseous state from the environment, unbalancing the interfacial equilibria at the printed layer surface and by displacing the vapor pressure point through transference of heat by convection.

The gradient of solvent from the gas/liquid interface at the top of the ink layer down to the ink/substrate interface, as well as the gradient of temperature in the ink layer can generate the flow of matter. The flow of solvent may displace functional particles. Depending on the heat, the vapor pressure over the drying film and the ink formulation, the kinetics of drying may favor the uniformity of physico-chemical characteristics of the ink layer or the contrary. Coffee ring defects typically appearing in ink-jet printing is an example of mismatch between the drying kinetics and the ink formulation, among other parameters.

Cracks and solvent-extraction chimneys forming in the ink layer are typical indications of overly hot and/or fast kinetic drying. On the other hand, while such drying may represent a defect for a conductive layer, it could be an advantage for a gas diffusion layer.

Drying may also use chemical mechanisms triggered by dryers emitting a part of the electromagnetic spectrum that allows selective interaction with matter to cure the inks, as summarized in Table 4.4.

### 4.6.2 Layer Functionalization

Each dried printed layer and each system of connected layers should fulfill a list of functions. This requirements sheet should include not only electro-active properties, but also target for physico-chemical characteristics such as layer cohesion, adequate porosity and permeability, significant network percolation with the targeted interface, good film morphology.

Ideally, all these properties may be attained through drying on the press after solidification of the layers. Nonetheless, achieving all the target performances frequently require additional treatment, sometimes called post-processing or post-drying. For instance, additional thermal-treatment can be applied to melt the particles and fuse them together at their points of contact allowing densification of the network and/or grain growth. This sintering process usually requires high energy with a temperature exceeding the temperature of stability of most of the commercial printing substrates (105°C for paper, $T_g$ of plastics). This may be mitigated with radiation-sintering processes to target the functional material and exclude the substrate or through the use of substrates with high temperature resistance such as polyethylene naphthalate (PEN) or polyimide (PI). Lamination, die-cutting, etc. are other post-processing steps that will provide additional-value to the dried film.

## 4.7 Printing Processes for PE Manufacturing

Prototyping aims to manufacture printed layers and applications with the best possible performance to prove the feasibility of a technology. Production is an optimization of application manufacturing at market scale with adapted pieces of equipment, methods, and supply chain, to manufacture at higher speeds and for longer durations, while having an adapted product life cycle that minimizes the impact on environment and society. At this stage, technical solutions may be implemented through the production chain to reduce the overall cost of production.

Therefore, a 100 or 1-million-unit manufacturing run may be considered either as prototyping or as production. Moreover, there is no process dedicated to prototyping or production, but there may be pieces of equipment integrating printing processes that are dedicated to one use or the other.

However, processes can be categorized depending on the level of technical expertise required to properly operate industrial equipment at various scales to obtain targeted results. Where ink-jet and flat-bed screen-printing presses are commonly found in labs working on additive manufacturing, it is less common for an electronics laboratory to acquire a flexographic lab-proofer or even a press with a gravure or a rotary screen unit.

In all cases, while it is very challenging to scale-up and industrialize a technology developed on a lab-scale device to an industrial system equipped with the same printing process, the industrialization becomes even more arduous when a change of printing process is required.

### 4.7.1 Versatile Processes

#### 4.7.1.1 Principle of Ink-Jet Printing

The ink jet is a technology where small ink droplets (with a volume about $10^{-9}$–$10^{-12}$ L) are ejected from a small ink chamber through a nozzle onto a substrate without any physical contact between the inking system and the substrate. Figure 4.10 exemplifies the two main techniques used to deposit ink onto the substrate:

- Continuous-Ink-Jet (CIJ), where the ink droplets are frequently ejected. Some droplets are directed toward the substrate by electrostatic deviation while the nonrequired drops are reused.
- or Drop-On-Demand (DOD) ink-jet. The ink droplets are generated when required.

The volume of the droplets ejected from the inking system is controlled by two main drop generation systems:

- Piezoelectric system. A small contraction of the volume of the ink chamber leads to the ejection of an ink droplet through the nozzle.

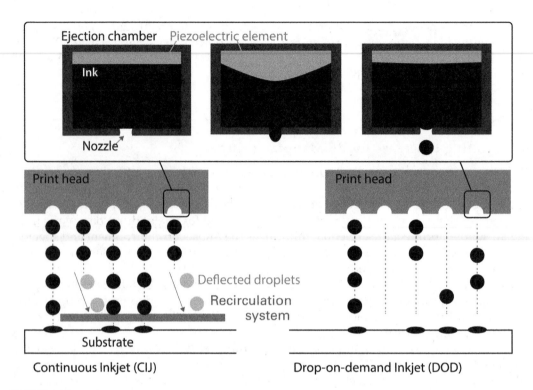

**FIGURE 4.10**
Ink-jet principle.

- Thermal system: A vapor bubble is formed by locally heating the ink. The bubble pushes the ink through the nozzle until a droplet of ink is formed. When the vapor bubble bursts, the ink droplet detaches from the nozzle.

Although this process equipped many research laboratories because it is a contact-less and masterless process that allows rapid customization of the printed pattern and change of substrate, its ink formulation and nozzle ejection parameters are particularly complex and as specific as any other roll-to-roll process. The scalability of such technique can be more challenging than flat-bed screen printing, for example, and it will require a complete reformulation of the ink for every piece of equipment. Extensive technical training on formulation and printability is mandatory to fully benefit from the many advantages of this process.

### 4.7.1.2 Principle of Flat-Bed Screen Printing

Screen printing involves forcing a pasty ink to go through the openings of a meshed screen with a squeegee. The mesh is covered by a stencil that blocks the non-printed area. Figure 4.11 gives details on the different steps that should be optimized to transfer the ink with accuracy and repeatability.

Flat-bed screen printing uses a flat screen in a discontinuous process that is well adapted for transferring high thickness in a single layer with not only low resolution and low pressure on various 2D and 3D substrates such as textile, cellulosic materials with high roughness, but also glass bottles.

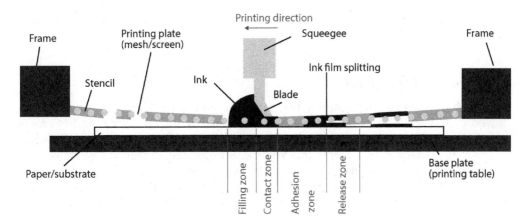

**FIGURE 4.11**
Flat-bed screen-printing principle. (Adapted from Kipphan, 2001.)

In electronics, screen printing is widely used for solder paste transfer using a stencil rather than a mesh screen for higher thickness deposition, but also for very high-resolution printing for circuit boards, membrane switches, legend marking, and touch screens.

This process, already widely integrated into graphics chain and electronics supply chain, is one of the most developed for PE applications.

### 4.7.2 Specialized Rotary Processes

#### 4.7.2.1 Principle of Rotary Screen Printing

Based on the same principle as described in Section 7.1.2, rotary screen printing uses a mesh screen covered by a stencil that is blocked in the non-printed areas. This screen is rolled to form a hollow cylinder and in its cavity are placed the squeegee and the inking systems. The cylinder turns in synchronization with the substrate, and, in the nip zone, the static squeegee forces the ink to transfer from the cavity to the substrate as described in Figure 4.12a.

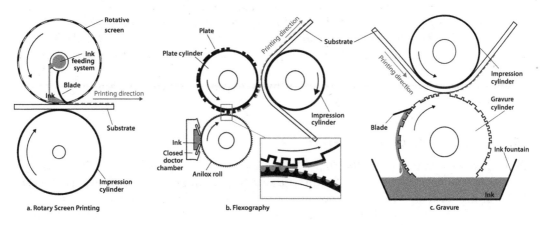

**FIGURE 4.12**
Rotary processes, cross view of a printing unit. a. Rotary screen printing b. Flexography and c. Gravure.

At this time, rotary screen-printing is mostly used to print on fabrics web or wallpaper web.

### 4.7.2.2 Principle of Flexography

Figure 4.12b clarifies the flexographic principle. The ink is put in contact with a micro-engraved cylinder (anilox), filling its cells, and is scraped with a doctor blade to assure that the cells are filled with a constant, controlled volume of ink. Aniloxes are defined by their theoretical volume. In Europe, this volume is expressed in metric system as $cm^3$ of ink contained for 1 $m^2$ of cells ($cm^3/m^{-2}$). In North America, because of the habit of using imperial system, this volume is expressed as $10^9$ $\mu m^3$ of ink contained in 1 $in^2$ of cells or billion cubic microns per square inch, typically written as $BCM/in^2$, BCM/sq in, or BCM. 1 $BCM/in^2 = 1.55$ $cm^3/m^2$.

The ink is then transferred onto a printing form which defines the pattern information. The inked pattern is then pressed against the substrate in the nip zone between the printing form and an impression cylinder. The ink transfer onto the substrate is performed with the lowest possible pressure. This particularity has given the nickname "kiss printing" to the flexography.

Thanks to the low pressure applied in the nip, the flexography is able to transfer ink onto:

- fragile substrates such as corrugated board that could be flattened during printing, and self-adhesive labels, that contain a layer made of adhesive and
- smooth substrates such as: plastics, foils, or glass.

### 4.7.2.3 Principle of Gravure Printing

Gravure printing (also known as heliogravure or rotogravure), described by Figure 4.12c, is a rotary process based on the use of a dual-purpose cylinder which contains the printed and non-printed area information, but also defines the potential quantity of ink which may be transferred onto the substrate.

Depending on the engraving technology, the cells have variable depth and/or variable area. The cells, which define the printed features, are filled with ink from a reservoir in which the cylinder is rotating and is scraped by a doctor blade, which removes excess ink from the non-printed area and controls the volume of ink in the cells. The ink is then put in contact with the substrate in a nip zone between the printing form and a printing cylinder.

Typical gravure printed products are of high quality, produced for very high-volume markets such as periodicals, magazines, and catalogs.

### 4.7.3 A Comparison of Printing Processes

Table 4.5 gives a summary of the characteristics, potential, and limitations of the printing processes previously discussed. Prototyping or industrializing PE applications consists of dealing with the possibilities and limitations of the ink/process/substrate systems by adjusting the parameters that can be adapted and bypassing those that are constrained. Often, one ink/process/substrate combination is adapted to prototyping an application,

**TABLE 4.5**

A Comparison of Printing Processes for PE

| | Ink jet | Screen Printing | Flexography | Gravure |
|---|---|---|---|---|
| Particle size limitation* | Nozzle diameter | Mesh opening | Cell size | Cell size |
| Type of ink and viscosity | Liquid<br>< 0.001 Pa.s<br>−~0.04 Pa.s | Fluid Paste<br>0.1–10 Pa.s | Liquid<br>0.01–0.1 Pa.s | Liquid<br>0.01–0.4 Pa.s |
| Liquid ink film thickness for PE | 20 nm to 1 μm | 20 nm (electronics) – 2 μm (graphics) to 100 μm | 300 nm to 5 μm (solvent)<br>> 50 μm (UV varnish) | 100 nm to 20 μm (solvent)<br>5 to 30 μm (UV) |
| Printing form | No printing form | Stencil / screen<br>$ | Relief<br>$ | Engraved Cylinder<br>$$$ |
| Control of ink transfer | Ink droplet volume | Mesh size and emulsion thickness | Anilox volume<br>$$ | |
| Printing definition potential** | Dependent on the droplet volume | High thickness (and low definition)<br>Or low thickness and high definition | Good definition | Very high definition |
| Information transfer | No printing form | Stencil / screen<br>$ | Relief<br>$ | Cell characteristics<br>$$$ |
| Common printing defect that impacts PE use | Coffee ring | Irregular edges and lines | 1. Halo around patterns<br>2. Marbling | Irregular edges around printed dots |
| Techniques for optimization | Film formation agents, kinetic drying | Orientation of the line vs. the mesh threads | 1. Pressure<br>2. Ink/substrate affinity | Engraving technique Orientation of the line vs. engraving angle |
| Use in PE | Mature for prototyping and small run production | Mature for PE (flatbed)<br>In development in rotary | Not adequate for prototyping<br>In development | In development |
| Industrialization challenges | Limited potential for scaling-up compared to roll-to-roll processes | Scaling up from flat-bed to rotary screen printing requires adaptation of formulation and printing conditions | Lab-proofing results difficult to scale-up | Not adequate for prototyping<br>In development for high-definition applications<br>Lab-proofing results difficult to scale-up<br>Printing cylinder very expensive<br>Special cylinder for nanoparticles |

\* *Rule of Thumb*: Largest particle dimension × 4 < printing processes size limitation parameter.
\*\* Strongly depends on the process/ink/substrate optimization.

but must be changed when considering the larger market, changing for a lower-cost substrate or when the process must be placed outside a clean-room or when temperature and humidity are no longer controlled.

## 4.8 Industrialization Challenges

In conclusion, Table 4.6 proposes a roadmap for the manufacture of PE applications by industrial printing processes. Currently, PE customers are very sensitive about the cost and the sustainability of the applications, while PE manufacturers validate primarily the volume capacity and run time before transferring the technology. PE industrialization must address these four pillars to successfully propose new applications on larger volume markets.

**TABLE 4.6**

PE Industrialization Challenges

| | Ink Formulation | Design | Printing and Drying | Hybrid Integration |
|---|---|---|---|---|
| **Volume** considerations Increase volume of production | High volume supply Mixing and dispersing techniques for high volume optimized for AR Ink adapted for high speed printing | Optimized positioning for speed and use of substrate Registration and quality control marks for multi-process line | High performance, driers Manufacture of functional multilayer and interfaces | Process selection for high volume Complementary processes for high volume |
| **Cost** considerations Reduce costs | Ink and material disposal and recovery after printing Low-energy drying Less material for same performance | Reduction of the number of layers or their dimensions | Selection of the type of printing process Ink/substrate/ process compatibility | Integration on web In-line characterization equipment |
| **Time** considerations Increase of running speed and production run time | Ink stability in storage Ink stability on press Press management Reusability of the ink Minimization of drier temperature | Patterns designed to help registration and overprinting Patterns adapted to printing defects and definition | Ink transfer stability Drying kinetic stability | Supplies durability Low-complexity integration |
| **Sustainability** considerations Product life cycle and sustainability | Selection of sustainable raw materials with high performance Materials' end of life compatible with end of life of the products Ink formulation adapted to product life cycle No VOC | Designed for low material consumption and high performance Multilayer selection to lower global material use Additive manufacturing process Less waste Low energy requirements High shelf-life | | |

## Acknowledgments

The authors wish to express their gratitude for the support of the Natural Sciences and Engineering Research Council of Canada (NSERC), the Ministère de l'Enseignement et de l'Éducation Supérieur du Québec (MEES) and the Collège Ahuntsic, Montréal, Québec, Canada.

ICI's team would like to thank the editor and publisher of this book for the opportunity to share its experience and knowledge. The sections on topics related to ink formulation, manufacturing, and PE high-scale industrialization condense the extensive skills developed empirically in the printing industry in Europe and North America. This huge amount of information has been mutualized and rationalized in the past decade by a multidisciplinary team of industrial and academic collaborators who work on the technological transfer of PE applications manufacturing at the Printability and Graphic Arts Institute, Montreal Québec, Canada.

Most of the printing process engineering knowledge summarized in this chapter was provided by H. Kippan, in the *Handbook of Printed Media*, ISBN 3-540-67326-1, and by the book "De la fibre à l'imprimé" in four volumes (ISBN 979-10-90188-00-6, ISBN 979-10-90188-01-3, ISBN 979-10-90188-02-0, ISBN 979-10-90188-03-7) edited in French by the School of Engineering, Grenoble-INP Pagora, Grenoble, France the only university to provide dedicated high-level theoretical courses on printing and paper-making sciences and academic research in the Laboratory of Pulp and Paper Science and Graphic Arts (LGP2 CNRS UMR 5518).

The authors wish that this book will support the efforts of the printed electronics community to train highly qualified personnel on sustainable manufacturing and greener applications development.

## Reference

Kippan, H. (2001). *Handbook of Printed Media*, Heidelberg, Springer.

# 5

## Carbon Nanotube-Based Flexible Electronics

Jianshi Tang

*Tsinghua University*

**CONTENTS**

## 5.1 Introduction

Economically driven by the Moore's law, the continuous scaling of silicon transistors has steadily boosted the performance of personal electronics and supercomputers over the past few decades. However, such a trend is getting more and more difficult to maintain as the transistor feature size is being shrunk toward the fundamental physical limit of just a few nanometers. Various nanomaterials with atomically thin body, such as one-dimensional carbon nanotube (CNT) [1–3] and two-dimensional materials like molybdenum disulfide ($MoS_2$) [4–6], have been extensively studied as potential channel replacements for silicon in the ultra-scaled logic technology. One attractive benefit of using these nanometer-thick materials is that they are inherently immune to the so-called short-channel effect, so that more aggressive scaling in the channel length can be achieved [7]. Also, CNT has many other intrinsic advantages over silicon, such as higher saturation velocity and carrier mobility, longer mean free path, and lower operation voltage [2, 3], which are highly favorable for high-speed, low-power electronics [8]. Despite their superb electrical properties, most demonstrations of nanomaterials-based logic so far are limited to the level of single device or small-scale circuits. In practice, there are still many technical challenges, in both material and device engineering (interestingly, some of those challenges arise from the ultrathin thin body itself), to deliver a viable logic technology with nanomaterials.

Taking CNT as an example again, Figure 5.1 briefly summarizes the key challenges for developing high-speed logic technology starting from raw materials: (1) the purification of semiconducting nanotubes, including the development of a reliable sorting technique and the accurate measurement of high purity beyond the detection limit of standard absorption spectrum; (2) the ultrahigh-density placement of nanotubes on the substrate with

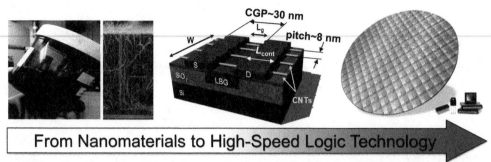

**FIGURE 5.1**
Challenges for developing high-speed logic technology using nanomaterials, from material preparation and placement to device scaling and circuit integration. Notations: "CGP" stands for "contact-gate pitch," "RO" stands for "ring oscillators," and "$\tau_{SD}$" stands for "stage delay."

precise pitch control down to about 8 nm or less while avoiding nanotube bundling; (3) the aggressive scaling of CNT transistor with a contact-gate-pitch (CGP) of 30 nm or less and low-resistance contacts; (4) the integration of millions or billions CNT transistors on a chip through proper management of variations (in threshold voltage $V_{th}$, nanotube diameter and length) and impurities or contaminations. To resolve those challenges and advance CNT-based logic technology, considerable progress has been made in the past 20 years in areas such as transistor scaling [2, 3], contact engineering [1, 9], improved semiconducting tube purity, and placement [10–14]. Nevertheless, the quality of materials and the integration level of circuits are still far away from meeting the extremely high standard in silicon industry. More work is needed in the future to demonstrate that a viable CNT-based logic technology could offer sufficient performance benefits on the circuit and system levels compared to its silicon counterpart [14, 15].

Despite the above-mentioned difficulties, it should be pointed out that the past research on nanomaterials is certainly not wasted. The related material development could enable some other applications that have much less stringent requirements on the material quality (e.g., semiconducting purity, diameter, and length variations) and device performance than high-speed logic technology while taking advantages of the unique merits of those nanomaterials, such as low-cost, large-area fabrication, and maybe more intriguingly, mechanical flexibility [16]. Such examples include emerging applications such as wearable electronics, real-time analytics, and Internet of Things (IoT), where high-performance logic circuits and sensors made on flexible or unconventional substrates are needed. Here, thin-film transistors (TFTs) with a channel length

**TABLE 5.1**

Comparison of Different Material Candidates for TFTs

| Metrics | Poly-Si | a-Si:H | Organic | Oxide | CNT |
|---|---|---|---|---|---|
| Mobility (cm²/Vs) | ~100 | ~1 | 0.1~1 | ~10 | 1–100 |
| Circuit type | CMOS | NMOS | PMOS | NMOS | CMOS |
| Stability | Good | Good | Poor | Good | Good |
| Cost | High | Low | Low | Low | Low |
| Process temperature | ~450°C | ~300°C | <100°C | ~300°C | <150°C |

of several microns (μm) or longer and an output current density over 10 microamperes per micron (μA/μm) are sufficient for most applications. More importantly, they can be easily fabricated using a much cheaper and simpler process over a large area. In literature, a variety of channel materials have been extensively studied for TFTs, including amorphous silicon [17], organic/oxide semiconductors [18, 19], CNTs [20], and two-dimensional layered materials [21]. Table 5.1 compares the key metrics of different material candidates for TFTs. As we can see, CNT stands out as an attractive candidate for making low-cost high-performance TFTs due to its high mobility, excellent stability, CMOS capability, and low-temperature processing [16, 22].

In this chapter, we will focus on CNT-based flexible electronics and sensors. We will first review the challenges and recent progress in making high-performance flexible CNT TFTs. Then, we will discuss different approaches, mainly through charge-transfer doping, to make *n*-type TFTs so that low-power complementary metal-oxide-semiconductor (CMOS) logic can be built. Recent advances in high-speed flexible logic circuits will be reviewed. After that, we will discuss the application of CNT TFTs for flexible sensors, especially integrated pressure sensors with TFT-based active matrix. Finally, we will give a perspective on future flexible electronics and sensors, intending to paint a picture where more functional components, including logic, sensor, memory, battery, and data transmitter, can be integrated on a single flexible substrate to realize more powerful flexible nanosystems.

## 5.2 CNT-Based Flexible TFTs

The fabrication of flexible CNT TFTs usually adopts a similar process flow as those fabricated on rigid substrates, e.g., silicon wafer or glass. Also, a rigid substrate is often employed as the handling wafer in the entire fabrication process, after which the flexible film with fabricated TFT devices is then carefully peeled off or delaminated. The use of flexible substrates, such as polyimide (PI), polyethylene naphthalate (PEN), or polyethylene terephthalate (PET) films, however, raises many process challenges, including substrate flatness and cleanness, thermal expansion, and limitations on process temperature and lithography resolution, which would affect the final device performance. Most flexible TFTs have either bottom-gate (Figure 5.2a) or top-gate (Figure 5.2b) transistor structure, in which a self-assembled monolayer (SAM, e.g., chloro(dimethyl)octadecylsilane)) is usually used for the deposition of CNT thin films in order to improve the density and uniformity. Considering the limitation on lithography resolution and

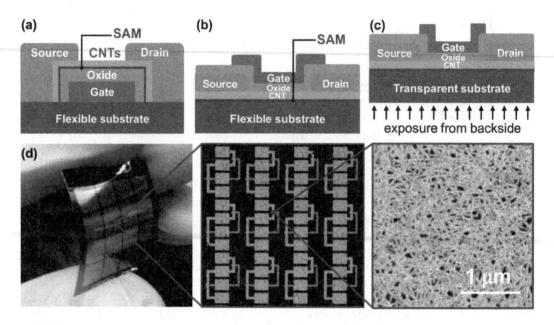

**FIGURE 5.2**
Structures and images of flexible CNT TFTs. Typical device structures of flexible TFTs: (a) Bottom-gate transistor, (b) Top-gate transistor, (c) Top-gate transistor with self-aligned gate by photoresist exposure from the backside using source/drain contacts as the mask in the photolithography process. (d) Images of typical CNT TFTs fabricated on a flexible substrate. (Reproduced with permission from Ref. [25].)

alignment accuracy, there is usually a relatively large overlap between gate and source/drain contacts, where the parasitic capacitance could limit the operation speed of flexible circuits. To minimize the parasitic capacitance, self-aligned gate structure is preferred but remains challenging to be reliably made on flexible substrates. One possible approach is to use the highly directional metal deposition and play with the thickness difference between gate and source/drain contacts [23]. Another interesting idea, as illustrated in Figure 5.2c, is to expose the photoresist from the backside of the substrate, which needs to be transparent to UV light, using source/drain contacts as the mask in the gate photolithography process [24].

Figure 5.2d illustrates typical flexible CNT TFTs after fabrication. While high-resolution e-beam lithography is usually not employed (also economically undesirable) here, the fabricated flexible CNT TFTs by conventional optical lithography or inkjet printing tend to have a relatively long channel length, in the range from a few microns up to hundreds or thousands of microns. Such channel length is longer than the typical length of solution-processed CNTs (in the range of 0.5~2 µm). As a result, the TFT channel is usually a network of randomly oriented CNTs and consists of many nanotube junctions. Those junctions, rather than the source/drain contacts, could dominate the transport property of CNT TFTs. It should be pointed out that such configuration somewhat relaxes the requirement on the semiconducting purity of CNT solution, but the improvement in purity (and density) is still beneficial for enhancing the TFT performance upon channel length scaling (see Figure 5.3a) [25]. As shown in Figure 5.2d, the deposited CNT thin film usually shows a good uniformity over a large scale; however, inhomogeneity clearly exists on the microscopic scale, which largely contributes to the device variations especially for devices with a short channel length of a few microns.

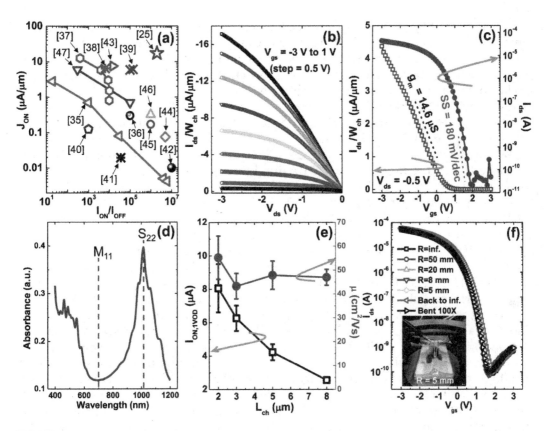

**FIGURE 5.3**

Flexible CNT TFTs with state-of-the-art performance. (a) Comparison of current density $J_{ON}$ versus $I_{ON}/I_{OFF}$ ratio for some representative flexible CNT TFTs reported in the literature. (b) $I_{ds}$-$V_{ds}$ curves from state-of-the-art flexible CNT TFT with a 2-μm long channel, achieving a large current density of 17.2 μA/μm. (c) The corresponding $I_{ds}$-$V_{gs}$ curve at $V_{ds} = -0.5$ V shows $I_{ON}/I_{OFF} > 10^6$, subthreshold slope of $SS = 180$ mV/dec, and mobility of μ = 52.6 cm²/Vs. (d) UV-vis-NIR absorption spectrum of the polymer-sorted CNT solution, indicating a high semiconducting purity above 99%. (e) Extracted ON current under 1 V gate overdrive ($I_{ON,1VOD}$, left) and carrier mobility (right) for CNT TFTs with different channel lengths. (f) $I_{ds}$-$V_{gs}$ transfer curves at $V_{ds} = -0.5$ V of a typical CNT flexible TFT bent under different radii $R$, showing no degradation on the device performance with $R$ down to 5 mm (inset). (Reproduced with permission from Ref. [25].)

Besides randomly connected CNT networks, aligned CNTs have also been used as the channel for CNT TFTs in the hope of further improving the device performance and reducing variations. There are two main approaches to prepare aligned CNTs. One is directly growing them by the chemical vapor deposition (CVD) technique, usually followed by a process to transfer them onto the target substrate [26, 27]. This approach produces high-quality CNTs with excellent structural integrity, but it requires the selective removal of metallic CNTs, which could induce a large variation in CNT density [28, 29]. The other approach is to purify the CNTs dispersed in solution and then align them on the target substrate during deposition using a variety of techniques, such as Langmuir-Blodgett assembly [30], Langmuir-Schaefer assembly [31], template-assisted chemical self-assembly [13, 14], dielectrophoretic assembly [32], floating evaporative self-assembly [33], and vacuum filtration [34]. The key challenge remains as the preparation of large-area monolayer aligned CNTs with a high enough density, considering that a thick CNT film would have adverse impacts on the effective gate modulation and contact [31, 34]. So far, nearly all attempts to use aligned CNTs as the channel have been made on rigid substrates rather than flexible substrates, mainly because such aligned channel configuration benefits more for short-channel devices, i.e., channel length smaller than nanotube length, which can be easily made on rigid substrates but much more difficult on flexible ones.

In the evaluation of the TFT performance, carrier mobility ($\mu$), current ON/OFF ratio ($I_{ON}/I_{OFF}$), and ON current density ($J_{ON}$) are usually considered as the three most important metrics. In general, flexible CNT TFTs usually $p$-type and exhibit lower performance ($J_{ON}$ in the range of 0.01~10 $\mu$A/$\mu$m [35–47]) than those made on rigid substrates ($J_{ON}$ in the range of 1~100 $\mu$A/$\mu$m [48, 49]). One critical challenge is to improve $J_{ON}$ without sacrificing $I_{ON}/I_{OFF}$ for CNT TFTs. In the presence of metallic tubes (even small percentages), $J_{ON}$ and $I_{ON}/I_{OFF}$ usually have an undesired trade-off with channel length in the case where the semiconducting purity is not high enough, which makes it challenging to improve TFT performance by simple scaling of the channel length [35]. As a comparison, Figure 5.3a plots $J_{ON}$ versus $I_{ON}/I_{OFF}$ for some representative flexible CNT TFTs reported in the literature. Here, as shown in Figures 5.3b-c, the best performed flexible CNT TFTs show $J_{ON} = 17.2$ $\mu$A/$\mu$m, $I_{ON}/I_{OFF} \approx 2\times10^6$ and $\mu \approx 55$ cm$^2$/Vs, which is on par with the best reported values from rigid CNT TFTs with similar channel lengths [48–50]. They are made using a highly purified CNT solution with a semiconducting purity above 99.99% [25], which is beyond the detection limit of commonly used optical absorption spectrum as shown in Figure 5.3d. It is noted that such high purity value is quantified by electrically measuring more than 20,000 CNT transistors, and it is among the highest purities ever quantified for a CNT solution [51]. Even though this solution-based purification process can be repeated many times to further improve the purity, a fast and convenient approach to accurately quantify the actual purity is still lacking.

To understand the carrier transport in CNT TFTs and the performance dependence on the device dimension, Figure 5.3e plots the ON current and field-effect mobility as a function of channel length. Considering the device variation in the threshold voltage $V_t$, for a fair comparison, here the device ON current is extracted at a constant gate overdrive (e.g., $|V_{gs} - V_t| = 1$ V, denoted $I_{ON,1VOD}$). In the evaluation of carrier mobility, the gate capacitance is usually calculated using the parallel plate model. However, it should be noted that this model is only valid for relatively dense CNT thin films where the tube spacing is smaller than the gate oxide thickness [52]. Figure 5.3e shows that $I_{ON,1VOD}$ decreases with increasing channel length, which is consistent with percolation transport in CNT TFTs [35]. In contrast, the carrier mobility shows a relatively weak dependence on the channel length and similar behavior has been observed in rigid CNT TFTs [49]. When the channel length is further scaled down to submicron, i.e., close to the length of individual CNT, both the current density and mobility would increase dramatically as the device approaches the direct-contact transport regime [2, 31, 33].

In addition, bending experiments are usually carried out in order to test the flexibility and durability of the fabricated CNT TFTs. Figure 5.3f plots the transfer curves of a typical flexible CNT TFT under different bending radii down to 5 mm, showing no performance degradation. This result implies that these flexible CNT TFTs would still be functional if wrapped around human finger, highlighting their potential applications as wearable electronics. A typical failure mechanism of flexible TFTs upon bending is the breaking of gate dielectric. Thinner flexible substrate could be used to reduce the bending-induced strain and enable the device to work under extremely small bending radius close to 100 $\mu$m [53].

Additional efforts can also be made by using an atomically smooth planarization coating and a hybrid encapsulation stack that places the transistors in the neutral strain position.

## 5.3 CNT-Based Flexible CMOS Circuits

Without passivation, as-fabricated CNT TFTs are naturally PFETs mainly due to water/oxygen absorption on the CNT surface from ambient [54]. However, CMOS logic is highly preferred for low-power applications; therefore, robust NFETs are still desired in the fabrication of flexible CNT circuits. For short-channel transistors where individual CNT is in contact with both source and drain contacts, the contact work function filters the carrier type being injected into the channel and, hence, determines the transistor polarity, as illustrated in Figure 5.4a. As a result, metals with high and low work functions are used

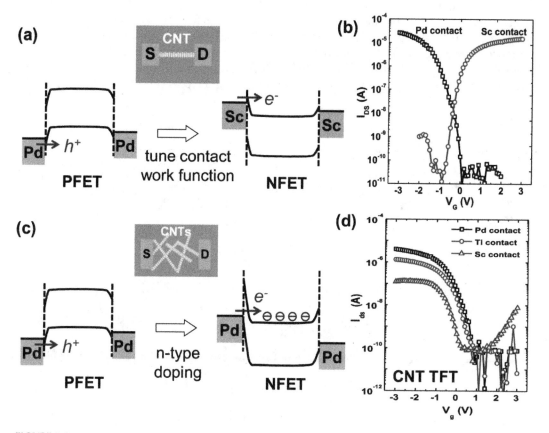

**FIGURE 5.4**
Strategies to make CNT NFETs. (a) Energy band diagrams of CNT PFET and NFET, where the work function of source/drain contacts in a short-channel transistor determines the transistor polarity. (b) Typical transfer characteristics of CNT PFET with Pd contacts and NFET with Sc contacts. (c) The tube-to-tube junctions rather than contacts dominate the carrier transport in CNT TFTs, where additional *n*-type channel doping is needed to make NFET. (d) Typical transfer characteristics of CNT TFTs with different metal contacts, showing that the use of low work function metal Sc alone in CNT TFTs is not effective to make unipolar NFETs.

as direct contacts to make PFETs (e.g., Pd and Au) and NFETs (e.g., Sc and Er) [14, 55–57], respectively, as shown in Figure 5.4b. In contrast, the carrier transport in CNT TFTs, whose channel length is typically much longer than tube length, is dominated by tube-to-tube junctions rather than contacts (i.e., percolation transport), as illustrated in Figure 5.4c. In this case, the use of low work function metal alone in CNT TFTs is usually not effective to make unipolar NFETs, as shown in Figure 5.4d (although exceptions are reported in literature [58]). As a result, additional *n*-type doping layer is usually adopted to convert PFETs into NFETs by shifting the threshold voltage, for which a variety of organic molecules (e.g., benzyl viologen) and dielectric films (e.g., $SiN_x$, $Al_2O_3$, $HfO_2$, and MgO) have been investigated in literature [42, 59–67].

As an example of charge transfer doping from organic molecules, Figure 5.5a shows the transfer characteristics before and after benzyl viologen (BV) doping in CNT TFTs, which successfully converts the PFET into NFET [59]. Considering the poor stability and process compatibility of organic molecule, a dedicated patterning and passivation scheme is used here. It starts with capping the BV layer with poly(methyl methacrylate) (PMMA) resist, and then patterns the channel doping region using hydrogen silsesquioxane (HSQ) negative resist, and finally passivates the sample with $Al_2O_3$ by atomic-layer deposition (ALD). With this, the fabricated NFETs with BV doping show pretty good stability in air over time. Similar *n*-type doping effect can be achieved using $Si_3N_4$ deposited by plasma-enhanced chemical vapor deposition (PECVD), in which the fixed positive charges arising from $^+Si\equiv N_3$ dangling bonds induce a field-effect doping in the TFT channel [61]. In the experiments shown in Figure 5.5b, it is found that higher plasma power (e.g., 450 W) is

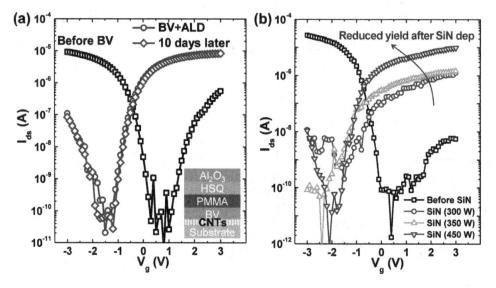

**FIGURE 5.5**

Different *n*-type doping methods in CNT TFTs. (a) Transfer characteristics before (black curve) and after (blue curve) benzyl viologen (BV) doping in CNT TFTs, showing successful conversion of PFET into NFET. Dedicated patterning and passivation scheme, including a multilayer capping with PMMA and HSQ resists followed by ALD $Al_2O_3$, is used to improve the stability (red curve) of BV doping. (b) Transfer characteristics before (black curve) and after (red, green and blue curves) $Si_3N_4$ deposition on CNT TFTs with different plasma power. Higher plasma power is needed to enhance the performance of the converted NFET; however, it also induces significant damage on CNTs and, hence, reduces the final device yield by more than 50%.

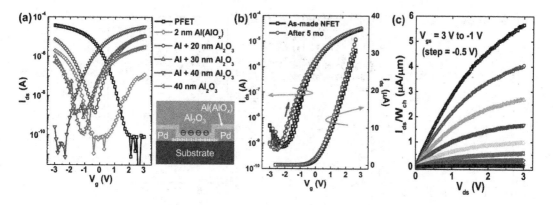

**FIGURE 5.6**

Robust *n*-type doping in CNT TFTs from bilayer oxide doping. (a) $I_{ds}$–$V_g$ transfer characteristics of CNT TFTs at $V_{ds} = 0.5$ V after the deposition of various oxide doping layers. The schematic inset shows the device structure. The starting 2-nm-thick Al layer, which is oxidized in air into $AlO_x$, acts as seeding layer for the subsequent ALD of $Al_2O_3$ and also provides *n*-type doping to the CNT TFT channel. (b) Small hysteresis (~100 mV) and excellent stability over 5 months of the fabricated NFET. (c) A set of $I_{ds}$–$V_{ds}$ output curves of an *n*-type CNT TFT after the bilayer doping of 2-nm-thick thermally evaporated Al ($AlO_x$) and 40-nm-thick ALD $Al_2O_3$. (Reproduced with permission from Ref. [25].)

needed to enhance the performance of the converted NFET, which in turn induces significant damage on CNTs and, hence, reduces the final device yield [59].

Besides the aforementioned two methods, ALD oxides on CNT TFTs could provide a convenient and harmless approach for effective *n*-type doping arising from oxygen vacancies [62–64], and it is readily process-compatible for building CMOS logic on flexible substrate. Figure 5.6a summarizes the experimental results of the doping effect of various ALD layers on flexible CNT TFTs. The *n*-type doping effect increases with the thickness of ALD $Al_2O_3$, which can be used to adjust the threshold voltage of CNT TFTs. However, the results here show that the ALD $Al_2O_3$ layer alone does not provide sufficient doping to fully convert PFET to unipolar NFET. The best doping recipe appears to be the combination of 2-nm-thick thermally evaporated Al and 40-nm-thick ALD $Al_2O_3$. Here, the thin Al layer is readily oxidized in air into $AlO_x$ [63], which serves as the seeding layer for subsequent ALD process to ensure better coverage on CNTs and also contributes to the *n*-type doping. With the $AlO_x$/$Al_2O_3$ bilayer doping, the NFET shows a comparable device performance as the PFET with symmetric threshold voltages. As shown in Figure 5.6b, such *n*-type doping layer also serves as an effective passivation layer to largely reduce the hysteresis and achieve excellent stability over time. Figure 5.6c shows the typical output characteristics of an individual NFET, where the linear $I_{ds}$–$V_{ds}$ relation at small $V_{ds}$ suggests that the sharp Schottky barrier at source/drain contacts is not limiting the electron tunneling transport.

With the demonstrated PFETs and NFETs, high-performance CMOS logic gates (such as inverters, NAND, and NOR gates) and circuits can be fabricated on flexible substrates with a high yield. Figure 5.7a shows the output characteristics of a flexible CNT CMOS inverter, exhibiting abrupt switching from $V_{DD} = +3$ V to $V_{SS} = -3$ V and a high inverter gain of 25. Besides the high gain, it also shows a large inverter noise margin of about 2.2 V, more than 36% of $|V_{DD}–V_{SS}|$. Both factors are critical for the demonstration of cascade logic circuits. In the evaluation of circuit performance and technology maturity, ring oscillator (RO) is a

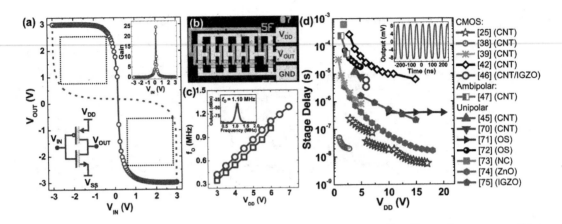

**FIGURE 5.7**

Flexible CNT CMOS logic and circuits. (a) Output characteristic of a flexible CNT CMOS inverter, showing abrupt switching from $V_{DD} = +3$ V to $V_{SS} = -3$ V. The bottom inset shows the schematic diagram of CMOS inverter. The top inset shows an inverter gain of about 25. (b) Microscope image of a 5-stage flexible CNT CMOS RO with an additional inverter stage as the output buffer. (c) Dependence of oscillation frequency $f_O$ on the supply voltage $V_{DD}$ for two exemplary CNT ROs with $L_{ch} = 3$ µm. The inset shows the output frequency spectrum with $f_O = 1.10$ MHz at $V_{DD} = 6$ V. (d) Comparison of stage delay among representative flexible ROs made with different nanomaterials, including CNT, organic semiconductors (OS), oxide semiconductors, nanocrystals (NC), and hybrid nanomaterials. The inset shows the output waveform with $f_O = 17.6$ MHz at $V_{DD} = 17$ V from the fastest flexible CNT CMOS RO, representing the smallest stage delay down to 5.7 ns. (Reproduced with permission from Ref. [25].)

widely used benchmark circuit that consists of a chain of odd number of inverters, where the output of the last inverter is fed back into the input of the first one [14]. Figure 5.7b shows a 5-stage flexible CNT CMOS RO, where an additional stage of inverter is used as the output buffer for electrical measurements [25]. Figure 5.7c plots the voltage dependence of the oscillation frequency, which increases almost linearly with the supply voltage $V_{DD}$. The inset shows the output frequency spectrum at a supply voltage of $V_{DD} = 6$ V, featuring an oscillation frequency of $f_O = 1.10$ MHz and stage delay of $\tau_{SD} = 90.9$ ns. Such sub-100 ns stage delay is already one of the fastest CNT ROs reported on flexible substrates [38, 39, 45–47, 68, 69]. The stage delay can be further reduced by scaling down the channel length, reducing parasitic capacitance, and increasing the supply voltage. The fastest flexible CNT RO achieved so far has a stage delay of only $\tau_{SD} = 5.7$ ns (inset of Figure 5.7d) [25], which is comparable to the best demonstrations on rigid substrates [50]. As a comparison, Figure 5.7d plots the supply voltage-dependent stage delay from flexible ROs made with different nanomaterials, including CNTs, organic semiconductors, oxide semiconductors, nanocrystals, and hybrid nanomaterials [42, 47, 70–75]. It clearly shows that CNT remains as the top candidate for making high-speed and low-power flexible integrated circuits.

## 5.4 CNT-Based Flexible Sensors

Besides flexible TFTs, CNTs have also been used to make a variety of flexible sensors to sense mechanical (e.g., pressure, strain, shear force, etc.) and biochemical signals [76–78]. For example, various composite materials by mixing CNTs with elastomers (e.g., polydimethylsiloxane) have been extensively investigated as pressure sensor and strain gauge [79, 80]. In particular, flexible pressure sensor is of great interest to mimic the human tactile sense [81]. Such artificial electronic skin, with the ability of accurately sensing arbitrarily shaped objects, is highly desired for the development of smart robotics and prosthetic solutions [82–86]. To make such an artificial skin, a critical challenge is the fabrication of high-performance large-area flexible TFT arrays as the active matrix, for which CNT is a superior candidate [87–91]. Here, besides the aforementioned TFT performance (including mobility, ON/OFF ratio, current density, operation voltage, and yield), pressure sensitivity, spatial resolution, accuracy, response speed, as well as mechanical flexibility are all considered as important metrics for a useful artificial electronic skin.

Figure 5.8 shows a fully integrated flexible pressure sensor built on an active matrix of $16 \times 16$ CNT TFTs spanning over a 4-inch area [92]. Each pixel consists of a flexible CNT TFT (with a channel dimension of $20 \times 75 \ \mu m^2$) connected to a pressure sensitive element as the load resistor, as illustrated in Figures 5.8a-b. For pressure mapping, the 256-pixel CNT TFT active matrix is scanned sequentially through the gate and drain lines, where a large sensing margin ($I_{Pressed}/I_{Released} > 10^4$) is achieved, as shown in Figure 5.8c. Besides, the threshold voltage of the CNT TFT needs to be carefully tuned to minimize crosstalk between the selected pixel and inactivated pixels during the measurements. Overall, the fully integrated pressure sensor system can operate within a small voltage range of 3 V, and exhibits superb performance with high spatial resolution of 4 mm, faster response than human skin (<30 ms), and excellent accuracy in sensing complex objects on both flat and curved surfaces, as shown in Figure 5.8d. It is noted that here a commercial pressure-sensitive rubber (PSR, ZOFLEX® ZL45.1) is used to sense the applied pressure. For a more

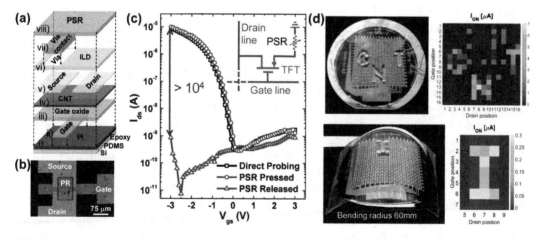

**FIGURE 5.8**
Fully integrated flexible pressure sensor with CNT TFTs active matrix. (a) Schematic of each pixel in the flexible pressure sensor by integrating a CNT TFT with pressure-sensitive rubber (PSR). (b) Microscope image of a CNT TFT with channel dimension of $W_{ch}/L_{ch} = 75 \ \mu m/20 \ \mu m$. (c) $I_{ds}-V_{gs}$ transfer curves at $V_{ds} = -1$ V for an individual pixel when directly probing through the source pad (black), when the PSR is pressed (red, applied pressure ~ 5 kPa), and when the PSR is released (blue), respectively. The inset shows the circuit diagram for a single pixel in the pressure sensor. (d) Pressure mapping of complex objects on flat and curved surfaces. Top panel: current mapping for the pressure sensing of word "CNT" made by PDMS. The defective pixels are labeled in gray color. Bottom panel: pressure sensing on curved surface with a bending radius of 60 mm. In both mappings, the applied pressure is about 6.8 kPa. (Reproduced with permission from Ref. [92].)

compact system with higher spatial resolution, the abovementioned CNT/elastomer composites may be used as the pressure-sensing element that may be integrated by microfabrications (e.g., spin coating and patterning) along with the CNT TFT active matrix.

## 5.5 Conclusions and Outlook

After nearly 20 years of research on CNT transistors [93], there has been tremendous progress in both material development and device engineering, which has dramatically boosted the individual device performance of CNT transistor (especially PFET) to be competitive compared to its silicon counterpart. There are still many remaining technical challenges, in particular further improving semiconducting purity and CNT placement [94, 95], to be resolved in order to develop a viable CNT-based logic technology. As the silicon industry continues marching toward deep sub-10 nm technology nodes, it is difficult to predict whether CNT (or any other candidates such as two-dimensional materials) is ever going to be able to replace silicon for the ultra-scaled high-speed logic technology in the not-so-distant future. Notably, the unique advantage of low-temperature and substrate-agnostic fabrication of CNT transistors (as well as emerging nonvolatile memories such as resistive random-access memory) may be taken to build logic circuits (and memory) in the back-end-of-line (BEOL) process, which facilitates the fabrication of monolithic three-dimensional system-on-chip (3D SoC) [96]. More importantly, the material development along the way enables other interesting applications for which CNT may be better suited, such as flexible TFTs and sensors. The excellent mechanical flexibility and strength of CNT, along with its superb electrical properties, makes it an outstanding candidate for high-performance flexible CMOS circuits. Recent advances in the material, device, and processing aspects have brought their performance to the same level as those made on rigid substrates. They have also been explored in broad applications beyond flexible logic circuits, such as fully integrated flexible sensors.

Besides the abovementioned pressure sensor, many other types of sensors can be made on flexible substrates as well, such as temperature sensors, bio- and chemical sensors [97], flow sensors [98, 99], magnetic sensors [100, 101], etc. An interesting scenario would be integrating different types of sensors on the same flexible substrate as a multiplex wearable sensor system, and one example is shown in Figure 5.9a. Here, the multiplex biosensors collect vital signals from sweat (e.g., PH value, temperature, etc.), which are converted to electrical signals using integrated circuit (IC) chips (e.g., trans-impedance amplifiers, transistors, resistors, and capacitors), and then they are transmitted to peripheral equipment for data collection and processing. A similar but more advanced system has been built for multiplexed in situ perspiration analysis with wireless data transmission [97]. However, most, if not all, flexible sensor systems reported in literature so far have only sensors on a flexible substrate while the peripheral circuits for signal processing are built externally or using discrete chips soldered on flexible PCB boards (as in the case shown in Figure 5.9a). As demonstrated earlier, flexible CNT TFTs can be used to make CMOS circuits with decent performance and low cost. Therefore, it is of great interest to replicate the functions of those silicon-based peripheral circuits using flexible CNT TFTs, so that they can be fabricated on the same substrate along with the flexible sensors. Moreover, besides sensors and data processors, nonvolatile memory [102], battery [103], solar cell [104], signal transmitters [50],

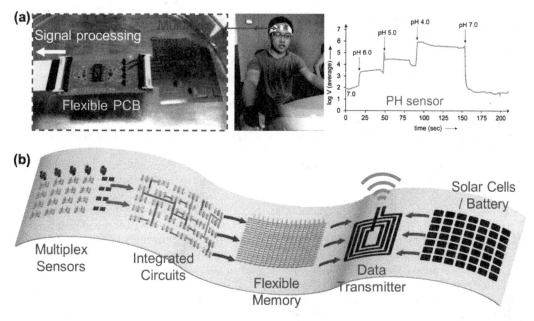

**FIGURE 5.9**
Fully integrated flexible nanosystems. (a) An example of fully integrated wearable multiplex sensor platform. The multiplex biosensors (e.g., PH sensor) collect vital signals from sweat, which are converted to electrical signals using IC chips soldered on a flexible PCB board and are then transmitted to peripheral equipment for data collection and processing. (b) Schematic illustration of a fully integrated nanosystem that may include many functional elements such as sensors and data processors, memory, data transmitters, battery, and solar cells on the same flexible substrate.

and many other functional elements may all be integrated on a single substrate to build a powerful flexible nanosystem [105], as illustrated in Figure 5.9b. One of the benefits of having such a highly integrated and compact system is that many data-processing tasks can now be done in situ (or locally) and such near-sensor computations could maximize data compression and reduce latency and power consumption. It is hence believed to have potential applications in areas such as healthcare (e.g., neural probe, human-machine interface), edge computing, and IoT.

## References

[1] Cao, Q., Han, S.J., Tersoff, J., Franklin, A.D., Zhu, Y., Zhang, Z., Tulevski, G.S., Tang, J., and Haensch, W. (2015) End-bonded contacts for carbon nanotube transistors with low, size-independent resistance. *Science*, **350** (6256), 68–72.

[2] Cao, Q., Tersoff, J., Farmer, D.B., Zhu, Y., and Han, S. (2017) Carbon nanotube transistors scaled to a 40-nanometer footprint. *Science*, **356** (6345), 1369–1372.

[3] Qiu, C., Zhang, Z., Xiao, M., Yang, Y., Zhong, D., and Peng, L.M. (2017) Scaling carbon nanotube complementary transistors to 5-nm gate lengths. *Science*, **355** (6322), 271–276.

[4] Radisavljevic, B., Radenovic, A., Brivio, J., Giacometti, V., and Kis, A. (2011) Single-layer MoS$_2$ transistors. *Nat. Nanotechnol.*, **6** (3), 147–150.

[5] Desai, S.B., Madhvapathy, S.R., Sachid, A.B., Llinas, J.P., Wang, Q., Ahn, G.H., Pitner, G., Kim, M.J., Bokor, J., Hu, C., Wong, H.-S.P., and Javey, A. (2016) MoS$_2$ transistors with 1-nanometer gate lengths. *Science*, **354** (6308), 99–102.

[6] Sarkar, D., Xie, X., Liu, W., Cao, W., Kang, J., Gong, Y., Kraemer, S., Ajayan, P.M., and Banerjee, K. (2015) A subthermionic tunnel field-effect transistor with an atomically thin channel. *Nature*, **526** (7571), 91–95.

[7] Ferain, I., Colinge, C.A., and Colinge, J.P. (2011) Multigate transistors as the future of classical metal-oxide-semiconductor field-effect transistors. *Nature*, **479** (7373), 310–316.

[8] Tang, J., and Han, S.J. (2018) Carbon nanotube logic technology, in *Advanced Nanoelectronics* (eds. Hussain, M.M.), Wiley, Weinheim, Germany, pp. 119–149.

[9] Tang, J., Cao, Q., Farmer, D.B., Tulevski, G., and Han, S. (2016) Carbon nanotube complementary logic with low-temperature processed end-bonded metal contacts. *IEDM Tech. Dig.*, 5.1.1–5.1.4.

[10] Nish, A., Hwang, J.Y., Doig, J., and Nicholas, R.J. (2007) Highly selective dispersion of single-walled carbon nanotubes using aromatic polymers. *Nat. Nanotechnol.*, **2** (10), 640–646.

[11] Mistry, K.S., Larsen, B.A., and Blackburn, J.L. (2013) High-yield dispersions of large-diameter semiconducting single-walled carbon nanotubes with tunable narrow chirality distributions. *ACS Nano*, **7** (3), 2231–2239.

[12] Tulevski, G.S., Franklin, A.D., and Afzali, A. (2013) High purity isolation and quantification of semiconducting carbon nanotubes via column chromatography. *ACS Nano*, **7** (4), 2971–2976.

[13] Park, H., Afzali, A., Han, S.J., Tulevski, G.S., Franklin, A.D., Tersoff, J., Hannon, J.B., and Haensch, W. (2012) High-density integration of carbon nanotubes via chemical self-assembly. *Nat. Nanotechnol.*, **7** (12), 787–791.

[14] Han, S.J., Tang, J., Kumar, B., Falk, A., Farmer, D., Tulevski, G., Jenkins, K., Afzali, A., Oida, S., Ott, J., Hannon, J., and Haensch, W. (2017) High-speed logic integrated circuits with solution-processed self-assembled carbon nanotubes. *Nat. Nanotechnol.*, **12** (9), 861–865.

[15] Shulaker, M.M., Hills, G., Patil, N., Wei, H., Chen, H.Y., Wong, H.S.P., and Mitra, S. (2013) Carbon nanotube computer. *Nature*, **501** (7468), 526–530.

[16] Franklin, A.D. (2015) Nanomaterials in transistors: From high-performance to thin-film applications. *Science*, **349** (6249), aab2750–aab2750.

[17] Nathan, A., Kumar, A., Sakariya, K., Servati, P., Sambandan, S., and Striakhilev, D. (2004) Amorphous silicon thin film transistor circuit integration for organic LED displays on glass and plastic. *IEEE J. Solid-State Circuits*, **39** (9), 1477–1486.

[18] Nomura, K., Ohta, H., Takagi, A., Kamiya, T., Hirano, M., and Hosono, H. (2004) Room-temperature fabrication of transparent flexible thin-film transistors using amorphous oxide semiconductors. *Nature*, **432** (7016), 488–492.

[19] Forrest, S.R. (2004) The path to ubiquitous and low-cost organic electronic appliances on plastic. *Nature*, **428** (6986), 911–918.

[20] Hu, L., Hecht, D.S., and Grüner, G. (2010) Carbon nanotube thin films: Fabrication, properties, and applications. *Chem. Rev.*, **110** (10), 5790–5844.

[21] Akinwande, D., Petrone, N., and Hone, J. (2014) Two-dimensional flexible nanoelectronics. *Nat. Commun.*, **5**, 5678.

[22] Cai, L., and Wang, C. (2015) Carbon nanotube flexible and stretchable electronics. *Nanoscale Res. Lett.*, **10** (1), 320.

[23] Javey, A., Guo, J., Farmer, D.B., Wang, Q., Yenilmez, E., Gordon, R.G., Lundstrom, M., and Dai, H. (2004) Self-aligned ballistic molecular transistors and electrically parallel nanotube arrays. *Nano Lett.*, **4** (7), 1319–1322.

[24] Haensch, W.E.A., and Liu, Z. (2013) Graphene and nanotube/nanowire transistor with a self-aligned gate structure on transparent substrates and method of making same. US8569121B2, issued 2013.

[25] Tang, J., Cao, Q., Tulevski, G., Jenkins, K.A., Nela, L., Farmer, D.B., and Han, S.J. (2018) Flexible CMOS integrated circuits based on carbon nanotubes with sub-10 ns stage delays. *Nat. Electron.*, **1** (3), 191–196.

[26] Cao, Q., and Han, S. (2013) Single-walled carbon nanotubes for high-performance electronics. *Nanoscale*, **5**, 8852–8863.

[27] Kang, S.J., Kocabas, C., Ozel, T., Shim, M., Pimparkar, N., Alam, M. a., Rotkin, S. V., and Rogers, J.A. (2007) High-performance electronics using dense, perfectly aligned arrays of single-walled carbon nanotubes. *Nat. Nanotechnol.*, **2** (4), 230–236.

[28] Jin, S.H., Dunham, S.N., Song, J., Xie, X., Kim, J.J., Lu, C., Islam, A., Du, F., Kim, J.J., Felts, J., Li, Y., Xiong, F., Wahab, M.A., Menon, M., Cho, E., Grosse, K.L., Lee, D.J., Chung, H.U., Pop, E., Alam, M.A., King, W.P., Huang, Y., and Rogers, J.A. (2013) Using nanoscale thermocapillary flows to create arrays of purely semiconducting single-walled carbon nanotubes. *Nat. Nanotechnol.*, **8** (5), 347–355.

[29] Hills, G., Zhang, J., Mackin, C., Shulaker, M., Wei, H., Wong, H.-S.P., and Mitra, S. (2013) Rapid exploration of processing and design guidelines to overcome carbon nanotube variations. *Proc. 50th Annu. Des. Autom. Conf. - DAC '13*, **34** (7), 1.

[30] Li, X., Zhang, L., Wang, X., Shimoyama, I., Sun, X., Seo, W.-S., and Dai, H. (2007) Langmuir-Blodgett assembly of densely aligned single-walled carbon nanotubes from bulk materials. *J. Am. Chem. Soc.*, **129** (16), 4890–4891.

[31] Cao, Q., Han, S., Tulevski, G.S., Zhu, Y., Lu, D.D., and Haensch, W. (2013) Arrays of single-walled carbon nanotubes with full surface coverage for high-performance electronics. *Nat. Nanotechnol.*, **8** (3), 180–186.

[32] Cao, Q., Han, S., and Tulevski, G.S. (2014) Fringing-field dielectrophoretic assembly of ultrahigh-density semiconducting nanotube arrays with a self-limited pitch. *Nat. Commun.*, **5**, 5071.

[33] Brady, G.J., Way, A.J., Safron, N.S., Evensen, H.T., Gopalan, P., and Arnold, M.S. (2016) Quasi-ballistic carbon nanotube array transistors with current density exceeding Si and GaAs. *Sci. Adv.*, **2** (9), e1601240.

[34] He, X., Gao, W., Xie, L., Li, B., Zhang, Q., Lei, S., Robinson, J.M., Hároz, E.H., Doorn, S.K., Wang, W., Vajtai, R., Ajayan, P.M., Adams, W.W., Hauge, R.H., and Kono, J. (2016) Wafer-scale monodomain films of spontaneously aligned single-walled carbon nanotubes. *Nat. Nanotechnol.*, **11** (7), 633–638.

[35] Chandra, B., Park, H., Maarouf, A., Martyna, G.J., and Tulevski, G.S. (2011) Carbon nanotube thin film transistors on flexible substrates. *Appl. Phys. Lett.*, **99** (7), 072110.

[36] Tian, B., Liang, X., Yan, Q., Zhang, H., Xia, J., Dong, G., Peng, L., and Xie, S. (2016) Wafer scale fabrication of carbon nanotube thin film transistors with high yield. *J. Appl. Phys.*, **120** (3), 034501.

[37] Wang, C., Chien, J.C., Takei, K., Takahashi, T., Nah, J., Niknejad, A.M., and Javey, A. (2012) Extremely bendable, high performance integrated circuits using semiconducting carbon nanotube networks for digital, analog, and radio-frequency applications. *Nano Lett.*, **12** (3), 1527–1533.

[38] Zhang, H., Xiang, L., Yang, Y., Xiao, M., Han, J., Ding, L., Zhang, Z., Hu, Y., and Peng, L.-M. (2018) High-performance carbon nanotube complementary electronics and integrated sensor systems on ultrathin plastic foil. *ACS Nano*, **12** (3), 2773–2779.

[39] Zhang, H., Liu, Y., Yang, C., Xiang, L., Hu, Y., and Peng, L. (2018) Wafer-scale fabrication of ultrathin flexible electronic systems via capillary-assisted electrochemical delamination. *Adv. Mater.*, **30** (50), 1805408.

[40] Cao, Q., Kim, H., Pimparkar, N., Kulkarni, J.P., Wang, C., Shim, M., Roy, K., Alam, M.A., and Rogers, J.A. (2008) Medium-scale carbon nanotube thin-film integrated circuits on flexible plastic substrates. *Nature*, **454** (7203), 495–500.

[41] Lau, P.H., Takei, K., Wang, C., Ju, Y., Kim, J., Yu, Z., Takahashi, T., Cho, G., and Javey, A. (2013) Fully printed, high performance carbon nanotube thin-film transistors on flexible substrates. *Nano Lett.*, **13** (8), 3864–3869.

[42] Zhao, Y., Li, Q., Xiao, X., Li, G., Jin, Y., Jiang, K., Wang, J., and Fan, S. (2016) Three-dimensional flexible complementary metal-oxide-semiconductor logic circuits based on two-layer stacks of single-walled carbon nanotube networks. *ACS Nano*, **10** (2), 2193–2202.

[43] Honda, W., Arie, T., Akita, S., and Takei, K. (2015) Mechanically flexible and high-performance CMOS logic circuits. *Sci. Rep.*, **5**, 15099.

[44] Wang, H., Wei, P., Li, Y., Han, J., Lee, H.R., Naab, B.D., Liu, N., Wang, C., Adijanto, E., Tee, B.C.K., Morishita, S., Li, Q., Gao, Y., Cui, Y., and Bao, Z. (2014) Tuning the threshold voltage of carbon nanotube transistors by n-type molecular doping for robust and flexible complementary circuits. *Proc. Natl. Acad. Sci. U. S. A.*, **111** (13), 4776–4781.

[45] Sun, D., Timmermans, M.Y., Tian, Y., Nasibulin, A.G., Kauppinen, E.I., Kishimoto, S., Mizutani, T., and Ohno, Y. (2011) Flexible high-performance carbon nanotube integrated circuits. *Nat. Nanotechnol.*, **6** (3), 156–161.

[46] Chen, H., Cao, Y., Zhang, J., and Zhou, C. (2014) Large-scale complementary macroelectronics using hybrid integration of carbon nanotubes and IGZO thin-film transistors. *Nat. Commun.*, **5** (1), 4097.

[47] Ha, M., Xia, Y., Green, A.A., Zhang, W., Renn, M.J., Kim, C.H., Hersam, M.C., and Frisbie, C.D. (2010) Printed, sub-3V digital circuits on plastic from aqueous carbon nanotube inks. *ACS Nano*, **4** (8), 4388–4395.

[48] Chen, B., Zhang, P., Ding, L., Han, J., Qiu, S., Li, Q., Zhang, Z., and Peng, L.-M. (2016) Highly uniform carbon nanotube field-effect transistors and medium scale integrated circuits. *Nano Lett.*, **16** (8), 5120–5128.

[49] Wang, C., Zhang, J., Ryu, K., Badmaev, A., De Arco, L.G., and Zhou, C. (2009) Wafer-scale fabrication of separated carbon nanotube thin-film transistors for display applications. *Nano Lett.*, **9** (12), 4285–4291.

[50] Yang, Y., Ding, L., Chen, H., Han, J., Zhang, Z., and Peng, L.M. (2018) Carbon nanotube network film-based ring oscillators with sub 10-ns propagation time and their applications in radio-frequency signal transmission. *Nano Res.*, **11** (1), 300–310.

[51] Lei, T., Shao, L.L., Zheng, Y.Q., Pitner, G., Fang, G., Zhu, C., Li, S., Beausoleil, R., Wong, H.S.P., Huang, T.C., Cheng, K.T., and Bao, Z. (2019) Low-voltage high-performance flexible digital and analog circuits based on ultrahigh-purity semiconducting carbon nanotubes. *Nat. Commun.*, **10** (1), 2161.

[52] Cao, Q., Xia, M., Kocabas, C., Shim, M., Rogers, J.A., and Rotkin, S. V. (2007) Gate capacitance coupling of singled-walled carbon nanotube thin-film transistors. *Appl. Phys. Lett.*, **90** (2), 023516.

[53] Sekitani, T., Zschieschang, U., Klauk, H., and Someya, T. (2010) Flexible organic transistors and circuits with extreme bending stability. *Nat. Mater.*, **9** (12), 1015–1022.

[54] Cao, Q., Han, S.J., Penumatcha, A.V., Frank, M.M., Tulevski, G.S., Tersoff, J., and Haensch, W.E. (2015) Origins and characteristics of the threshold voltage variability of quasiballistic single-walled carbon nanotube field-effect transistors. *ACS Nano*, **9** (2), 1936–1944.

[55] Javey, A., Guo, J., Wang, Q., Lundstrom, M., and Dai, H. (2003) Ballistic carbon nanotube field-effect transistors. *Nature*, **424** (6949), 654–657.

[56] Zhang, Z., Liang, X., Wang, S., Yao, K., Hu, Y., Zhu, Y., Chen, Q., Zhou, W., Li, Y., Yao, Y., Zhang, J., and Peng, L.M. (2007) Doping-free fabrication of carbon nanotube based ballistic CMOS devices and circuits. *Nano Lett.*, **7** (12), 3603–3607.

[57] Han, S.J., Oida, S., Park, H., Hannon, J.B., Tulevski, G.S., and Haensch, W. (2013) Carbon nanotube complementary logic based on Erbium contacts and self-assembled high purity solution tubes. *IEDM Tech. Dig.*, 19.8.1–19.8.4.

[58] Yang, Y., Ding, L., Han, J., Zhang, Z., and Peng, L.M. (2017) High-performance complementary transistors and medium-scale integrated circuits based on carbon nanotube thin films. *ACS Nano*, **11** (4), 4124–4132.

[59] Tang, J. Farmer, D., Bangsaruntip, S., Chiu, K.C. Kumar, B., and Han, S.J. (2017) Contact engineering and channel doping for robust carbon nanotube NFETs. *2017 Int. Symp. VLSI Technol. Syst. Appl.*, 1–2.

[60] Geier, M.L., McMorrow, J.J., Xu, W., Zhu, J., Kim, C.H., Marks, T.J., and Hersam, M.C. (2015) Solution-processed carbon nanotube thin-film complementary static random access memory. *Nat. Nanotechnol.*, **10** (11), 944–948.

[61] Ha, T., Chen, K., Chuang, S., Yu, K.M., Kiriya, D., and Javey, A. (2015) Highly uniform and stable n-type carbon nanotube transistors by using positively charged silicon nitride thin films. *Nano Lett.*, **15** (1), 392–397.

[62] Li, G., Li, Q., Jin, Y., Zhao, Y., Xiao, X., Jiang, K., Wang, J., and Fan, S. (2015) Fabrication of air-stable n-type carbon nanotube thin-film transistors on flexible substrates using bilayer dielectrics. *Nanoscale*, **7** (42), 17693–17701.

[63] Wei, H., Chen, H.Y., Liyanage, L., Wong, H.S.P., and Mitra, S. (2011) Air-stable technique for fabricating n-type carbon nanotube FETs. *IEDM Tech. Dig.*, 23.2.1–23.2.4.

[64] Zhang, J., Wang, C., Fu, Y., Che, Y., and Zhou, C. (2011) Air-stable conversion of separated carbon nanotube thin-film transistors from p-type to n-type using atomic layer deposition of high-κ oxide and its application in CMOS logic circuits. *ACS Nano*, **5** (4), 3284–3292.

[65] Kojima, A., Shimizu, M., Chan, K., Kamimura, T., Maeda, M., and Matsumoto, K. (2005) Air stable n-type top gate carbon nanotube filed effect transistors with silicon nitride insulator deposited by thermal chemical vapor deposition. *Jpn. J. Appl. Phys.*, **44** (10), L328–L330.

[66] Moriyama, N., Ohno, Y., Kitamura, T., Kishimoto, S., and Mizutani, T. (2010) Change in carrier type in high-k gate carbon nanotube field-effect transistors by interface fixed charges. *Nanotechnology*, **21** (16), 165201.

[67] Franklin, A.D., Koswatta, S.O., Farmer, D.B., Smith, J.T., Gignac, L., Breslin, C.M., Han, S.J., Tulevski, G.S., Miyazoe, H., Haensch, W., and Tersoff, J. (2013) Carbon nanotube complementary wrap-gate transistors. *Nano Lett.*, **13**, 2490–2495.

[68] Ha, M., Seo, J.W.T., Prabhumirashi, P.L., Zhang, W., Geier, M.L., Renn, M.J., Kim, C.H., Hersam, M.C., and Frisbie, C.D. (2013) Aerosol jet printed, low voltage, electrolyte gated carbon nanotube ring oscillators with sub-5 μs stage delays. *Nano Lett.*, **13** (3), 954–960.

[69] Xiang, L., Zhang, H., Dong, G., Zhong, D., Han, J., Liang, X., Zhang, Z., Peng, L.M., and Hu, Y. (2018) Low-power carbon nanotube-based integrated circuits that can be transferred to biological surfaces. *Nat. Electron.*, **1** (4), 237–245.

[70] Sun, D.M., Timmermans, M.Y., Kaskela, A., Nasibulin, A.G., Kishimoto, S., Mizutani, T., Kauppinen, E.I., and Ohno, Y. (2013) Mouldable all-carbon integrated circuits. *Nat. Commun.*, **4** (1), 2302.

[71] Myny, K., Steudel, S., Smout, S., Vicca, P., Furthner, F., van der Putten, B., Tripathi, A.K., Gelinck, G.H., Genoe, J., and Dehaene, W. (2010) Organic RFID transponder chip with data rate compatible with electronic product coding. *Org. Electron.*, **11** (7), 1176–1179.

[72] Zschieschang, U., Ante, F., Yamamoto, T., Takimiya, K., Kuwabara, H., Ikeda, M., Sekitani, T., Someya, T., Kern, K., and Klauk, H. (2010) Flexible low-voltage organic transistors and circuits based on a high-mobility organic semiconductor with good air stability. *Adv. Mater.*, **22** (9), 982–985.

[73] Kim, D.K., Lai, Y., Diroll, B.T., Murray, C.B., and Kagan, C.R. (2012) Flexible and low-voltage integrated circuits constructed from high-performance nanocrystal transistors. *Nat. Commun.*, **3**, 1216.

[74] Zhao, D., Mourey, D.A., and Jackson, T.N. (2010) Fast flexible plastic substrate ZnO circuits. *IEEE Electron Device Lett.*, **31** (4), 323–325.

[75] Kim, Y.H., Heo, J.S., Kim, T.H., Park, S., Yoon, M.H., Kim, J., Oh, M.S., Yi, G.R., Noh, Y.-Y., and Park, S.K. (2012) Flexible metal-oxide devices made by room-temperature photochemical activation of sol-gel films. *Nature*, **489** (7414), 128–132.

[76] Chang, Y.T., Huang, J.H., Tu, M.C., Chang, P., and Yew, T.R. (2013) Flexible direct-growth CNT biosensors. *Biosens. Bioelectron.*, **41**, 898–902.

[77] Lee, S., Reuveny, A., Reeder, J., Lee, S., Jin, H., Liu, Q., Yokota, T., Sekitani, T., Isoyama, T., Abe, Y., Suo, Z., and Someya, T. (2016) A transparent bending-insensitive pressure sensor. *Nat. Nanotechnol.*, **11** (5), 472–478.

[78] Schroeder, V., Savagatrup, S., He, M., Lin, S., and Swager, T.M. (2019) Carbon nanotube chemical sensors. *Chem. Rev.*, **119** (1), 599–663.

[79] Yogeswaran, N., Tinku, S., Khan, S., Lorenzelli, L., Vinciguerra, V., and Dahiya, R. (2015) Stretchable resistive pressure sensor based on CNT-PDMS nanocomposites. *2015 11th Conf. Ph.D. Res. Microelectron. Electron.*, 326–329.

[80] Kanoun, O., Müller, C., Benchirouf, A., Sanli, A., Dinh, T., Al-Hamry, A., Bu, L., Gerlach, C., and Bouhamed, A. (2014) Flexible carbon nanotube films for high performance strain sensors. *Sensors*, **14** (6), 10042–10071.

[81] Sundaram, S., Kellnhofer, P., Li, Y., Zhu, J.Y., Torralba, A., and Matusik, W. (2019) Learning the signatures of the human grasp using a scalable tactile glove. *Nature*, **569** (7758), 698–702.

[82] Hammock, M.L., Chortos, A., Tee, B.C.K., Tok, J.B.H., and Bao, Z. (2013) 25th anniversary article: The evolution of electronic skin (e-skin): A brief history, design considerations, and recent progress. *Adv. Mater.*, **25** (42), 5997–6038.

[83] Bauer, S., Bauer-Gogonea, S., Graz, I., Kaltenbrunner, M., Keplinger, C., and Schwödiauer, R. (2014) 25th anniversary article: A soft future: From robots and sensor skin to energy harvesters. *Adv. Mater.*, **26** (1), 149–162.

[84] Dahiya, R.S., Mittendorfer, P., Valle, M., Cheng, G., and Lumelsky, V.J. (2013) Directions toward effective utilization of tactile skin: A review. *IEEE Sens. J.*, **13** (11), 4121–4138.

[85] Wang, X., Dong, L., Zhang, H., Yu, R., Pan, C., and Wang, Z.L. (2015) Recent progress in electronic skin. *Adv. Sci.*, **2** (10), 1500169.

[86] Zang, Y., Zhang, F., Di, C., and Zhu, D. (2015) Advances of flexible pressure sensors toward artificial intelligence and health care applications. *Mater. Horiz.*, **2** (2), 140–156.

[87] Endoh, H. (2015) Development of carbon nanotube TFT for sheet electronic device by printing technology. *J. Photopolym. Sci. Technol.*, **28** (3), 349–352.

[88] Takahashi, T., Takei, K., Gillies, A.G., Fearing, R.S., and Javey, A. (2011) Carbon nanotube active-matrix backplanes for conformal electronics and sensors. *Nano Lett.*, **11** (12), 5408–5413.

[89] Wang, C., Hwang, D., Yu, Z., Takei, K., Park, J., Chen, T., Ma, B., and Javey, A. (2013) User-interactive electronic skin for instantaneous pressure visualization. *Nat. Mater.*, **12** (10), 899–904.

[90] Yeom, C., Chen, K., Kiriya, D., Yu, Z., Cho, G., and Javey, A. (2015) Large-area compliant tactile sensors using printed carbon nanotube active-matrix backplanes. *Adv. Mater.*, **27** (9), 1561–1566.

[91] Lee, W., Koo, H., Sun, J., Noh, J., Kwon, K.-S., Yeom, C., Choi, Y., Chen, K., Javey, A., and Cho, G. (2016) A fully roll-to-roll gravure-printed carbon nanotube-based active matrix for multi-touch sensors. *Sci. Rep.*, **5** (1), 17707.

[92] Nela, L., Tang, J., Cao, Q., Tulevski, G., and Han, S.J. (2018) Large-area high-performance flexible pressure sensor with carbon nanotube active matrix for electronic skin. *Nano Lett.*, **18** (3), 2054–2059.

[93] Editorial (2018) 20 years of nanotube transistors. *Nat. Electron.*, **1** (3), 149–149.

[94] Shulaker, M.M., Hills, G., Park, R.S., Howe, R.T., Saraswat, K., Wong, H.S.P., and Mitra, S. (2017) Three-dimensional integration of nanotechnologies for computing and data storage on a single chip. *Nature*, **547** (7661), 74–78.

[95] Zhao, M., Chen, Y., Wang, K., Zhang, Z., Streit, J. K., Fagan, J. A., Tang, J., Zheng, M., Yang, C., Zhu, Z., et al. (2020) DNA-directed nanofabrication of high-performance carbon nanotube field-effect transistors. *Science*, **368** (6493), 878–881.

[96] Sun, W., Shen, J., Zhao, Z., Arellano, N., Rettner, C., Tang, J., Cao, T., Zhou, Z., Ta, T., Streit, J.K., et al. (2020). Precise pitch-scaling of carbon nanotube arrays within three-dimensional DNA nanotrenches. *Science*, **368** (6493), 874–877.

[97] Gao, W., Emaminejad, S., Nyein, H.Y.Y., Challa, S., Chen, K., Peck, A., Fahad, H.M., Ota, H., Shiraki, H., Kiriya, D., Lien, D.-H., Brooks, G.A., Davis, R.W., and Javey, A. (2016) Fully integrated wearable sensor arrays for multiplexed in situ perspiration analysis. *Nature*, **529** (7587), 509–514.

[98] Petropoulos, A., Pagonis, D.N., and Kaltsas, G. (2012) Flexible PCB-MEMS flow sensor. *Procedia Eng.*, **47**, 236–239.

[99] Liu, P., Zhu, R., and Que, R. (2009) A flexible flow sensor system and its characteristics for fluid mechanics measurements. *Sensors*, **9** (12), 9533–9543.

[100] Wang, Z., Wang, X., Li, M., Gao, Y., Hu, Z., Nan, T., Liang, X., Chen, H., Yang, J., Cash, S., and Sun, N.-X. (2016) Highly sensitive flexible magnetic sensor based on anisotropic magneto-resistance effect. *Adv. Mater.*, **28** (42), 9370–9377.

[101] Melzer, M., Mönch, J.I., Makarov, D., Zabila, Y., Cañón Bermúdez, G.S., Karnaushenko, D., Baunack, S., Bahr, F., Yan, C., Kaltenbrunner, M., and Schmidt, O.G. (2015) Wearable magnetic field sensors for flexible electronics. *Adv. Mater.*, **27** (7), 1274–1280.

[102] Kim, S., Jeong, H.Y., Kim, S.K., Choi, S.-Y., and Lee, K.J. (2011) Flexible memristive memory array on plastic substrates. *Nano Lett.*, **11** (12), 5438–5442.

[103] Wan, J., Xie, J., Kong, X., Liu, Z., Liu, K., Shi, F., Pei, A., Chen, H., Chen, W., Chen, J., Zhang, X., Zong, L., Wang, J., Chen, L.-Q., Qin, J., and Cui, Y. (2019) Ultrathin, flexible, solid polymer composite electrolyte enabled with aligned nanoporous host for lithium batteries. *Nat. Nanotechnol.*, 10.1038/s41565-019-0465-3.

[104] Lungenschmied, C., Dennler, G., Neugebauer, H., Sariciftci, S.N., Glatthaar, M., Meyer, T., and Meyer, A. (2007) Flexible, long-lived, large-area, organic solar cells. *Sol. Energy Mater. Sol. Cells*, **91** (5), 379–384.

[105] Nathan, A., Ahnood, A., Cole, M.T., Sungsik Lee, Suzuki, Y., Hiralal, P., Bonaccorso, F., Hasan, T., Garcia-Gancedo, L., Dyadyusha, A., Haque, S., Andrew, P., Hofmann, S., Moultrie, J., Daping Chu, Flewitt, A.J., Ferrari, A.C., Kelly, M.J., Robertson, J., Amaratunga, G.A.J., and Milne, W.I. (2012) Flexible electronics: The next ubiquitous platform. *Proc. IEEE*, **100**, 1486–1517.

# 6

## Flexible Sensor Sheets for Healthcare Applications

**Kuniharu Takei**

*Osaka Prefecture University*

### CONTENTS

## 6.1 Introduction

Daily and home-use medical and healthcare diagnoses are of great interest due to the worldwide increase of the elder population and the technological progress in integrating multiple sensors. In terms of wearable sensors, continuous real-time health condition monitoring may be a potential target. If continuous monitoring of multiple health conditions can be realized by simply attaching a comfortable noninvasive device onto the skin, preventive medical care, or diagnosis of diseases in the early stage can be realized by detecting trends in the health condition changes. It should be noted that the absolute values of vitals such as body temperature, blood pressure, glucose level, etc., are not so important for this application. Additionally, this concept may be easier to build a robust device market because most applications do not require the approval for medical uses since the exact values corresponding to the diagnosis are not indicated. For wearable healthcare applications, real-time continuous trend changes of the health conditions should be useful information to predict disease, which may be a way to spread flexible healthcare sensor sheets to the market.

Another useful target for healthcare applications is a feedback system between patients and doctors via the internet (Figure 6.1) (Honda et al. 2014). Briefly, patients or users can monitor their real-time health condition changes by tracking trends and changes. If an artificial intelligence (AI) or deep learning concept can be integrated into the wearable sensor system, a drug or curing agent may be delivered to the user automatically based on

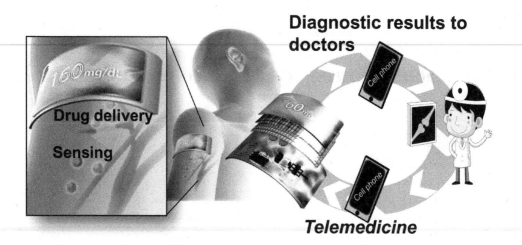

**FIGURE 6.1**
Concept of wearable healthcare patches with interactions between a user and a doctor. (Reproduced with permission from Honda et al. (2014). Copyright 2014, John Wiley and Sons.)

the signals. However, this medical action capability will depend on the laws in a country. Regardless of the drug delivery function, the health condition trends provide useful information for doctors and users. Doctor can treat patients based on the condition changes during the patients' daily life. Furthermore, users can make decisions to keep their physical and mental conditions without having to consult medical doctors. This health condition monitoring may have an extremely high impact on future medical and healthcare platforms.

Although this concept and platform are promising for not only academic research but also big markets, almost no practical devices have been commercialized. There are several challenges toward the realization of this platform. The first one is the integration of multiple sensors on a flexible film. The second is the integration of signal processing, a wireless system, and battery without sacrificing mechanical flexibility and increasing the cost. The third one is determining which information is required to monitor and diagnose health conditions.

To decide which kinds of sensors to integrate, a lot of data must be analyzed by developing different combinations of sensors, although some may prove unnecessary for the healthcare applications. Once the above-mentioned challenges are addressed, wearable healthcare devices should create a huge market and change how people check their health conditions. Herein, the progress of flexible sensor sheets attached onto the human skin to monitor health conditions is reviewed.

## 6.2 Flexible Physical Healthcare Device Patches

To address the aforementioned challenges, many studies have investigated flexible and/or stretchable multifunctional sensor sheets (Takei et al. 2015, Xu, Lu, and Takei 2019, Gao et al. 2019). Research has included sensors (Bariya, Nyein, and Javey 2018, Martin et al. 2017,

Yan, Wang, and Lee 2015, Takei et al. 2010, Kim, Lu, Ghaffari, et al. 2011, Kim, Lu, Ma, et al. 2011), circuits (Wang et al. 2018, Someya et al. 2005, Cao et al. 2008, Honda et al. 2015), and power sources (Kim et al. 2016, Hu et al. 2009, Yoon et al. 2008). Here, flexible healthcare sensors are introduced to understand the fundamental characteristics and applications as a flexible sensor sheet based on device developments in Takei's laboratory at Osaka Prefecture University in Japan.

Many groups, including Takei, have proposed noninvasive monitoring of the surface of the skin using attachable bandage-type sensor sheets (Figure 6.2) (Gao et al. 2019, Yamamoto et al. 2016). Takei's group's idea involves a device that consists of at least two layers (Figure 6.2b). The first layer is disposable and the second is reusable.

The first layer is a sheet in direct contact with the skin. This layer is disposable because direct contact with the skin is necessary for precise monitoring, but also creates hygiene concerns. To realize a disposable sheet, a low-cost fabrication process is required. For the disposable sheet, Takei's group printed flexible sensors to make electrocardiogram (ECG) sensor, skin temperature sensor, and three-axis acceleration sensor (motion sensor) on a polyethylene terephthalate (PET) film.

The other sheet is integrated with expensive components such as signal-processing circuits, wireless systems, and battery. Hence, it is reusable due to the high fabrication costs. To minimize device costs, the sheet with the more expensive components should not come into direct contact with the skin. Thus, this sheet is assembled on top of the disposable sheet.

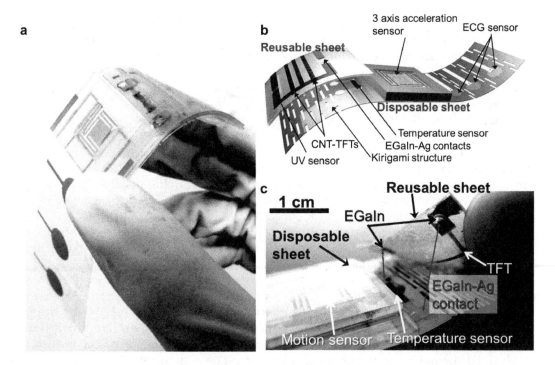

**FIGURE 6.2**
(a) Photo and (b) schematic of a flexible healthcare patch integrated with a three-axis acceleration sensor, skin temperature sensor, ECG sensor, UV sensor, and switching transistors. (c) Photo of the flexible connection region between the reusable and disposable sheets. (Reproduced with permission from Yamamoto et al. (2016). Copyright 2016, American Association for the Advancement of Science.)

Another critical feature is creating an electrical connection between the sheets that does not affect the mechanical flexibility. One approach is to use a human-friendly (biocompatible) liquid metal of eutectic gallium-indium metal alloy (Figure 6.2c). For the details about this material, including its structure and stability, please refer to the literature (Harada et al. 2015).

## 6.2.1 Printed Flexible Three-Axis Acceleration Sensors

For wearable healthcare applications, activity monitoring is important not only to collect vital signs but also to diagnose conditions. This is because vital signs such as ECG, respiration rate, and skin temperature strongly depend on the body's movements such as resting state and exercise state. Two types of sensors have been reported to detect human motion. One monitors bending of extremities (e.g., arm or leg) by attaching a strain sensor at the joints (Yamada et al. 2011). The other detects acceleration of human motion (Yamamoto et al. 2016).

A platform that incorporates acceleration monitoring is better suited to precisely monitor human motion and activity. Joint bending monitoring may misclassify the movement of an object as the sensor detects strain. For example, the sensor can detect strain while running as well as when bending the arm without moving the rest of the body. To prevent this misclassification, herein a flexible acceleration sensor (Yamamoto et al. 2016) is discussed.

To realize an acceleration sensor, there are two types of structures inspired from standard microelectromechanical system (MEMS) devices. One is a capacitive type to measure the electrode movements caused by acceleration, which corresponds to the capacitance change due to a distance change between interdigitated electrodes. The other is a resistive type, which measures the resistance change of a piezoresistive sensor. For a flexible and printed acceleration sensor, the resistive type sensor may be easier to fabricate because many flexible piezoresistive strain sensors have been reported (Takei et al. 2014, Lipomi et al. 2011, Harada et al. 2014). Here, the resistive three-axis acceleration sensor reported by the author is discussed (Yamamoto et al. 2016).

Figure 6.3 describes a flexible and printed acceleration sensor using printed piezoresistive strain sensors. Three strain sensors are integrated to distinguish the acceleration directions (x-, y-, and z-directions). The fabricated beam structures in the PET film (Figure 6.3b) detect applied strain due to bending caused by acceleration. An acrylic plate is mounted at the center of the sensor as a mass for effective movements of the structure against the acceleration with spacers on top and bottom, where the sensors are placed (Figure 3c). A finite element method (FEM) simulation confirms the dependence of the strain distribution on the acceleration direction.

Figure 6.3d shows that all four beams, including strain sensors, have identical strain when the strain is applied in the z-axis acceleration. In contrast, when applying acceleration in the y-axis direction, the strain distribution becomes nonidentical. Beams #1 and #3 (strain sensors) have larger strain than beam #2. By monitoring the strain distribution from each beam using the printed strain sensors, the acceleration direction can be distinguished. Furthermore, the amplitude of strain corresponds to the amplitude of acceleration. Based on these, the direction and amplitude of acceleration can be readily monitored using a fully printed sensor sheet.

After designing the device structure and confirming the strain distributions, the device was experimentally measured by applying different directions of acceleration

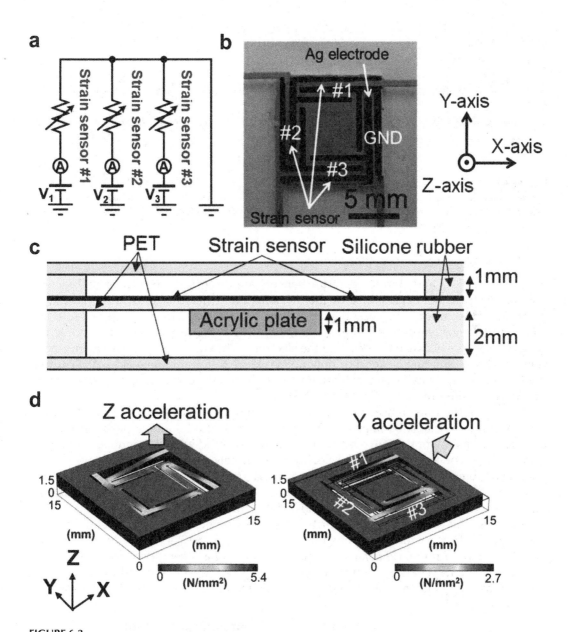

**FIGURE 6.3**
(a) Equivalent circuit diagram and (b) photo of the printed three-axis acceleration sensor. (c) Cross-sectional image of the acceleration sensor. (d) Strain distribution calculated by FEM simulations when applying z-axis (left) and y-axis (right) acceleration. (Reproduced with permission from Yamamoto et al. (2016). Copyright 2016, American Association for the Advancement of Science.)

(Figure 6.4). As expected by the FEM simulation results, the beam vibration due to acceleration depends on the direction. In the z-axis direction, all strain sensors detect a vibration. However, only strain sensor #2 detects acceleration in the x-axis direction, whereas strain sensors #1 & #3 detect acceleration in the y-axis direction. These results show that this sensor platform can monitor human motion like a conventional MEMS-based acceleration sensor.

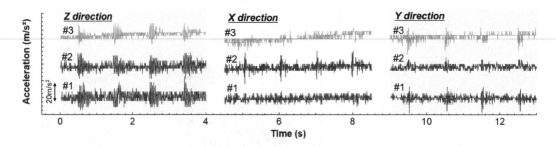

**FIGURE 6.4**
Acceleration detection calculated from the resistance change of the strain sensors with $z$-axis, $x$-axis, and $y$-axis acceleration. (Reproduced with permission from Yamamoto et al. (2016). Copyright 2016, American Association for the Advancement of Science.)

### 6.2.2 ECG and Skin Temperature Sensors

Next, flexible ECG and skin temperature sensors are discussed. For ECG signal recording, the electrical signal generated by the movement of the cardiac muscle is detected by using electrodes around the chest region. To record the signal with a high signal-to-noise (SN) ratio, the most important parameter is electrical impedance, which corresponds to the contact resistance between the skin and the electrode. The equivalent measurement circuit consists of a resistor and a capacitor in parallel (Figure 6.5) (Yamamoto et al. 2017). If this impedance is high, the ECG signal is attenuated at the interface, resulting in a low SN ratio. The ECG signal is only tens of mV through skin.

To decrease the impedance, conformal contact and better adhesion between the electrode and the skin must be improved. Ultrathin film electrodes and adhesive electrodes are often discussed for this purpose. In this chapter, an adhesive type conductive polymer is introduced. For conventional ECG electrodes, ionic gels are often used. However, when the sensor comes into contact with water or sweat, swelling may occur, which leads to unstable signal recording. This is one of the bottlenecks to apply gel-type electrodes in wearable applications because skin is almost always in a moist condition.

To address this issue, a sticky conductive polymer composed of a mixture of a carbon nanotube (CNT) as a conductor, poly(dimethylsiloxane) (PDMS) as a polymer, and ethoxylated polyethylenimine (PEIE) as an adhesive material has been developed (Yamamoto et al. 2017). The impedance at 100 Hz is about 600 kΩ, but it varies according to the composition ratio of CNTs. Increasing of the CNT content decreases the impedance. However,

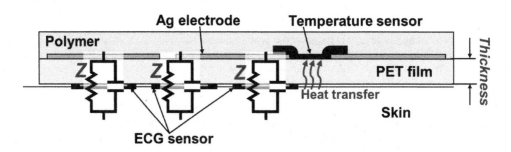

**FIGURE 6.5**
Schematic of the device structure with an equivalent circuit for the ECG sensor and heat transfer through a PET film from the skin to the temperature sensor. (Reproduced with permission from Yamamoto et al. (2017). Copyright 2017, John Wiley and Sons.)

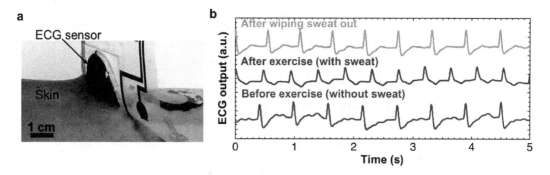

**FIGURE 6.6**
(a) Photo of a sticky conductive polymer electrode for ECG recording. (b) ECG output with and without sweat between the skin and the ECG sensor. (Reproduced with permission from Yamamoto et al. (2017). Copyright 2017, John Wiley and Sons.)

a composition >10% in PDMS polymer makes the film less adhesive due to the limitation of continuous polymer formation. The impedance of a conventional ionic gel is ~100 kΩ. Hence, the impedance of a polymer-based conductive ECG electrode must be decreased if a similar SN ratio is required in the future. Although the impedance is still higher than that of the conventional electrodes, the advantage of this sensor is that this can be used under sweat conditions. Because the polymer-based conductive polymer does not swell much when in contact with water or sweat, the ECG signal can be continuously and stably recorded (Figure 6.6). It should be noted that under sweat conditions, the amplitude of the ECG signal is slightly attenuated most likely because sweat between the skin and the electrode increases the impedance. However, after wiping sweat from the surfaces of the electrode and skin, the amplitude returns to the initial value, suggesting that this material does not absorb sweat. This is an advantage compared to an ionic gel for the wearable healthcare applications.

Next, the temperature difference caused by the environmental temperature is discussed for efficient and precise monitoring. To measure the targeted temperature (skin temperature in this case), the temperature must diffuse through the film, where the temperature sensor is formed (Figure 6.5). However, the environmental temperature is also diffused through the passivation film to the temperature sensor.

To precisely monitor the skin temperature, the temperature difference between the sensor output and the real surface temperature was investigated experimentally (Yamamoto et al. 2017) by measuring both the surface of the hotplate and PET films with various thicknesses as the temperature difference $\Delta T$ (Figure 6.7a) under an environmental temperature of ~25°C. Figure 6.7b clearly depicts that $\Delta T$ is larger at higher temperatures of the hotplate due to the low environmental temperature compared to the hotplate. More importantly, this difference can be reduced by using a thinner PET film due to the thermal transfer process through the materials.

Because both films on the top and the bottom of the temperature sensor are polymer materials with similar thermal conductivities, the temperature changes linearly though the polymer between the surfaces with different temperature. Using a 25-µm-thick PET film results in a temperature error of a few degrees Celsius, which is critical error for the wearable healthcare applications. To address this issue, a thermal management structure should be developed to support using thermal conductivity materials with a large difference. Based on our preliminary study (data not shown), an ultra-thin film temperature

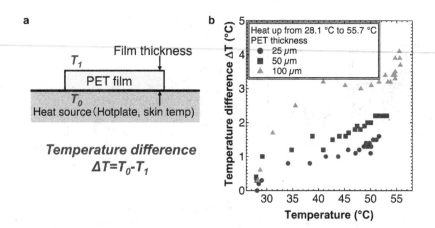

**FIGURE 6.7**

(a) Schematic to define the temperature difference. (b) Measured temperature difference at different temperatures as a function of PET film thickness. (Reproduced with permission from Yamamoto et al. (2017). Copyright 2017, John Wiley and Sons.)

sensor less than 100 μm in total thickness may be unsuitable to monitor the temperature precisely from the surface for wearable and other applications. To address this issue, a thermal management structure should be developed to support using thermal conductivity materials with a large difference.

### 6.2.3 Multifunctional Sensor Sheet Demonstration

Although several issues remain to realize stable and precise monitoring, multiple sensor integration is demonstrated. A proof-of-concept has been demonstrated by forming three-axis acceleration sensor, skin temperature sensor, and ECG sensor on a PET film as a disposable sheet while integrating a CNT transistor to switch functions and a UV sensor on another film as the reusable sheet (Figure 6.2) (Yamamoto et al. 2016). The sensor sheet can be attached on a human body like in Figure 6.8a. It should be noted that all sensors are connected to the equipment to monitor information even though Figure 6.8a does not show electrical wiring to the sensor sheet. Due to the wired connection between the sensor sheet and equipment, walking and running were simulated with a volunteer while all sensor outputs were recorded. All sensors successfully detect information of the UV light in the room, skin temperature, ECG, and human motion corresponding to acceleration (Figure 6.8b). Especially, for the human motion, the spikes correspond to the steps. The spike pitch and amplitude are related with speed and strength of steps, respectively.

The resistance changes without observing spikes indicates a change in the volunteer position from lying down to sitting up. It is well known that the heartbeat frequency extracted from the ECG R-R peak intervals strongly depends on the activity. Using this sensor sheet, the status can be monitored by detecting human motion while relaxing or being active (Figure 6.8c) as well as lying down or sitting up (Figure 6.8d). As expected, Figures 6.8c–d clearly show that right after an activity causes a higher beating speed (i.e., short R-R interval). This result suggests that the acceleration sensor to detect human activity precisely is an important component for wearable healthcare applications. Without detecting motion, health conditions detected from the sensors cannot be diagnosed accurately.

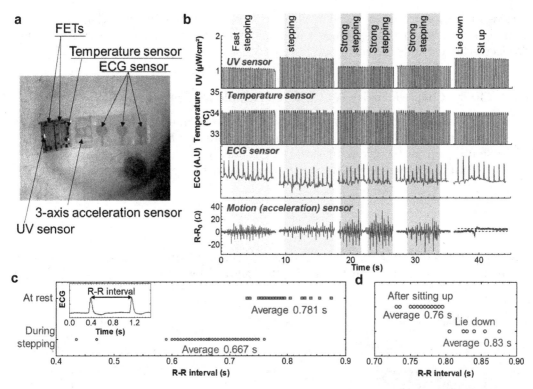

**FIGURE 6.8**
(a) Photo of a multifunctional flexible healthcare patch attached on a volunteer. (b) Real-time condition detections (UV, skin temperature, ECG, and motion from top to bottom). R-R intervals, which correspond to heartbeat rates, when (c) the volunteer is active and at rest and (d) moving from lying down to sitting up. (Reproduced with permission from Yamamoto et al. (2016). Copyright 2016, American Association for the Advancement of Science.)

## 6.3 Flexible Chemical Healthcare Devices

Section 6.2 introduced physical monitoring conditions such as motion, skin temperature, and ECG signals detected from the surface of the skin. These vital signs are important information, but alone are typically insufficient to diagnose health conditions accurately. At a hospital, blood is often taken in combination with checking vital signs. The blood sample checks several things inside the body. However, there is a high probability of causing infection if users self-collect blood samples at home. In addition, the process is inconvenient for real-time monitoring.

A viable alternative may be to detect the chemical contents in sweat or skin gas for real-time continuous monitoring as a wearable device. In fact, sweat or skin gas contains a variety of chemicals, including glucose, potassium, sodium, etc. For example, the glucose level in sweat is correlated with that in the blood (Tierney et al. 2000). Although the absolute value of the glucose level for diagnosis cannot be used, the trend of glucose level changes is more important for wearable healthcare applications. Such trend monitoring is only available for methods using continuous real-time monitoring like a wearable device.

Here, a highly sensitive chemical flexible sensor is introduced. In particular, a pH sensor is discussed because pH detection is a fundamental method for chemical detection using an electrochemical mechanism.

### 6.3.1 Charge-Coupled Device (CCD)-Based Flexible pH Sensor

Figure 6.9 shows the concept of a pH-monitoring device integrated with a temperature sensor (Nakata et al. 2018). Figure 6.9c shows photos of the device. To prevent direct contact of the solution on the flexible transistors, an extended gate electrode structure was used. For the electrochemical method, the maximum sensitivity is express by Nernst theory as

$$E = E_{REF} + \frac{RT}{nF} \ln \alpha_{H^+}$$

where $E_{REF}$ is the reference voltage, $R$ is the gas constant, $T$ is the measured temperature, $n$ is the number of electrons transferred, $F$ is the Faraday constant, and $\alpha_{H^+}$ is the hydrogen

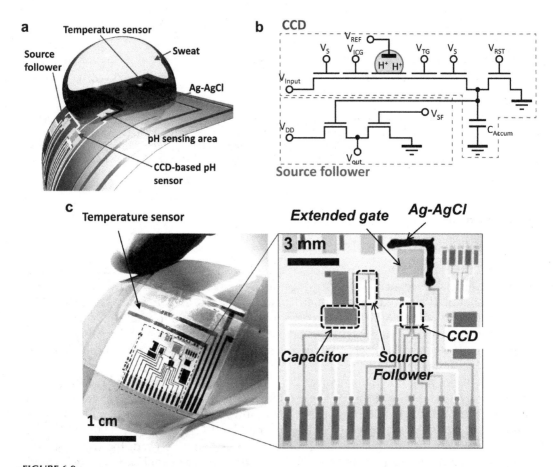

**FIGURE 6.9**
(a) Schematic, (b) equivalent circuit, and (c) photo of a highly sensitive CCD-based flexible pH sensor. (Reproduced with permission from Nakata et al. (2018). Copyright 2018, Nature Publishing Group.)

ion concentration. At room temperature (298 K), the pH sensitivity is limited to ~59 mV/pH, which is sufficient for pH detection. However, a higher sensitivity may be necessary to precisely and accurately monitor other chemicals like the glucose level in sweat.

To address the theoretically low-sensitivity limitation, the electrically amplified method was proposed using a CCD method, where electrons, which depend on the pH level, are transferred and accumulated in a capacitor. By repeating this process, the sensitivity can be enhanced without increasing the noise level or fabricating a complicated analog circuit. The circuit diagram on a flexible film is shown in Figure 6.9b, and the detailed band diagram to explain the mechanism is described in Figure 6.10a. $V_{ICG}$ and $V_{TG}$ work to transfer the electrons injected from $V_{Input}$. The number of electrons to transfer is defined by the well depth of the channel between $V_{ICG}$ and $V_{TG}$ and corresponds to the pH level (Figure 6.10a(2)). Unlike a conventional Si-based CCD, $V_s$ is also applied to control the Schottky barrier height to efficiently inject electron. It should be noted that the band diagram for $V_s$ is omitted from Figure 6.10a. Schottky junctions are used in this device because

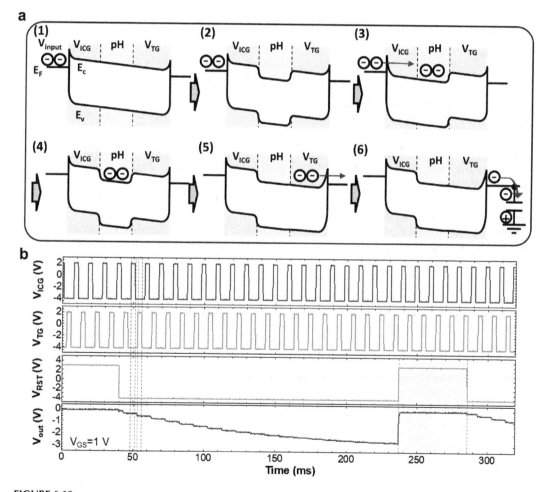

**FIGURE 6.10**

(a) Band diagrams to explain the CCD-based pH sensing mechanism. (b) Output voltage under transfer, accumulation, and reset processes. (Reproduced with permission from Nakata et al. (2018). Copyright 2018, Nature Publishing Group.)

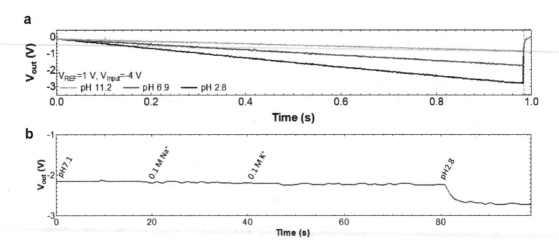

**FIGURE 6.11**

(a) Output voltage at different pH solutions up to 100-cycle transfers and accumulations. (b) Selectivity test conducted by adding sodium and potassium ions over the sensor. (Reproduced with permission from Nakata et al. (2018). Copyright 2018, Nature Publishing Group.)

the formation of a p-n junction is difficult due to limitations of the thermal budget of the doping process for flexible devices.

By turning $V_{ICG}$ on, electrons are injected from $V_{Input}$ to fill the well depth of the pH region (Figure 6.10a(3)). After filling the electron in the well, $V_{ICG}$ returns to the off-state (Figure 6.10a(4)). Next, $V_{TG}$ is turned on to extract electrons from the well (Figure 6.10a(5)), which results in charging of all electrons in the integrated capacitor (Figure 6.10a(6)). This accumulation process is repeated to increase the output voltage corresponding to the sensitivity. After reading the output voltage, the reset voltage, $V_{RST}$, is applied to remove all electrons from the capacitor and to reset the signal.

Figure 6.10b displays the output voltage as a function of accumulation cycles. As described in Figure 6.10a, when $V_{TG}$ is applied, the output voltage negatively increases stepwise. After $V_{RST}$, the output voltage returns to zero, which allows it to continuously measure the pH level in real-time. By changing pH level, the output voltage difference is larger by increasing the transfer and accumulation processes, as described in Figure 6.11a. In this case, after 100 cycles, the pH sensitivity is ~240 mV/pH, which is roughly a sensitivity four times higher than the theoretical limit (i.e., 59 mV/pH).

Another important parameter for the chemical sensor is selectivity between different chemicals. Due to the inorganic oxide-based membrane ($SiO_x$) used in this sensor, the sensor only detects the hydrogen ion concentration (i.e., pH level) in a solution. Different chemical solutions (sodium and potassium) with pH 7.1 were also dropped over the sensor (Figure 6.11b). Specifically, 0.1 M sodium and potassium solutions were used. The results clearly indicate that the sensor is only sensitive to the pH level in the solution.

### 6.3.2 Highly Sensitive Real-Time pH Monitoring

Real-time monitoring of the pH level was conducted by adding different pH solutions into a solution with pH 2.6. Figure 6.12a shows that the CCD-based flexible pH sensor can monitor the pH level continuously. In the experiment, the transfer and accumulation process was repeated 100 times. Each cycle consisted of transfer and accumulation and a

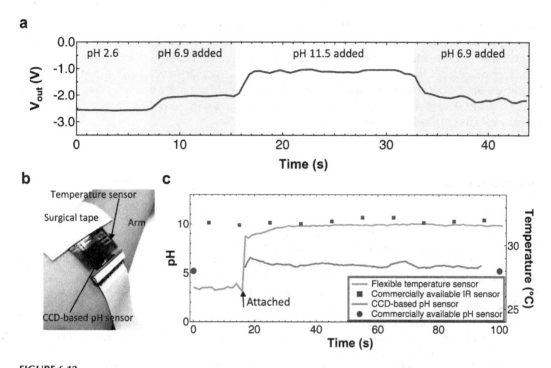

**FIGURE 6.12**

(a) Real-time pH monitoring using the CCD-based pH sensor. (b) Photo and (c) the real-time monitoring of sweat pH and skin temperature detections by attaching sensors onto the skin. (Reproduced with permission from Nakata et al. (2018). Copyright 2018, Nature Publishing Group.)

subsequent reset voltage to reset the accumulation. For a more practical demonstration, the sensor was integrated with a flexible temperature sensor and attached on a human skin with sweat (Figures 6.12b-c). The sensor can monitor the pH level in real human sweat precisely similar to a commercially available stick-type pH meter. In addition, the integrated temperature sensor can monitor the skin temperature and help calibrate the output voltage of the pH sensing. This is because the electrochemical sensors and transistors are temperature dependent, which is critical because the measured temperature is readily changed in wearable applications due to the environmental temperature.

## 6.4 Summary and Outlook

This chapter introduces flexible sensors applied to the health condition monitoring attached onto the human skin. After discussing the fundamental characteristics of a three-axis acceleration sensor, an ECG sensor, and a temperature sensor as a physical sensor sheet, real-time condition monitoring is conducted. In addition to the physical sensor sheets, a chemical sensor based on CCD-structure to enhance the sensitivity is explained.

For the practical applications, system integration such as a battery, signal processing, and wireless circuits must be addressed in the future. To reduce the power consumption, wireless communications to a smartphone should be minimized by decreasing the amount of data transmitted. For this purpose, it may be necessary to integrate edge computing with AI on a wearable sensor patch. Moreover, how to collect big data from real-time datasets for healthcare monitoring and addressing medical diagnosis simultaneously must be resolved. Once these requirements are met, this platform has a high potential to create a paradigm shift in the fields of medical and healthcare applications and to create a huge market. These should be achieved by collaborating with many researchers and engineers bridging multidisciplinary fields.

# References

Bariya, M., H. Y. Y. Nyein, and A. Javey. 2018. "Wearable sweat sensors." *Nat. Electron.* 1:160–171.

Cao, Q., H. S. Kim, N. Pimparkar, J. P. Kulkarni, C. Wang, M. Shim, K. Roy, M. A. Alam, and J. A. Rogers. 2008. "Medium-scale carbon nanotube thin-film integrated circuits on flexible plastic substrates." *Nature* 454:495–500. doi: 10.1038/nature07110.

Gao, W., H. Ota, D. Kiriya, K. Takei, and A. Javey. 2019. "Flexible electronics toward wearable sensing." *Acc. Chem. Res.* 52:523–533.

Harada, S., W. Honda, T. Arie, S. Akita, and K. Takei. 2014. "Fully printed, highly sensitive multifunctional artificial electronic whisker arrays integrated with strain and temperature sensors." *ACS Nano* 8:3921–3927.

Harada, S., T. Arie, S. Akita, and K. Takei. 2015. "Highly stable liquid-solid metal contact toward multilayered detachable flexible devices." *Adv. Electron. Mater.* 1:1500080.

Honda, W., S. Harada, S. Ishida, T. Arie, S. Akita, and K. Takei. 2015. "High-performance, mechanically flexible, and vertically integrated 3D carbon nanotube and InGaZnO complementary circuits with a temperature sensor." *Adv. Mater.* 27:4674–4680.

Honda, W., S. Harada, T. Arie, S. Akita, and K. Takei. 2014. "Wearable, human-interactive, health-monitoring, wireless devices fabricated by macroscale printing techniques." *Adv. Funct. Mater.* 24:3299–3304.

Hu, L., J. W. Choi, Y. Yang, S. Jeong, F. La Mantia, L. F. Cui, and Y. Cui. 2009. "Highly conductive paper for energy-storage devices." *Proc. Natl. Acad. Sci. U.S.A.* 106:21490–21494.

Kim, D. H., N. Lu, R. Ghaffari, Y. S. Kim, S. P. Lee, L. Xu, J. Wu, R. H. Kim, J. Song, Z. Liu, J. Viventi, B. de Graff, B. Elolampi, M. Mansour, M. J. Slepian, S. Hwang, J. D. Moss, S. M. Won, Y. Huang, B. Litt, and J. A. Rogers. 2011. "Materials for multifunctional balloon catheters with capabilities in cardiac electrophysiological mapping and ablation therapy." *Nat. Mater.* 10:316–323.

Kim, D. H., N. Lu, R. Ma, Y. S. Kim, R. H. Kim, S. Wang, J. Wu, S. M. Won, H. Tao, A. Islam, K. J. Yu, T. I. Kim, R. Chowdhury, M. Ying, L. Xu, M. Li, H. J. Chung, H. Keum, M. McCormick, P. Liu, Y. W. Zhang, F. G. Omenetto, Y. Huang, T. Coleman, and J. A. Rogers. 2011. "Epidermal electronics." *Science* 333:838–843.

Kim, J., G. A. Salvatore, H. Araki, A. M. Chiarelli, Z. Xie, A. Banks, X. Sheng, Y. Liu, J. W. Lee, K.-I. Jang, S. Y. Heo, K. Cho, H. Luo, B. Zimmerman, J. Kim, L. Yan, X. Feng, S. Xu, M. Fabiani, G. Gratton, Y. Huang, U. Paik, and J. A. Rogers. 2016. "Battery-free, stretchable optoelectronic systems for wireless optical characterization of the skin." *Sci. Adv.* 2:e1600418.

Lipomi, D. J., M. Vosgueritchian, B. C. Tee, S. L. Hellstrom, J. A. Lee, C. H. Fox, and Z. Bao. 2011. "Skin-like pressure and strain sensors based on transparent elastic films of carbon nanotubes." *Nat. Nanotechnol.* 6:788–792. doi: 10.1038/nnano.2011.184.

Martin, A., J. Kim, J. F. Kurniawan, J. R. Sempionatto, J. R. Moreto, G. Tang, A. S. Campbell, A. Shin, M. Y. Lee, X. Liu, and J. Wang. 2017. "Epidermal microfluidic electrochemical detection system: Enhanced sweat sampling and metabolite detection." *ACS Sens.* 2:1860–1868.

Nakata, S., M. Shiomi, Y. Fujita, T. Arie, S. Akita, and K. Takei. 2018. "A wearable pH sensor with high sensitivity based on a flexible charge-coupled device." *Nat. Electron.* 1:596–603.

Someya, T., Y. Kato, T. Sekitani, S. Iba, Y. Noguchi, Y. Murase, H. Kawaguchi, and T. Sakurai. 2005. "Conformable, flexible, large-area networks of pressure and thermal sensors with organic transistor active matrixes." *Proc. Natl. Acad. Sci. U.S.A.* 102:12321–12325.

Takei, K., W. Honda, S. Harada, T. Arie, and S. Akita. 2015. "Toward flexible and wearable human-interactive health-monitoring devices." *Adv. Healthcare Mater.* 4:487–500.

Takei, K., T. Takahashi, J. C. Ho, A. Ko, A. G. Gillies, P. W. Leu, R. S. Fearing, and A. Javey. 2010. "Nanowire active-matrix circuitry for low-voltage macroscale artificial skin." *Nat. Mater.* 9(10):821–826.

Takei, K., Z. Yu, M. Zheng, H. Ota, T. Takahashi, and A. Javey. 2014. "Highly sensitive electronic whiskers based on patterned carbon nanotube and silver nanoparticle composite films." *Proc. Natl. Acad. Sci. U.S.A.* 111:1703–1707.

Tierney, M. J., H. L. Kim, M. D. Burns, J. A. Tamada, and R. O. Potts. 2000. "Electroanalysis of glucose in transcutaneously extracted samples." *Electroanalysis* 12:666–671.

Wang, S., J. Xu, W. Wang, G. N. Wang, R. Rastak, F. Molina-Lopez, J. W. Chung, S. Niu, V. R. Feig, J. Lopez, T. Lei, S. K. Kwon, Y. Kim, A. M. Foudeh, A. Ehrlich, A. Gasperini, Y. Yun, B. Murmann, J. B. Tok, and Z. Bao. 2018. "Skin electronics from scalable fabrication of an intrinsically stretchable transistor array." *Nature* 555:83–88.

Xu, K., Y. Lu, and K. Takei. 2019. "Multifunctional skin-inspired flexible sensor systems for wearable electronics." *Adv. Mater. Technol.* 4:1800628.

Yamada, T., Y. Hayamizu, Y. Yamamoto, Y. Yomogida, A. Izadi-Najafabadi, D. N. Futaba, and K. Hata. 2011. "A stretchable carbon nanotube strain sensor for human-motion detection." *Nat. Nanotechnol.* 6:296–301.

Yamamoto, Y., S. Harada, D. Yamamoto, W. Honda, T. Arie, S. Akita, and K. Takei. 2016. "Printed multifunctional flexible device with an integrated motion sensor for health care monitoring." *Sci. Adv.* 2:e1601473.

Yamamoto, Y., D. Yamamoto, M. Takada, H. Naito, T. Arie, S. Akita, and K. Takei. 2017. "Efficient skin temperature sensor and stable gel-less sticky ECG sensor for a wearable flexible healthcare patch." *Adv. Healthcare Mater.* 6:1700495.

Yan, C., J. Wang, and P. S. Lee. 2015. "Stretchable graphene thermistor with tunable thermal index." *ACS Nano* 9:2130–2137.

Yoon, J., A. J. Baca, S. I. Park, P. Elvikis, J. B. Geddes, III, L. Li, R. H. Kim, J. Xiao, S. Wang, T. H. Kim, M. J. Motala, B. Y. Ahn, E. B. Duoss, J. A. Lewis, R. G. Nuzzo, P. M. Ferreira, Y. Huang, A. Rockett, and J. A. Rogers. 2008. "Ultrathin silicon solar microcells for semi-transparent, mechanically flexible and microconcentrator module designs." *Nat. Mater.* 7:907–915.

# 7

## Controlled Spalling Technology

**Huan Hu**
*Zhejiang University*

**Renwei Mao**
*Zhejiang University*

**Katsuyuki Sakuma**
*IBM Thomas J. Watson Research Center*

## CONTENTS

## 7.1 Introduction

Wearable electronics have shown fast growing needs in entertainment, robotics, personalized healthcare, and medicine because they provide convenient and continuous monitoring of human or robotics that wear them. For entertainment, wearable electronics senses the human movement and even brain activities and feed into the gaming system in a convenient fashion and provides unprecedented substitution for players [1]. For robotics, wearable electronics can render robotics human-like senses and enables seamless interfacing with humans such as prosthetic limbs and other types of wearable robots [2]. For healthcare applications, wearable electronics provides long-term and continuous recording of physical and physiological signals that can either deepen understandings of the disease such as Parkinson's disease and enable rapid treatment by alerting immediately [3].

Current technology for producing wearable electronics can be mainly categorized into four types. The first type uses existing inorganic semiconductor materials and manufacturing processes as it is. The electronics are made into small chips and can be packaged into a wearable fashion like a watch or a bracelet. The second type also uses existing inorganic semiconductor materials but employs new manufacturing methods of thinning down these semiconductor materials into thin films or ribbons to achieve flexibility [4]. The third type is to use organic semiconductor materials and conductive polymers [5] that are intrinsically flexible. The fourth type uses 1-D or 2-D nanomaterials such as carbon nanotube [6], 2-D materials [7] nanowires [8, 9] liquid metal sealed in flexible forms [10, 11] (Figure 7.1).

Among the four types of wearable electronics technology, the second type uses existing semiconductor materials that own outstanding material properties such as fast carrier mobility and reliability compared to other materials. In addition, the manufacturing process of semiconductor materials is mature and is more likely to be commercialized into reliable product. The mainstream method within the second technology is to use silicon-on-insulator (SOI) substrate and transfer-suspended thin film of silicon to flexible substrates [12]. This method includes extra steps of etching top silicon and buried oxide for releasing, which reduce the effective area of devices and pose challenges for succeeding manufacturing procedures [13]. Another emerging method is controlled spalling technology (CST), which is to directly remove the top part of semiconductor substrate off using a layer of material with tensile stress [14, 15]. In this book chapter, we first introduce the history of the CST. Then, we will list various applications enabled by the CST. After that, we focus on the recent development of the CST in wearable electronics. Finally, we will comment on the future development trends.

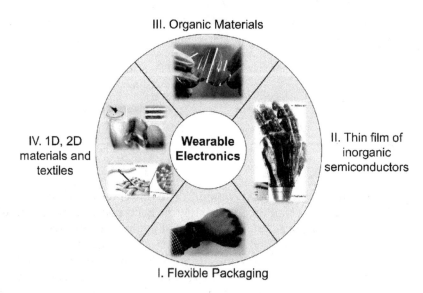

**FIGURE 7.1**
Four major types of wearable electronics technology.

## 7.2 History and Theoretical Basis of Controlled Spalling Technology

The CST, as a new novel and ingenious spalling layer transfer process, has been attracted more and more attentions from its first report [16]. In contrast to sophisticated and high-temperature experiment condition of other traditional layer release methods [17–19], the CST can work at room temperatures using a common laboratory [14, 15]. The crucial elements in the CST include a tensile stress layer (stressor layer) and crack initiation. The stressor layer in the CST mostly employs metal material such as nickel [14] and aluminum [20] prepared by physical sputtering or evaporation as well as electroplating or polymer materials [15, 16]. Laser cutting [21], temperature distribution [20], and handle layer assistance [8] have been demonstrated to introduce the initial crack site for spalling. The ability of properly spalling the entire thin film off its host substrate without cracking or damages requires the initial crack to occur in a single location and propagate as a single crack front. Actually, two main fracture failures can occur when the films are applied under tension condition: for the first case, some internal cracks would propagate spontaneously into the substrate if the film tension stress is too large, which will destroy the whole samples. Another case is that the internal initial crack occurs and propagates along the interface under the low enough toughness interface condition [22].

Figure 7.2 shows the procedures of a typical spalling experiment. A stressor film such as nickel is deposited on the surface of a brittle substrate to get the compression boundary conditions of the substrate. To initiate a crack, a small force exerted on the edge of the stressor film can induce a type I facture, which then would lead to the major type-II fracture, This type II fracture then propagates as a single fracture front by mechanically guiding the handle layer and completes a CST process.

Two main factors, film stress and thickness value, are the critical parameters to determine the spalled layer thickness. Suo and Hutchinson developed spalling mode fracture model in late 1980s [23], which offered a direct and simple means of predicting the relationship

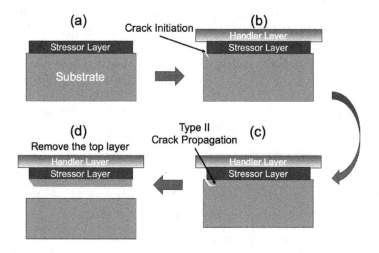

**FIGURE 7.2**

Schematics describing a typical procedure of the CST: (a) a tensile stressor layer is deposited on the substrate; (b) a flexible handle layer is then applied on the top of the stressor layer in order to form the initiation crack of type I at the edge of the substrate; (c) type II crack propagates along the interface under the low-enough toughness interface condition; (d) the top layer is removed from the substrate.

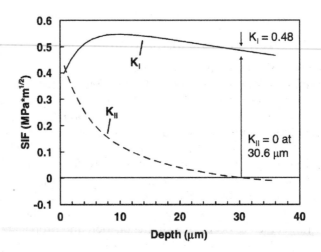

**FIGURE 7.3**
Calculated stress intensity factors $K_I$ and $K_{II}$ for a 6-μm-thick Ni layer with 500-MPa tensile stress on a Ge <1, 0, 0> wafer [37].

between critical loading conditions and crack depths. The resulting expressions for the $K_I$ and $K_{II}$ stress intensity factors were written by:

$$K_I = \frac{P}{\sqrt{2Uh}}\cos(\omega) + \frac{M}{\sqrt{2Vh^3}}\sin(\omega + \gamma) \tag{7.1}$$

$$K_{II} = \frac{P}{\sqrt{2Uh}}\sin(\omega) - \frac{M}{\sqrt{2Vh^3}}\cos(\omega + \gamma) \tag{7.2}$$

where $P$ is the load force applied on the edge and $M$ is the moment induced by deposited stressor layer, $U$, $V$, and $\gamma$ are dimensionless constants related to the elastic energy stored in the structure far behind the crack tip, $h$ is the stressor layer thickness, and $\omega$ is a dimensionless number that depends on the elastic dissimilarity of the stressor and the substrate and depends on crack depths. All of the dimensionless constants can be calculated from the stressor layer and substrate materials' properties, such as Young's Modulus and Poisson ratio. Combining with the fracture criterion $K_{II} = 0$, the steady-state cracking depth can be predicted from these results (Figure 7.3).

## 7.3 Application in Wearable Electronics

### 7.3.1 Materials and Fabrication Processes

Silicon spalling is a unique mode of brittle fracture whereby a tensile surface layer induces fracture to parallel the substrate interface. Although the phenomenon was discovered and the fracture mechanics analysis was developed by the late 1980s [23–25], it was till quite recently that the spalling technique was implemented to make a wearable sensor [26–28].

If we use the materials of conventional semiconductors, it could offer better performance and reliability in electronics for use in healthcare applications. However, the use of electronics in some biological applications requires contact with biological surfaces such as human skin. Compared to several methods to produce a sufficiently thin film of semiconductor material to withstand larger mechanical deformation, silicon spalling has advantages such as being CMOS (complementary metal-oxide-semiconductor) compatible, includes fewer processing steps, and uses low process temperature.

Figure 7.4 shows a fabrication process flow of forming a flexible piezoresistive sensor using a silicon substrate, where the region of the piezoresistors is selectively spalled. The silicon substrate can be entirely or selectively doped; thus, electronic devices and sensors can be fabricated on the same silicon substrate. Dopant materials can be p-type dopants (e.g., boron) or n-type dopants (e.g., phosphorus). Spin-coat processes were used to apply the first photoresist

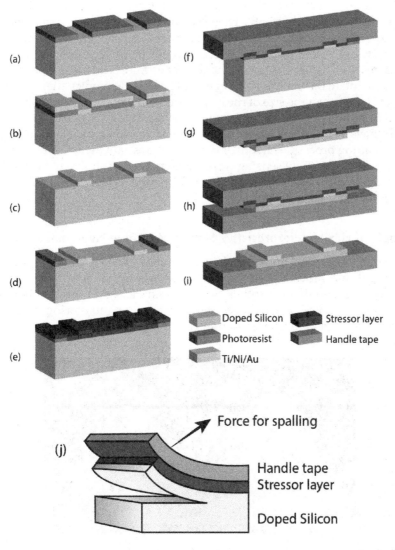

**FIGURE 7.4**
A schematic diagram of major spalling process of forming flexible electronics using doped silicon.

layer on the silicon substrate and the photoresist was patterned using the first lithography mask to define the regions of the pads (Figure 7.4a). Ohmic contact material such as Ti/Ni/Au was then sputtered on top of the structure (Figure 7.4b). The lift-off process was performed to remove the first photoresist (Figure 7.4c). A second photoresist was then applied and patterned using a second lithography mask to define the shape over the substrate regions to be spalled (Figure 7.4d). Then, a blanket metal film (e.g., 0.2-μm Ti) was deposited by sputtering to serve as a conductive seed layer for a subsequent electroplating process (Figure 7.4e). As a stressor layer, nickel was electroplated on the exposed conductive layer because nickel creates a large mechanical stress at the edge of the film to initiate a crack into the substrate. The thickness of the stressor layer can be adjusted to control the desired fracture depth. Then, a top flexible handle tape was bonded to the top of the stressor layer (Figure 7.4f) to enable spalling. A thin layer of silicon including the piezoresistive sensors was peeled off from the silicon substrate in regions unmasked by the second photoresist layer (Figure 7.4g and 7.4j). The spalled structure was then applied to the top of the bottom flexible handle tape with the device surface facing up (Figure 7.4h). Then, the top handle tape was debonded with ultraviolet (UV) exposure, followed by the removal of the second photoresist layer, the sputtered conductive layer, and the electroplated stressor layer by using organic solvents and chemical etchants (Figure 7.4i). The top and bottom handle tapes were used to provide fracture control and helped to maintain the integrity of thin film of flexible electronics after spalling. The size of the tape was larger than the size of the base silicon substrate. The bottom handle tape can be debonded from the spalled silicon with the assistance of UV exposure.

Figure 7.5 shows a 40-μm-thick flexible piezoresistive sensor using a doped silicon substrate by the spalling process. The sensor is sensitive to strain, temperature, and light. The temperature sensing and light sensing were characterized as shown in Figure 7.6 [26]. The results indicated that the spalling method could be used to fabricate a wide range of flexible wearable electronics application.

### 7.3.2 Example of Flexible Piezoresistive Sensor: Wearable Fingernail Sensor

Humans use their hands and fingers in their daily lives. When fingers are pressed directly against an object and moved left or right, this makes the fingernail deformation more pronounced on one side than the other. Sakuma and Heisig reported a fingernail-based wearable sensor, and also some processing methods for monitoring fingernail deformation to quantify and characterize hand activities at the time of daily movement [29].

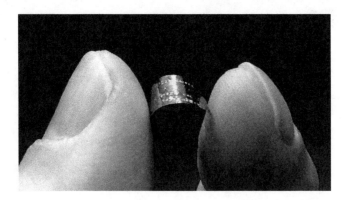

**FIGURE 7.5**
A photo of a flexible piezoresistive sensor formed by controlled spalling process bent by two fingers.

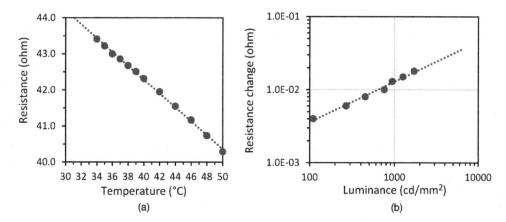

**FIGURE 7.6**
Temperature-sensing and light-sensing characterization results for spalled silicon sensors: (a) measured electrical resistances at different temperatures; (b) measured electrical resistances at different light intensities.

The fingernail-based wearable sensor can also be used for a new type of human-computer user interface [29, 31, 32]. They quantified the fingernail deformation when a fingertip pad was pressed and, thus, the displacement of fingernail was more than 130 µm and the strain change was more than 0.3%. A mechanical strain of the fingernail can be sensed as a change in electrical resistance in piezoresistive strain gauges. Using the piezoresistive property, a flexible strain gauge sensor attached to the fingernail can capture the deformation of the fingernail by using a strong, fast-acting, biocompatible adhesive.

Figure 7.7 shows a block diagram illustrating the simplified hardware architecture for a wearable fingernail deformation sensor. The system includes a flexible piezoresistive

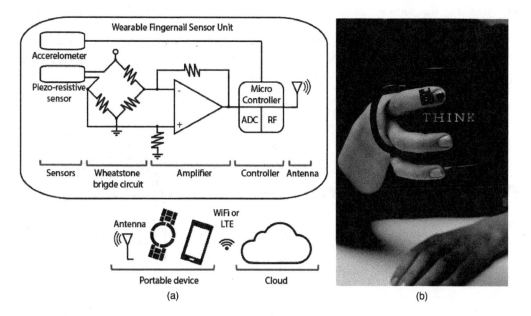

**FIGURE 7.7**
(a) A diagram illustrating a system employing the wearable wireless fingernail sensor; (b) a photo of a prototyped wearable fingernail sensor worn on a finger of a hand wrapping a mug.

strain gauge sensor, a three-axis accelerometer, Wheatstone bridge circuit, an amplifier, a micro controller, an analog-to-digital converter (ADC), a radio frequency (RF) controller, a network antenna, a portable device, and cloud-based machines running the analytics code [29, 30]. The fingernail deformation signal from the piezoresistive sensor goes through an offset compensation circuit in the form of a bridge network. The signals are then amplified before being connected to the ADC module of a microcontroller. The overall system also includes a three-axis accelerometer to monitor finger movement. The digital signals are transmitted via Bluetooth low energy (BLE) to one or more user devices, such as a smartphone, a smartwatch, tablet, laptop, etc. These signals are analyzed using machine learning and then processed for measuring the finger grip strength and the applied force. The raw data and analytics output from the user devices are sent to a cloud platform via Wi-Fi or long-term evolution (LTE) technology and can be saved in a database.

The most commonly used type of strain gauges includes metal alloys such as constantan, the piezoelectric strain gauge, and the doped semiconductor materials as described in Section 7.3.1. The strain gauge sensor is placed and glued on the fingernail via biocompatible adhesives. The sensor must remain securely attached for the duration of the data-collection process. The archived data was used for trend analysis, training of new models, and longitudinal studies. A metal foil strain gauge sensor was chosen for use in hand activity-detection experiment as shown in Figures 7.7(b) and 7.8. However, as an optimization, it could be replaced by a flexible piezoresistive semiconductor sensor that is formed by the spalling process as described in Section 7.3.1 to allow the fingernail to move more freely. The wearable, wireless device is designed to be worn on a finger and not get in the way with most normal hand activities (Figure 7.7(b)). The use of this technology was demonstrated as a human-computer interface, fingertip writing recognition, behavioral and biomechanical monitoring, and a clinical feature generator [29–32]. Figure 7.8 shows examples of fingernail strain responses during finger extension and flexion. This test demonstrates that the sensor system can even capture forces produced by movements that have no object interaction [29].

The ability to measure contour changes via the piezoresistive sensors has a variety of important clinical applications. For instance, patients with degenerative neurological conditions, such as Parkinson's disease, experience motor symptoms, such as tremors and rigidity.

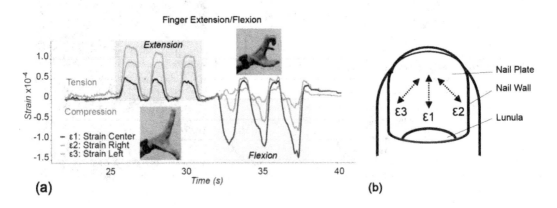

**FIGURE 7.8**
(a) Examples of fingernail strain response data: a set of index finger three extensions followed by three flexions and (b) position of three piezoresistive sensors on the fingernail (ε1: center, ε2: right, ε3: left).

When evaluating patients with such neurological conditions, it is important to have an efficient and effective way to evaluate their treatment effectiveness and the progression of their symptoms. The wearable fingernail sensor with piezoresistive sensors can be included as a Unified Parkinson's Disease Rating Scale (UPDRS) for daily neurological examinations, such as based on a finger-tapping test.

## 7.4 Other Applications of Controlled Spalling Technology

### 7.4.1 Application in Photovoltaic Technology

The earliest application for the CST is to produce thin film of photovoltaic devices. In photovoltaic (PV) industry, there is tremendous stimulus for reducing the silicon solar cell device thickness as well as contents. Moreover, in manufacturing, thinner wafer tend to cause issues in wafer handling and lead to yield loss. Rao et al. demonstrated successful spalling of 25-μm-thick thin film off a 6-inch crystalline silicon wafer with a measured efficiency of 12.5%. The final solar cell cost is predicated to be reduced more than 50% [33]. Bedell et al. demonstrated spalling III-V single-junction solar cells off Ge substrate and showed 17% efficiency close to same devices on a bulk substrate [15]. They further applied the CST to spall a thin-film InGaP/(In)GaAs/Ge tandem solar cells with a high-efficiency of 28.7% off a Ge substrate [34]. The solar cells have multiple different layers and heterojunctions. Tunneling electron microscopy (TEM) and electrical characterizations were used together to verify the structural integrity of the solar cells. The results indicated that the CST could maintain the integrity of all layers and thus the performance in a complicated solar cell. Jain et al. grew III-V solar cells directly on unpolished spalled Ge substrates. Although there is roughness on the spalled Ge substrates, the final performance of the III-V solar cells is almost equivalent [35].

Most studies report no significant degradation of solar cells after spalling from bulk substrates. For example, the open circuit voltage (Voc) increases by 0.2%, the short-circuit current density (Jsc) and the fill factor both decreases around 1% after spalling as reported by [36].

### 7.4.2 Application for Other Electronic Devices and Materials

CMOS circuits with 100 ring-resonators [37] as well as memory cells [38] have also been applied with the CST to construct flexible circuits. Shahrjerdi and Bedell applied the CST to an SOI wafer with only 60-A-thick top silicon. Although ~10-μm-thick bottom silicon was spalled together with the buried oxide and the top 60-A-thick silicon, the thick bottom silicon can be removed with tetramethylammonium hydroxide (TMAH) solutions to render an ultra-thin layer consisting of only buried oxide and the top thin silicon prefabricated with circuits. This methodology brings down the thickness of the thin film from conventionally several tens of micrometers to several tens of nanometers with the assistance of SOI wafer.

The CST has also been applied to spall a thin film of 3-μm-thick GaN off the substrate [39]. Multiquantum well (MQW)-based InGaN/GaN epitaxial LED layers grown on PSS substrates were spalled off the substrate and optical emission was measured on the spalled vertical LED devices.

In addition to Si, Ge, and III-V semiconductor materials, the CST has also played roles in producing high-quality mono-layer graphene from silicon carbide substrate [40]. He et al. demonstrated thin lead zirconate titanate (PZT) film spalled by the CST [41].

## 7.5 Outlook

To satisfy the growing demands of wearable electronics, current CST technology still faces two major issues. First, the spalled film thickness is still too thick (~20 μm) compared to thin films released from SOI wafer using other techniques (hundreds of nanometers thick). Thicker films indicate less flexibility and thus limited wearability. Second, the spalled film has rough backside surface and irregular boundaries, reducing the quality of the electronics and uniformity of devices. Following aspects may potentially solve the above-mentioned challenges. First, better crack initiation methods can be implemented. For example, larger discontinuous surface stresses or jump changes can be the first step to create sufficient strain gradient. Second, additional external stimuli might be added such as thermal stress, electrical field, magnetic field, laser, etc., to induce even larger stress to facilitate the spalling.

The advantages of the CST lie in its CMOS compatibility and its great potential in creating system-level wearable electronics integrating sensors, signal processing, communicating unit that can have minimum footprint and are easier to be worn.

## Acknowledgments

The authors would like to sincerely thank S. W. Bedell, B. Webb, S. Wright, and M. Agno for their contribution in developing spalling silicon sensor. The authors would like to sincerely thank S. Heisig, G. Blumrosen, A. Abrami, J. W. Ligman, J. Dicarlo, S. Lukashov, R. Narayanan, M. Stellitano, and V. Caggiano for their contribution in developing hardware, software, and system. Authors also would like to acknowledge management support and encouragement from D. McHerron, J. Knickerbocker, A. Royyuru, and T.C. Chen of IBM during this research. The book chapter writing is also supported by ZJUI and led by Prof. Huan Hu with funding from Tang Foundation.

## References

[1] J. J. S. Norton, *et al.*, "Soft, curved electrode systems capable of integration on the auricle as a persistent brain-computer interface," *Proceedings of the National Academy of Sciences of the United States of America*, vol. 112, pp. 3920–3925, Mar 31 2015.
[2] N. Lu and D-H. Kim, "Flexible and stretchable electronics paving the way for soft robotics," *Soft Robotics*, vol. 1, pp. 53–62, 2014.

[3] D. Son, et al., "Multifunctional wearable devices for diagnosis and therapy of movement disorders," *Nature Nanotechnology*, vol. 9, p. 397, 2014.

[4] J. A. Rogers, et al., "Materials and mechanics for stretchable electronics," *Science*, vol. 327, pp. 1603–7, Mar 26 2010.

[5] D. J. Lipomi, et al., "Skin-like pressure and strain sensors based on transparent elastic films of carbon nanotubes," *Nature Nanotechnology*, vol. 6, pp. 788–792, Dec 2011.

[6] D. J. Lipomi, et al., "Skin-like pressure and strain sensors based on transparent elastic films of carbon nanotubes," *Nature Nanotechnology*, vol. 6, pp. 788–792, Dec 01 2011.

[7] L. Gao, "Flexible device applications of 2D semiconductors," *Small*, vol. 13, p. 1603994, 2017.

[8] F. Xu, et al., "Controlled 3D buckling of silicon nanowires for stretchable electronics," *ACS Nano*, vol. 5, pp. 672–8, Jan 25 2011.

[9] J. M. Weisse, et al., "Fabrication of flexible and vertical silicon nanowire electronics," *Nano Lett*, vol. 12, pp. 3339–43, Jun 13 2012.

[10] W. Xi, et al., "Soft tubular microfluidics for 2D and 3D applications," *Proceedings of the National Academy of Sciences of the United States of America*, vol. 114, pp. 10590–10595, Oct 3 2017.

[11] H. Hu, et al., "Super flexible sensor skin using liquid metal as interconnect," *2007 IEEE Sensors*, vols. 1–3, pp. 815–817, 2007.

[12] E. Menard, et al., "A printable form of silicon for high performance thin film transistors on plastic substrates," *Applied Physics Letters*, vol. 84, pp. 5398–5400, 2004.

[13] A. M. Hussain and M. M. Hussain, "CMOS-technology-enabled flexible and stretchable electronics for internet of everything applications," *Advanced Materials*, vol. 28, pp. 4219–4249, Jun 8 2016.

[14] S. W. Bedell, et al., "Layer transfer of bulk gallium nitride by controlled spalling," *Journal of Applied Physics*, vol. 122, p. 025103, 2017.

[15] S. W. Bedell, et al., "Kerf-less removal of Si, Ge, and III–V layers by controlled spalling to enable low-cost PV technologies," *IEEE Journal of Photovoltaics*, vol. 2, pp. 141–147, 2012.

[16] M. Tanielian, et al., "A new technique of forming thin free standing single-crystal films," *Journal of the Electrochemical Society*, vol. 132, pp. 507–509, 1985.

[17] M. Bruel, "The history, physics, and applications of the smart-cut (R) process," *Mrs Bulletin*, vol. 23, pp. 35–39, Dec 1998.

[18] T. Yonehara, et al., "Epitaxial layer transfer by bond and etch back of porous Si," *Applied Physics Letters*, vol. 64, pp. 2108–2110, Apr 18 1994.

[19] M. Konagai, et al., "High-efficiency gaas thin-film solar-cells by peeled film technology," *Journal of Crystal Growth*, vol. 45, pp. 277–280, 1978.

[20] J. Hensen, et al., "Directional heating and cooling for controlled spalling," *IEEE Journal of Photovoltaics*, vol. 5, pp. 195–201, Jan 2015.

[21] J. Vaes, et al., "SLiM-cut thin silicon wafering with enhanced crack and stress control," *Next Generation (Nano) Photonic and Cell Technologies for Solar Energy Conversion*, vol. 7772, 2010.

[22] A. S. Argon, et al., "Intrinsic toughness of interfaces between SiC coatings and substrates of Si or C fibre," *Journal of Materials Science*, vol. 24, pp. 1207–1218, April 01 1989.

[23] Z. Suo and J. W. Hutchinson, "Steady-state cracking in brittle substrates beneath adherent films," *International Journal of Solids and Structures*, vol. 25, pp. 1337–1353, 1989.

[24] M. D. Thouless, et al., "The edge cracking and spalling of brittle plates," *Acta Metallurgica*, vol. 35, pp. 1333–1341, Jun 1987.

[25] A. G. Evans, et al., "The cracking and decohesion of thin-films," *Journal of Materials Research*, vol. 3, pp. 1043–1049, Sep-Oct 1988.

[26] K. Sakuma, et al., "CMOS-compatible wearable sensors fabricated using controlled spalling," *IEEE Sensors Journal*, vol. 19, pp. 7868–7874, Sep 15 2019.

[27] K. Sakuma, et al., "Flexible Piezoresistive Sensors Fabricated by Spalling Technique," in *2018 International Flexible Electronics Technology Conference (IFETC)*, 2018, pp. 1–2.

[28] N. Li, et al., "Single crystal flexible electronics enabled by 3D spalling," *Advanced Materials*, vol. 29, May 10 2017.

[29] K. Sakuma, *et al.*, "Wearable nail deformation sensing for behavioral and biomechanical monitoring and human-computer interaction," *Scientific Reports*, vol. 8, Dec 21 2018.

[30] K. Sakuma, *et al.*, "A wearable fingernail deformation sensing system and three-dimensional finite element model of fingertip," in *2019 IEEE 69th Electronic Components and Technology Conference (ECTC)*, 2019, pp. 270–276.

[31] K. Sakuma, *et al.*, "Turning the Finger into a writing tool," in *2019 41st Annual International Conference of the IEEE Engineering in Medicine and Biology Society (EMBC)*, 2019, pp. 1239–1242.

[32] G. Blumrosen, K. Sakuma, J. J. Rice, and J. Knickerbocker, "Back to finger-writing: fingertip writing technology based on pressure sensing," in *IEEE Access*, vol. 8, pp. 35455–35468, 2020, doi: 10.1109/ACCESS.2020.2973378.

[33] R. A. Rao, *et al.*, "A novel low cost 25μm thin exfoliated monocrystalline Si solar cell technology," in *2011 37th IEEE Photovoltaic Specialists Conference*, 2011, pp. 001504–001507.

[34] D. Shahrjerdi, *et al.*, "High-efficiency thin-film InGaP/InGaAs/Ge tandem solar cells enabled by controlled spalling technology," *Applied Physics Letters*, vol. 100, p. 053901, 2012.

[35] N. Jain, *et al.*, "III–V Solar cells grown on unpolished and reusable spalled Ge substrates," *IEEE Journal of Photovoltaics*, vol. 8, pp. 1384–1389, 2018.

[36] C. A. Sweet, *et al.*, "Controlled exfoliation of (100) GaAs-based devices by spalling fracture," *Applied Physics Letters*, vol. 108, Jan 4 2016.

[37] S. W. Bedell, *et al.*, "Layer transfer by controlled spalling," *Journal of Physics D-Applied Physics*, vol. 46, Apr 17 2013.

[38] D. Shahrjerdi and S. W. Bedell, "Extremely flexible nanoscale ultrathin body silicon integrated circuits on plastic," *Nano Letters*, vol. 13, pp. 315–320, Jan 2013.

[39] S. W. Bedell, *et al.*, "Vertical light-emitting diode fabrication by controlled spalling," *Applied Physics Express*, vol. 6, Nov 2013.

[40] J. Kim, *et al.*, "Layer-resolved graphene transfer via engineered strain layers," *Science*, vol. 342, pp. 833–836, Nov 15 2013.

[41] J. He, *et al.*, "Flexible heterogeneous integration of PZT film by controlled spalling technology," *Journal of Alloys and Compounds*, vol. 807, Oct 30 2019.

# 8

## *Flexible and Stretchable Liquid Metal Electronics*

**Dishit P. Parekh**
*IBM Research*

**Ishan D. Joshipura**
*Lawrence Livermore National Laboratory*

**Yiliang Lin**
*University of Chicago*

**Christopher B. Cooper**
*Stanford University*

**Vivek T. Bharambe**
*North Carolina State University*

**Michael D. Dickey**
*North Carolina State University*

**KEYWORDS:** *Liquid metals, EGaIn, Soft electronics, Gallium, Galinstan, Stretchable electronics, Gallium oxide, Flexible electronics*

## CONTENTS

## 8.1 Introduction

Nonplanar surfaces are ubiquitous, ranging from natural curved structures to artificial complex-shaped objects. Soft electronics have the capability to conform their structures on nonplanar surfaces, which significantly expands the applications of conventional rigid electronics in sensing, monitoring, and diagnosing. There are two key advantages of soft electronics: (1) the intimate contact between the electronic components and the nonplanar structures enables the collection of high-quality electrical signal transfer due to the elimination of any air gaps; (2) the similar mechanical properties largely prevent the damage to the final application environments of soft electronics such as human tissues (especially during their movement), allowing it to be used in long-term, continuous healthcare monitoring.

Although the terms flexible electronics and soft electronics commonly appear together in literature, flexible electronics is not necessarily soft electronics. From a materials perspective, flexible electronics refers to the bending capabilities of the device, while soft electronics requires them to not only be bendable but also stretchable and possess a relatively low modulus. From the perspective of material mechanics, the bending stiffness is inversely proportional to the third power of thickness.[1] Essentially, one could make a material flexible through decreasing the material's thickness, a common example of how aluminum can be transformed from a rigid plate to a flexible foil.

There are mainly two strategies to achieve stretchability in electronics: (1) make nonstretchable materials stretchable through proper design of structures to absorb applied strain; and (2) develop intrinsic stretchable materials applicable in electronic applications.

Various designs, including wave/wrinkle,[2] island-bridge,[3] origami,[4] kirigami,[5] and cracks and interlocks[6] have been developed to endow materials with stretchability that combines the conventional rigid silicon wafer-based devices and elastomeric polymers such as poly(dimethylsiloxane) – PDMS to be physically deformed together as shown in Figure 8.1. Structural design serves as a universal tool for stretchability enhancement and it is applicable to different types of materials — including organic or inorganic, lab-made or commercial materials. While ~30–50% strain is easy to achieve with simple fabrication strategies, larger strain would require more complex fabrication techniques. Therefore, the maximum strain to which the materials can be stretched is limited to some extent in practice.

Alternate way to achieve stretchability is to utilize intrinsic soft and stretchable materials as components in electronics. There are multiple candidate materials that have been applied for flexible/soft electronics, including hydrogels,[7] conductive polymers,[8] nanomaterials,[9] and liquid metals[10] as shown in Figure 8.2. While all the above-mentioned materials have been integrated into different applications, in some cases it requires the conductor to possess a relatively high electrical conductivity, even at extremely high strains/deformations (i.e., a few hundred percentage strain or more). Liquid metals have the unique combination of possessing both fluidic and metallic nature, namely, (1) their electrical conductivity is higher than other room-temperature conductors, such as ionic conductors; (2) their conductivity changes minimally depending on the cross-section under extreme deformation due to their fluidic properties, making them a suitable candidate for soft and stretchable electronics. Mercury is another commonly known liquid metal, but its toxicity limits the applications. Gallium and its alloys serve as promising liquid metals due to low toxicity and negligible vapor pressure. While the melting point of gallium is ~30 °C, alloying with other metals, e.g., indium and tin, lowers the melting point below room temperature. The two most popular examples are eutectic gallium-indium (EGaIn, 75.2 wt.% gallium and 24.8 wt.% indium) with a melting point of 15.5 °C and galinstan (composed of 67 wt.% of gallium, 20.5 wt.% of indium, and 12.5 wt.% of tin) with a melting point of ~10.7 °C.

## 8.2 Scope

The primary focus of this chapter is on liquid metal-based flexible and stretchable electronics because liquid metals are extremely soft while being intrinsically stretchable, in addition to having superior conductivity as compared to other soft materials.[13–15] Prototyping electronics such as field-effect transistors (FETs), displays, sensors, and actuators, using other classes of soft materials such as nanoparticle-based inks, hydrogels, etc., have been covered in literature[16,17] and the rest of the book chapters. There are impressive examples of soft electronics ionic devices[16,17] composed of hydrogels[18] (for ionic diodes,[19,20] soft photovoltaics,[21] transparent electrodes,[22] capacitors,[23] and electrocardiogram – ECG patches), ionic liquids/ionogels[24] (for stretchable electrodes[25]

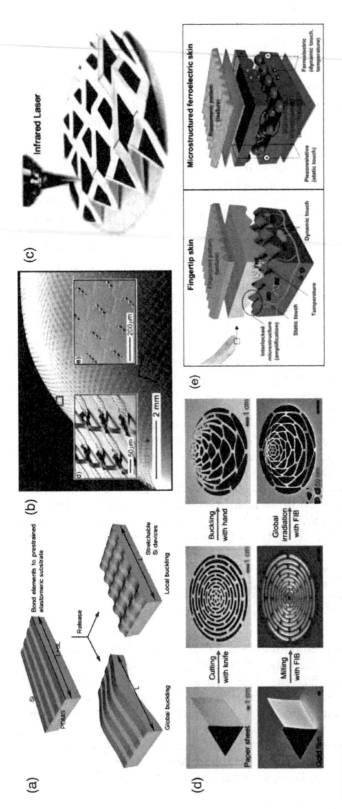

### FIGURE 8.1

Examples of different flexible and stretchable structures combining rigid and soft materials. (a) Schematic illustration of the process for fabricating buckled single crystal Si ribbons (yellow) on a PDMS (gray) substrate; bond Si elements to prestretched PDMS; local buckling; and global buckling due to release of prestretch.[2] (b) Colorized angled-view scanning electron microscope (SEM) images of a silicon circuit mesh on a PDMS transfer element on the surface of a plastic model of a human heart. Inset shows magnified views of the areas indicated by the red and blue box. The gray, yellow, and blue colors correspond to silicon, polyimide (PI), and PDMS, respectively.[3] (c) An optical photo of a fabricated origami structure after the direct laser-write patterning process (black) on a paper substrate (white).[4] (d) Camera images of the paper kirigami process of an expandable dome (top). SEM images of an 80-nm-thick gold film, a 2D concentric arc pattern, and a 3D microdome (bottom). The high-dose FIB milling corresponds to the 'cutting' process and the global low-dose FIB irradiation of the sample area (enclosed by the dashed ellipse) corresponds to the 'buckling' process in nano-kirigami. The inset shows a 3D feature size of 50 nm.[5] (e) Structural and functional characteristics of human fingertips (left). Fingertip skin consists of slow- and fast-adapting mechanoreceptors for static and dynamic touch, free nerve endings (FNE) for temperature, fingerprint patterns for texture, and epidermal/dermal interlocked microstructures for tactile signal amplification. Flexible and multimodal ferroelectric e-skin (right). The functionalities of human skin are mimicked by elastomeric patterns (texture) and piezoresistive (static pressure), ferroelectric (dynamic pressure and temperature), and interlocked microdome arrays (tactile signal amplification).[6]

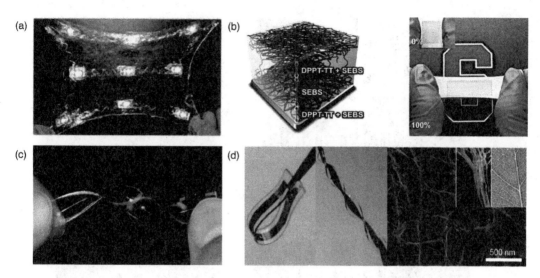

**FIGURE 8.2**
Strategies for intrinsically stretchable devices. (a) A hydrogel electronic device that encapsulates an array of light-emitting diodes (LEDs) lights connected by stretchable silanized titanium wire. The device is transparent and robust under multiple cycles of high stretch.[7] (b) A 3D illustration of the morphology of the CONPHINE-1 film (left)[11] and photographs of a CONPHINE-1 film (blue area) at 0% strain (top left inset) and stretched to 100% strain on a rubber substrate (center).[12] (c) Photograph of a stretchable liquid metal dipole antenna.[10] (d) Digital photographs of a bent and twisted SWCNT/PDMS stretchable conductor, showing their flexibility (left and center). SEM images of the as-grown single-walled carbon nanotube films (right). The insets show the comparison of the hierarchical reticulate structure features of the SWCNT film (left inset) and the leaf veins (right inset).[9]

and electrolytes[26]), and nanoparticle pastes/inks (for stretchable batteries,[27,28] conductors,[29,30] rubber-like conductors,[31–35] and fluidic conductors[36]). While outside the scope of this chapter, these materials have their own advantages; however, they do not provide the combination of mechanical fluidity and high electrical conductivity under large strains afforded by liquid metals.

## 8.3 The Need for Liquid Metals in Soft Electronics

Soft electronics are devices that can be bent, folded, stretched, or conformed regardless of their material composition, without losing the electronic functionality. In addition to adding smart functionality and compatible form factor to the conventional rigid electronics, these devices have the promise of being employed in healthcare – designing low-cost stretchable electronic skins or lightweight smart sensors conformal to human body for biomonitoring and energy harvesting applications.

One of the subsets of soft electronic devices is flexible electronics that embeds conductors in thin form factors inside a polymer matrix and retain their function while being bent. Intrinsically stiff materials, such as copper, can be rendered flexible by making them sufficiently thin. Examples of such electronics include large area flexible displays

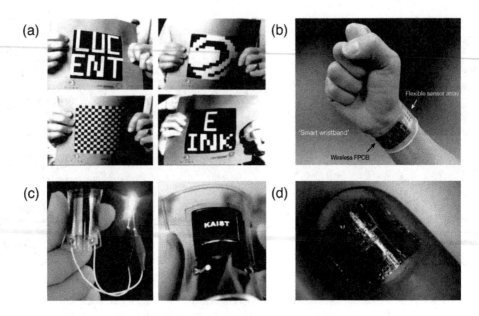

**FIGURE 8.3**
Flexible devices fabricated from low-thickness metal films. (a) Images of a sheet of electronic paper displaying images while being mechanically flexed where bending does not alter the performance of the display.[54] (b) Photograph of a wearable, flexible integrated sensor on a subject's wrist with the embedded multiplexed sweat sensor array and the wireless flexible PCB.[42] (c) Photographs of (left) a lithium-ion battery powering a blue LED in bent condition and (right) a picture of an all-in-one flexible LED system integrated with a bendable lithium ion battery.[49] (d) Image of an ultra-flexible, lightweight, and transparent thin film transistor on the top of a 1-μm-thick parylene film sticker on a finger nail.[50]

fabricated using plastic sheets—whose performance does not get altered upon bending[37–41], fully integrated and wearable sensor arrays for *in situ* perspiration analysis,[42–45] flexible antennas,[46,47] bendable inorganic thin-film batteries,[48,49] and transparent, lightweight electronics that can be transferred on another object, surface or biological tissues, i.e., human skin.[50–53] Some of these examples are shown in Figure 8.3.[42,49,50,54] One of the drawbacks of these devices is that they are not inherently stretchable as the conductors and the off-the-shelf electronic components used in the fabrication process are rigid and brittle by nature.

While flexible electronics can be bent, stretchable electronics can be elongated. Thus, stretchable electronics can be used in a wider application space while providing increased durability. To build a stretchable electronic device, in addition to having the backbone of a soft polymer matrix, it is required to pattern interconnects that are intrinsically stretchable. Using multiple patterning processes such as chemical vapor deposition, sputtering, soft lithography, and 3D printing, researchers have fabricated a variety of stretchable electronics such as optoelectronic skin for sensing and display,[55–57] soft neural implants that sustain millions of mechanical stretch cycles and assist in drug delivery,[58–60] stretchable batteries with wireless recharging capabilities,[61–63] stretchable displays,[64–66] soft silicon integrated circuits using wavy metal films,[67–70] and epidermal electronics for the skin.[71–73] Some of these examples are shown in Figure 8.4.[54,55,61,65]

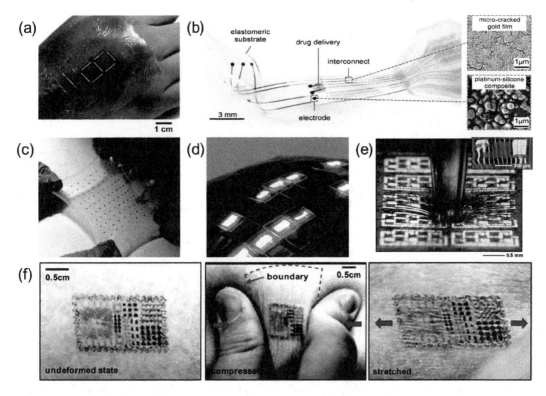

**FIGURE 8.4**
Stretchable devices with stretchable interconnects. (a) Photograph of an ultra-flexible and conformal red seven-segment PLED display on the surface of a human hand.[55] (b) Optical image of an e-dura (electronic dura meter) – a soft neural implant with the insets exhibiting scanning electron micrographs of the gold film and the platinum-silicone composite.[54] (c) Operation of a stretchable battery connected to red LEDs while being biaxially stretched to 300%.[61] (d) A demonstration of a stretchable display that can be spread over arbitrary curved surfaces.[65] (e) Image of a stretchable silicon circuit in a wavy geometry, compressed in its center by a glass capillary tube (main) and wavy logic gate built with two transistors (top right inset).[54] (f) A multifunctional epidermal electronic system (EES) on skin: undeformed (left), compressed (middle), and stretched (right).[54]

Despite their increasing demand, only a handful of these devices have been commercialized due to the lack of novel functional materials available along with the complex fabrication mechanisms needed to process them. Unlike conventional silicon-based microelectronics manufacturing which is limited to rigid wafers, stretchable electronics need to be incorporated onto plastics, paper, fibers, and even biological tissues – necessitating low-temperature processing. A brief comparison of Young's modulus of these materials is provided in Figure 8.5a.[14,15,74,75] According to Wagner & Bauer,[74] the two most important parameters for stretchable interconnects are having a high electrical conductance and a large critical strain at which the conduction is lost. Most of the 'true' soft electronics patterned today are based on conductive inks made with silver,[34,76,77] gold,[78] carbon nanotubes,[65,79] or conductive polymers.[80] These materials are inherently rigid: solid materials that are dispersed in colloidal solutions in the form of flakes, nanoparticles, nanotubes, or nanowires to form composite conductive inks. Once patterned, especially using direct-writing/3D printing, to obtain conductivity on the order

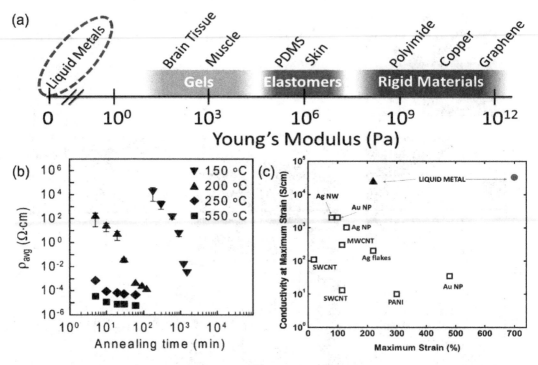

**FIGURE 8.5**

Key advantages of liquid metal alloys. (a) Young's modulus of liquid metals compared with materials found in the human body and common materials used for flexible electronics.[15] (b) A plot of electrical resistivity of direct-write printed flexible and stretchable silver microelectrodes as a function of annealing temperature and time.[54] (c) A plot comparing the conductivity at maximum strain for different composite conductors and liquid metals indicating that liquid metals have the highest conductivity at maximum strain values.[15]

of bulk materials, harsh post-processing treatments such as sintering and annealing at high temperatures need to be applied to burn away the viscous, toxic organic binders and obtain close-to-bulk electrical conductivities. An example of such dependence of the electrical conductivity on annealing time for printed silver nanoparticle inks[81] is provided in Figure 8.5b.[54] This limits the strain on the final electronics and the type of substrates used for patterning, thus, leading to the formation of cracks and fractures over time due to the mechanical mismatch between the soft substrate and the rigid conductor.

Consequently, there is a need for processing soft materials that are intrinsically stretchable and capable of maintaining bulk conductivity during reversible elongation (i.e., elastic deformation). A plot, as shown in Figure 8.5c[15] adapted from Matsuhisa et al.,[34] exhibits this quantitatively where the conductivity at maximum strain is compared for the different conventional conductors versus gallium-based liquid metals. To date, no conductive 'inks' offer the same combination of stretchability and conductivity as liquid metal.[74] The open squares represent composite approaches, whereas the triangle[82] and the circle[83] are select examples that use a gallium-based liquid metal embedded in elastomer.

In Section 8.4, we will discuss about the different low melting point metals available in nature and their properties and applications.

## 8.4 How to Pick a Liquid Metal?

Typically, metals are hard in their solid state and are known for their physical properties of high electrical and thermal conductivity in addition to being malleable (can be hammered into thin sheets), ductile (can be drawn out into thin wires), and fusible (can be melted to their molten liquid state).[84,85] In addition, metals are characterized by their high melting points; however, when liquefied, the conductive and fluidic nature of these metals allows them to be promising candidates for applications in microfluidics and soft, flexible, and stretchable electronics. Nevertheless, it is fairly difficult to process and work with high melting point alloys like bulk copper (1085°C) and silver (961.8°C). Therefore, researchers have focused their work on the other metals that are liquid at or near room temperature – colloquially, known as liquid metals. Table 8.1 lists these elements from the periodic table with their respective melting points (M.P.), uses, and concerns associated with them. These elemental liquid metals are discussed further below with their advantages and disadvantages.

### 8.4.1 Mercury (Hg)

Hg is perhaps the most well-known and extensively studied liquid metal. Notably, it has been utilized in thermometers and for electrochemical measurements (e.g., falling Hg

**TABLE 8.1**

Comparing the Properties of Different Liquid Metals from the Periodic Table[86–92]

| Element | Mercury | Cesium | Francium | Rubidium | Gallium |
|---|---|---|---|---|---|
| **Melting Point** | -39 °C | 29 °C | 27 °C | 40 °C | 30 °C |
| **Uses** | Electrochemical measurements, Dental fillings, and (formerly) thermometers | Atomic clocks, drilling fluids, spectroscopy | N/A | N/A | Low toxicity, Negligible vapor pressure |
| **Challenges** | Toxic | Radioactive, Violent reactivity | Radioactive, Short half-life | Radioactive, Violent reactivity | Oxidation, Corrosive to other metals |

drops, rotating Hg electrodes, etc.). In addition to the thermometer, mercury has been a key ingredient in dental amalgams for more than 165 years as it mixes easily with other metal powders forming a paste that hardens into an intermetallic alloy after a short time.[93–96] However, Hg is acutely toxic and forms harmful vapors.[97] To the best of our knowledge, Hg drops have only been utilized in soft electronics as micro-electromechanical (MEMS) devices that are hermetically sealed to prevent exposure to the environment.[98] Due to toxicity concerns, the utilization of Hg in flexible and wearable electronics has been limited otherwise.

### 8.4.2 Cesium, Francium, and Rubidium

Cesium (Cs), francium (Fr), and rubidium (Rb) are a few other metals that are liquid at or near room temperature. All three metals are radioactive, thus limited to certain applications. Nevertheless, for health and safety considerations, they are not currently utilized or considered in flexible electronics. To provide a broader perspective on the utility of these metals, we highlight the merits and prior use of these metals.

Cesium (Cs), with a low melting point of 29°C, has been previously utilized in atomic clocks,[99] drilling fluids in the oil industry,[100] and as an ion source for spectroscopic measurements.[101] Interestingly, a Cs alloy with sodium and potassium (41% Cs, 47% K, 12% Na) forms the lowest-known melting point for a metallic system (M.P.: –78°C). However, Cs is radioactive, pyrophoric (with an auto-ignition temperature of –116°C), and corrosive. In addition, Cs is soluble in water, making it particularly concerning for environmental and biological settings.[102] Due to its violent reactivity, radioactivity, and concerns with environmental contamination, it is not suitable for consumer electronics or other electronics that interface with humans or living systems.

Rubidium (Rb), with a melting point of 40°C, has no known uses in commercial applications. Due it being an alkali metal, it is violently reactive with water. Likewise, francium (Fr) has no known applications or commercial use; due to it having a very short half-time, it is concerned to be one of the rarest naturally occurring elements.

### 8.4.3 Gallium (Ga)

Out of all the liquid metals listed previously, gallium has the least amount of toxicity or hazards associated with it.[86] Unlike Hg, gallium has essentially no vapor pressure (can be heated up to ~2400°C before boiling) at room temperature, which implies that it can be handled outside of a chemical hood without concern for inhalation.[103] Like most metals, metallic gallium has negligible solubility in water and, therefore, could only feasibly enter the blood stream as a salt. Gallium salts (such as gallium nitrate) are an oxidized yet more soluble form of gallium and have been approved by the United States Food and Drug Administration (FDA) for magnetic resonance imaging (MRI) contrast agents and have some therapeutic value.[104] Recently, gallium-based liquid metals were shown to be an effective carrier for anti-cancer drugs.[105] Nevertheless, liquid metals should still be handled with caution as more is learned about their toxicity.[106] For example, a synthesized organo-metallic salt of gallium was found to be poisonous.[86] Due to its low toxicity, gallium is replacing mercury in many cases and is a widely used liquid metal in emerging soft electronic applications. Section 8.5 discusses gallium and its low melting point alloys in detail with a short historic background, followed by its properties and the current applications from the scope of this chapter.

## 8.5 Properties of Gallium-Based Liquid Metal Alloys

In 1875, gallium was first discovered by a French scientist named Paul-Émile Lecoq de Boisbaudran. It is not present in nature in its pure form, but instead, it is extracted as a byproduct of aluminum and zinc production.[107] It lies within the group 13 of the periodic table and its atomic weight is 69.716.[108] It is the element with the widest temperature range in which it is a liquid with the normal melting point being 29.83°C and the normal boiling point being 2402.85°C.[109] Gallium compounds are known in the periodic table due to their remarkable semiconducting properties with the elements of group 15, the most famous of them being, GaAs (gallium-arsenide) and GaN (gallium nitride).[110,111] In addition, gallium is characterized by its high affinity to mix with a wide range of other elements (both metals and nonmetals) under ambient conditions, forming halides, nitrides, oxides, and sulfides.[109,112]

Eutectic alloys are mixtures that form using a specific composition of binary or ternary elements and have a melting point lower than the melting point of the individual constituents. Gallium forms a couple of eutectic alloys when mixed with indium and tin. The most common examples are EGaIn – a eutectic alloy of gallium (Ga) and indium (In) in 75.2 and 24.8 wt.%, respectively, and galinstan – a eutectic alloy of gallium (Ga), indium (In), and tin (Sn) in 67, 20.5, and 12.5 wt.%, respectively. The physical properties of these alloys vary from the pure liquid metal gallium and, hence, we compare them with water as shown in Table 8.2.[103,109,113–123]

In Sub-section 8.5.1, we will review the properties of the oxide skin present on all gallium-based liquid metals that dominate the behavior of gallium in all applications.

### 8.5.1 Role of Oxide Skin on Gallium-Based Liquid Metals

Similar to aluminum, gallium-based liquid metals rapidly form a thin oxide layer on its surface when exposed to air or dissolved oxygen in ambient conditions[124] – even at ~1 ppm concentrations of oxygen.[125] The oxide skin is a critical feature for patterning liquid metals. The skin forms instantaneously, thus allowing the metal to adhere to the surface where it is deposited and adopt stable, non-equilibrium shapes[126] against the destabilizing effects of gravity and high surface tension (~10 × that of water).[114] An example of the such non-spherical shapes is as shown in Figure 8.6a[127] where cones

**TABLE 8.2**

Comparing the Physical Properties of Different Gallium-Based Liquid Metal Alloys with Water at Room Temperature[103,109,113–123]

| Physical Property | Gallium | EGaIn | Galinstan | Water |
|---|---|---|---|---|
| Melting point (°C) | 29.8 | 15.5 | 10.7 | 0 |
| Boiling point (°C) | 2402 | 2000 | >1300 | 100 |
| Density (gm/cm³) | 5.91 | 6.25 | 6.44 | 1 |
| Viscosity (mPa.sec) | 1.969 | 1.99 | 2.09 | 1 |
| Surface tension (mN/m) | 750 | 632 | 534.6 | 72.8 |
| Thermal conductivity (W/m.K) | 30.54 | 26.43 | 25.41 | 0.6 |
| Electrical resistivity (μΩ.cm) | 27.2 | 29.4 | 30.3 | $20 \times 10^8$ |

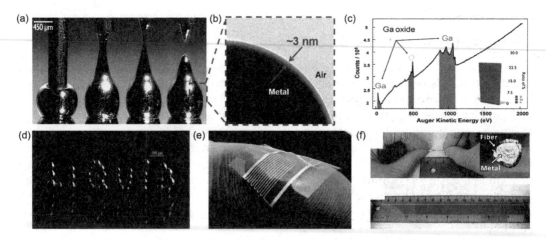

**FIGURE 8.6**
The importance of the oxide skin in gallium-based liquid metal alloys. (a) A series of photographs of the formation of a conical tip of EGaIn being manipulated using a syringe into contact with the bare, reflective surface of a silver film.[127] (b) TEM images of EGaIn nanoparticles synthesized in dodecane using sonication with an inset of the thickness measurement of the oxide skin. (Courtesy: Prof. Mohammed G. Mohammed) (c) Auger spectroscopy peaks of EGaIn surface oxide showing the composition of the skin is majorly dominated by gallium oxide.[113] (d) Image of the word "LIQUID" 3D printed using free-standing, out-of-plane liquid metal microdroplets dispensed from a 10-μm needle.[128] (e) Image of a biphasic traces created with liquid metal deposited onto solid gold trace that can be encased in soft elastomer to form stretchable interconnects.[136] (f) Images of an ultra-stretchable thermoplastic elastomer fiber with a hollow core filled with liquid metal that can be stretched to extreme values of strain while maintaining metallic conductivity.[83]

of liquid metal are formed using a stainless steel needle tip and Figure 8.6d[128] where multiple free-standing beads of the liquid metal EGaIn are stacked in 3D using a 10-μm glass capillary needle tip. The native oxide is very thin ~0.7-nm thick in a controlled environment under vacuum[129] and ~3 nm or thicker at ambient laboratory[130,131] conditions as seen in Figure 8.6b, which shows transmission electron microscope (TEM) images of the EGaIn nanoparticles synthesized in dodecane using sonication for oxide thickness measurements. The oxide passivates and does not grow significantly thicker over time in dry air in the absence of a driving force or physical perturbation.[129,132] The thin oxide 'skin' is composed primarily of gallium oxides as shown in Figure 8.6c.[113] The Auger spectroscopy analysis of the oxide exhibits that the amount of indium is very low as shown by the peaks in ambient conditions, due to the fact that indium oxidizes slowly unlike gallium.[133] Although the most stable form of the oxide is β-$Ga_2O_3$ (a wide-bandgap semiconductor),[134] the crystallinity of the skin is yet to be characterized (one study speculates it is amorphous or poorly crystallized).[129] In the presence of moisture, the gallium oxide can change to gallium oxide monohydroxide that is mechanically weaker and less passivating than gallium oxide.[135] Because these alloys are liquid, they are intrinsically stretchable as shown in Figure 8.6e and f;[83,136] for example, a liquid metal wire (a few millimeters in dimeter) on a stretchable substrate has been stretched 'down' to ~10 μm while maintaining electrical functionality.[137] Although other strategies exist to form stretchable metallic conductors, this magnitude of stretchability is not possible with other types of stretchable conductors. Because these conductors are liquid, they are also 'virtually soft' and, thus, form more conformal electrical interfaces. In addition, the use of such 'soft' conductors makes them more suitable for applications that interface with living tissues, such as skin and the brain, which are inherently soft.

In addition, Ga alloys may be useful in electronics due to their high thermal conductivity; as a result, these liquids can dissipate heat generated from integrated circuits and other components that produce heat (such as lasers). These alloys are advantageous for thermal applications because they have a negligible vapor pressure and do not evaporate or boil at elevated temperatures. Liquid metals are also known to supercool,[138] meaning that they can remain liquid below their freezing points. As a result, the use of these materials in thermal applications can be interesting because they can expand the operating range of devices to harsh climates (both hot and cold environments).

The mechanical strength of the oxide skin allows us to pattern monolithic architectures seen in Figure 8.6d. The same oxide film can be removed using aqueous acid (e.g., HCl) at pH < 3 as shown in Figure 8.7[139] or base (e.g., NaOH) at pH > 10, according to the Pourbaix diagram.[140] Once the oxide skin is removed, the bare liquid metal, that has a large surface tension, beads up and dewets most surfaces. The same oxide skin can also be removed electrochemically.[141] The ability to deposit and remove the oxide electrochemically offers new opportunities to reconfigure the shape of the metal using low voltages.[142]

To understand the role of the oxide skin better, a parallel-plate rheometer has been used in the past to quantify the mechanical properties of the oxide skin that forms on the surface of the liquid metal. EGaIn is utilized for this experiment as it is a liquid at room temperature and a less complex binary alloy system than galinstan. The rheometer instrument consists of two disks that sandwich the metal as shown in Figure 8.8a in an 'oreo-cookie' configuration where the bottom disk remains stationary while the top disk rotates back and forth with the desired applied torque as shown in Figure 8.8b.[143]

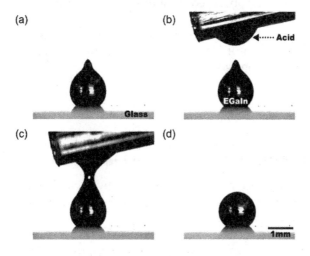

**FIGURE 8.7**
Removal of the surface oxide skin on gallium-based liquid metal alloys. (a) Image of a liquid metal droplet in the shape of a cone stabilized by the surface oxide.[139] (b-c) Acid (1 M HCl) removes the oxide. In the absence of the oxide, the metal beads up due to the large surface tension of the bare metal.

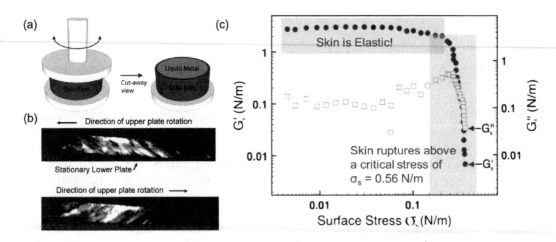

**FIGURE 8.8**
Rheological characterization of gallium-based liquid metal alloys. (a) Illustration showing the 'oreo-cookie' configuration of liquid metal with the oxide skin between the top and the bottom rheometer plate. (b) Side view images of EGaIn buckling in the rheometer showing the top rotating and the bottom stationary plate, the oxide is stretched as the rheometer rotates left and right.[143] (c) A plot of viscous and elastic modulus of the oxide skin as a function of the surface stress.[113]

The measurements show that the oxide skin is elastic and its yield stress dominates the mechanical response to the applied stress. Figure 8.8c[113] plots the elastic modulus (G') and the viscous modulus (G") of EGaIn as a function of the surface stress ($\sigma_s$). The plot shows that the skin is elastic and holds a constant value of G' until it reaches a critical (yield) stress of ~560 mN/m. Below this stress, the oxide is stable and the metal will not flow; but at higher stresses, the oxide ruptures and the metal flows like a true liquid.[113] Similar behaviors have been observed for gallium and galinstan using a Du Noüy ring setup.[144,379]

Section 8.6 briefly discusses the different patterning techniques available to manipulate liquid metals at room temperature that defines its wide range of applications in the field of microfluidics and soft electronics.

## 8.6 Techniques to Pattern Gallium-Based Liquid Metals

Liquid metals can be patterned at room temperature by taking advantage of the mechanically stiff yet elastic oxide skin. There are multiple ways to pattern liquid metals[128] and they can be divided into four categories:

1. **Lithography-assisted:** Conventional photolithography is utilized to create molds, stencils, and other guiding elements for patterning liquid metals into predetermined features, geometries, and traces.

2. **Injection:** Liquid metal is injected into hollow cavities (fibers, microfluidic channels). Pattern dimensions are predetermined based on shape and size of the cavity.

3. **Additive:** Liquid metal is selectively deposited at the desired location, often using the same equipment or techniques as other additive manufacturing techniques.

4. **Subtractive:** Liquid metal is selectively removed to form the desired pattern.

We note that many of the examples presented in this section may belong to multiple categories. Thus, these categories primarily serve for the sake of organization. In addition, any given patterning technique may be enabled by one or several phenomena or behaviors, such as adhesion or wetting, surface tension, etc. We will briefly review the different patterning techniques, highlighting the applications in the field of microfluidics and soft electronics.

### 8.6.1 Lithography-Assisted

Conventional microfabrication techniques for patterning metals typically involve depositing thin films of solid metals that are subsequently patterned by lithography and etching. It is challenging to create thin, uniform films of the liquid metal. In addition, etching removes the oxide and causes the metal to flow due to capillary forces.[145] Although photolithography is poorly suited for direct-patterning of liquid metals, researchers in the past have used masks such as molds, stamps, and stencils using lithography to fabricate liquid metal electronics.

There are different techniques used in lithography-assisted patterning such as imprinting thin films of liquid metals using a silicone mold,[146] stencil lithography to enable the spreading of liquid metals on a receiving substrate,[147–149] selective wetting on pre-patterned substrates,[150] and lift-off where the liquid metal is spread onto substrates that are covered by photoresist and then dissolving away the photoresist to produce high resolution features on the order of tens of microns using conventional photolithography processes.[151] Similar to conventional photolithography, it is also possible to deposit gallium via sputter-coating or e-beam evaporation; a stencil or mask can be utilized to determine the pattern. Recently, this approach has been utilized to create stretchable conductors by coating gallium (Ga) onto gold (Au); the two metals alloy to form a biphasic-stable (liquid/solid) pattern that remains conductive upon stretching, flexing, or deforming out of plane (i.e., twist or torsion).[136,152] A recent work has also shown similar biphasic alloy formation between gallium-indium (GaIn) and nickel (Ni).[153] In addition, the liquid metal traces can serve as templates for other metals that can be deposited via electroplating or galvanic replacement.[154,155] This method allows other metals (that would still be liquid at room temperature) to be deposited without the need for vacuum-based processes or at high temperatures. Likewise, this approach may be utilized to form electrically conductive encapsulating layers or even patterned traces of intermetallic compounds.

Unlike conventional photolithography processes, the ability of a metal to be a liquid at room temperature enables new patterning methods (assisted by lithography to determine the pattern and feature size). For example, Figure 8.9c illustrates an approach where photolithography is utilized to pattern a template feature of Au. Next, the liquid metal selectively wets the Au as it is spread over the substrate. Thereafter, the liquid metal feature is encapsulated to finalize the patterning process. Similarly, it is possible to form 3D liquid metal structures by assembling micro-droplets of liquid metal with the aid of electric forces, specifically dielectrophoresis.[156] This approach requires photolithography to make a patterned electrode pad, determining the lateral shape of liquid metal.

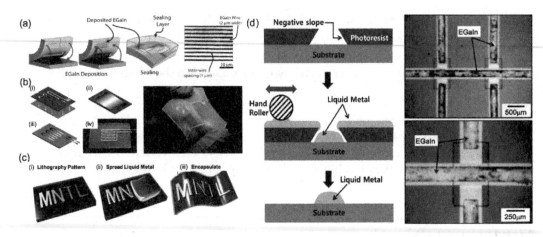

**FIGURE 8.9**
Lithography-assisted methods for patterning of liquid metals. (a) Schematic for imprinting-assisted lithography (left) and a microscopic image of the high-resolution liquid metal traces patterned (right).[146] (b) Fabrication of a soft circuit with stencil lithography.[149] (i) A stencil is produced by laser engraving a thin-film mask and placed on an elastomer. (ii) Galinstan is spread over the stencil. (iii) The stencil is removed and copper wires are inserted before sealing. (iv) Unsealed galinstan heater on an elastomer produced with stencil lithography. (right) Image of the galinstan heater being stretched after being embedded in a soft silicone elastomer. (c) Fabrication of selectively plated stretchable liquid metal wires for transparent electronics.[162] (i) Au is patterned onto a substrate using photolithography techniques. (ii) Thereafter, liquid metal is spread onto the substrate and selectively wets the patterned Au. (iii) Finally, the substrate is encapsulated and delaminated from the substrate to complete the patterning process. (d) Schematic illustration of the lift-off patterning process (left) and the liquid metal patterns obtained (right) that are aligned to prefabricated micropatterns.[151]

Moreover, this approach is interesting because it is possible to create 2D or 3D 'monoliths' of liquid metal from droplets with the aid of external forces or stimuli. Alternatively, one can create molds (inverse of desired shape and pattern) of silicones and then press the mold against a puddle of liquid metal; the metal fills the ingress of the molds to form the feature.[146,157,158] One drawback to this approach is that it is generally difficult to pattern liquid metals as interconnects between other components. To overcome this challenge, one can utilize photolithography to create stencils. The liquid metal is then sprayed or spread over these stencils to form a desired pattern.[150,159–161] As a result, the metal can be selectively deposited without contacting or damaging other areas of the substrate.[162,163] To date, stencils have been utilized to create patterns with pitch resolutions of ~20 μm.[164] Some of these examples are shown in Figure 8.9.[146,149,151,162]

In general, the biggest advantage of liquid metals is the ability to deposit metal outside clean-room settings, without the need for vacuum-based processes, and at room temperature (i.e., no thermal sintering or curing steps required). Furthermore, lithography-enabled patterning is useful for creating liquid metal patterns where the surface is exposed that could be utilized as soft and conformal electrodes or encapsulated if used as a conductive trace. In most cases, the resolution of pattern is based on the resolution or limitation of the lithography technique. To date, traces with resolutions down to three to five microns have been achieved using lithography-enabled techniques.[162,165] At this length-scale, it is possible to form features that are visually imperceptible to the naked eye; thus, one can form stretchable and transparent conductors when patterned onto transparent substrates (such as elastomers or gels). Nevertheless, there remain opportunities to form submicron resolution features using lithography tools.

Finally, this approach overlaps with other approaches highlighted in this section. For example, in some cases, the metal is forced into ingresses and cavities, much like injection-based approaches. In addition, many of the methods rely on the adhesion or wetting of the oxide-skin of the liquid metal, which may be important for all three other approaches. Likewise, additive approaches can also form exposed electrical contacts.

### 8.6.2 Injection

The easiest way to pattern the metal precisely on the sub-mm length scale is to inject it into microfluidic channels. Due to its low viscosity, liquid metal can readily fill channels, assuming a sufficient pressure is applied to yield the oxide layer to fabricate soft reconfigurable antennas.[10,113] The pressure required for filling microchannels scales inversely with the diameter of the channels due to the need to yield the oxide.[113] Atmospheric pressure (~15 psi) is sufficient to inject the metal into channels with diameters as small as 10 μm using a handheld syringe. The metal maintains its shape in the microchannel due to the oxide skin that mechanically stabilizes it. The use of posts ('Laplace barriers') can direct the metal into complex channels[166] and it is also possible to fill multilayered channels.[167] Liquid metals have been injected into capillaries as small as 150 nm using large pressures.[168] By injecting the metal, the feature sizes are predefined by the size of the void spaces; as a result, this approach can create patterns from macro-scale down to nanoscale.[169,170] In addition, the metal is exposed at the injection point (and potentially in other locations) that allows the pattern to be electrically addressable. Similarly, one may backfill the metal into a substrate, which is in contact with a nonwetting surface. After injection, the substrate can be removed to form 'open-to-air' and electrically addressable patterns.[171] These patterns can be transferred onto other surfaces, including curved surfaces (similar to a stamp).[172–174]

However, injecting the metal by hand can lead to void-spaces at sharp corners and turns. Recently, this challenge was overcome by utilizing vacuum to drive the metal into void spaces.[175] Unlike hand-injection technique, this method does not require an outlet hole or terminal to displace the gas or fluid within the cavity. A major advantage with this approach is the ability to completely fill void-spaces, as shown in Figure 8.10c.[175] Furthermore, metal can be injected by vacuum even if the encapsulating substrate is not porous.[176]

The limiting factor with injection-based patterning is that the applied pressure must overcome the surface tension of the metal as well as the yield stress of the oxide skin.[113] The energetic barrier to infiltrate the liquid metal can be lowered by simply prefilling the void space with an aqueous solution.[135] The presence of water weakens mechanical properties of the surface oxide and, thus, allows the metal to be injected with less pressure.[135] In addition, applying an oxidative voltage across an aqueous electrolyte and liquid metal can lower the interfacial tension to nearly zero, allowing the metal to be infiltrated spontaneously.[141] Some of the soft electronics and multilayered microfluidics fabricated using this method are shown from literature in Figure 8.10.[10,15,166,175]

### 8.6.3 Additive

The advantage of using additive patterning processes is to directly deposit the liquid metal in the desired areas. As a result, there is virtually no material waste generated or loss of material from the substrate during the process. Inkjet printing is a conventional additive

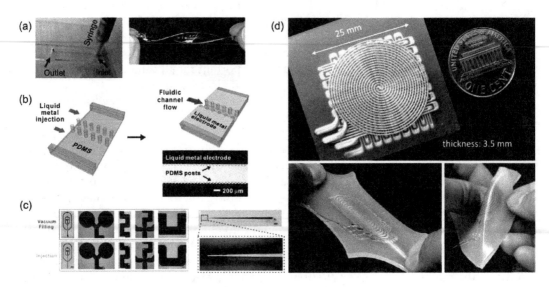

**FIGURE 8.10**

Injection methods for patterning of liquid metals. (a) Injecting liquid metal into microchannels using a hand-held syringe (left) is a simple method to create well-defined and high-resolution structures along with a photograph (right) showing a reconfigurable dipole antenna.[10] (b) Inherently aligned liquid metal electrodes in microchannels.[166] Cut-away depiction of the process used to fabricate liquid metal microelectrodes in contact with a central fluidic channel. Two parallel rows of posts separate the liquid metal electrodes from the fluidic channel. Top-down, backlit optical micrograph of the fluidic channel flanked by two liquid metal electrodes. The posts prevent the metal from entering the fluidic channel. (c) Comparison of vacuum filling and injection (the latter, into microchannels with an outlet located at the top of the image). Photograph (left, scale bar 3 mm) and optical micrographs (right, scale bar 200 μm). A linear, tapered microfluidic channel and its inset taken by top-down optical microscopy (so the metal appears shiny), scale bars 1 mm (main image) and 10 μm (zoomed in image), respectively.[175] (d) Soft artificial skin prototype containing multilayered microchannels filled with injected liquid metal inside an elastomeric matrix.[15]

patterning technique, yet it is difficult to use conventional inkjet printing to pattern liquid metal due to the surface oxide and large surface tension. It is, however, possible to inkjet print liquid metals in the form of colloidal solutions that can be sintered mechanically at room temperature[177] or by using an acidic environment to break the liquid metal stream into droplets based on the principle of Rayleigh-Plateau instability.[178,179] The adhesion of the oxide to the substrate is the most crucial factor in enabling additive methods. The presence of the surface oxide allows the metal to be 3D printed, both out-of-plane[126] and in-plane[180,181] via direct-writing with features stabilized by the oxide. The smallest features using this approach are ~100 μm.

The simplest approach is to simply spread or paint the metal onto the substrate using either a draw bar, paint brush, or a roller. It is also possible to transfer liquid metal using microcontact printing[149] (by contacting liquid metal with an elastomeric stamp and transferring it to a targets substrate) or by loading it in a ball-point pen.[182] This can form smooth films of the metal with a native surface oxide in a single step. Because the surface oxide is thin, it acts as a dielectric material between the metal and surrounding environment and, thus, exhibits capacitive behavior.[183] Accordingly, this approach forms electrodes with a self-healing or regenerating dielectric coating in a single step which can be useful for capacitive sensors as well as applications that utilize electrowetting-on-a-dielectric. Further resolution can be improved by selectively wetting the metal using either rough surfaces or solvents and wet surfaces that prevent adhesion.[171,172,184–188] In general, these techniques are versatile because they can be utilized as stamps to transfer the metal onto other substrates. Likewise, this technique is also useful for creating surfaces bearing only the native oxide skin; the oxide can serve as a template for other materials, such as chalcogenides and 2D materials.[189–193]

Much like other liquids and inks, it is possible to spray-coat liquid metal to directly pattern it.[160] This approach is simplistic in nature and works well for creating conductive traces or films that are thin. The use of carrier fluids, such as organic solvents, can improve this process. However, it is limited with patterning resolution and accuracy (i.e., line-width-roughness). The use of stencils can greatly improve the patterning resolution of spray-coated liquid metals, though such a method would not be strictly 'additive' in fashion. Alternatively, one can spray or cast a dispersion of liquid metal (in a solvent or 'oligomerized' ink) into a smooth film. As processed, the film itself is not electrically conductive but does exhibit high thermal conductivity; the latter property may be useful for thermal management applications. This film can be rendered electrically conductive by sintering the liquid metal particles via mechanical pressure[177,194,195] or laser sintering.[196–198]

Among additive approaches, direct-write techniques are perhaps the most appealing for their accuracy and the ability to form complex shapes, including 3D and out-of-plane structures.[126,180,199–204,380,381] To date, it is possible to pattern liquid metal lines to ~2 µm widths using direct-write techniques, making direct-writing one of the most accurate methods.[199] This method is often referred to as direct-ink-write (DIW), 3D printing (3DP), as well as 2D plotting (when limited to planar patterns). In general, this approach involves extruding/dispensing liquid metal using a syringe needle (via pneumatic or volumetric displacement) onto a substrate; where the nozzle-head or underlying stage can be moved by motion stages, computer numerically controlled (CNC) routers, or other motorized equipment. A computer-aided design (CAD) model can be used to generate a pattern. The resolution of printed parts is dependent on a variety of parameters, such as nozzle diameter, stage velocity, distance between needle and substrate, applied pressure, and the environmental conditions. In addition, chemical properties of the liquid, namely viscosity and surface tension, are some important parameters to consider when printing liquid metals. When combined, these parameters influence whether drops, jets, or fibers are extruded from the nozzle, ultimately impacting the shape, pattern, and resolution of the printed structures. Moreover, it is possible to perform multi-material printing by extruding polymer inks containing liquid metal particles or by utilizing co-axial needles that continuously dispense the polymer-coated (outside needle) liquid metal (inside needle) fibers.[205–207] Most recently, a multi-material printing platform has been exploited to create shape-changing lattice structures of elastomers with innervated liquid metal; as a result, these composite lattices behave as reconfigurable antennas in response to temperature.[208]

The mechanical properties of the oxide play an important role for direct-write techniques. The oxide provides mechanical support to the liquid metal structures; accordingly, it is possible to print liquid metals into 3D and nonequilibrium shapes. Such structures are not possible with conventional inks (composed of water or organic solvents) for direct-write techniques. Because the oxide skin exhibits a yield stress, it is possible to print the metal by applying shear forces to the surface (i.e., without any applied pressure); either moving the stage or nozzle-head can apply shear forces to extrude the liquid metal. Finally, the wetting or adhesion of the oxide substrate plays a crucial role in determining whether liquid metal will print onto a given substrate. Recent work has demonstrated tuning adhesion of the oxide based on the velocity of the motion stages to pattern liquid metal lines then partially lift-off to form out-of-plane structures such as loops.[199] Similar to direct-write techniques, micro-contact printing can pattern liquid metal features without need for extrusion.[149,209] This approach utilizes a PDMS tip coated with liquid metal that comes in contact with a substrate; the adhesion of the oxide onto the substrate to enable the printing. Similarly, one can fill roller-ball pens to directly pattern the metal by hand.[182]

Examples of additively patterned liquid metal soft electronics are shown in Figure 8.11.[54,177,180,199]

Finally, there are several important considerations for utilizing additive patterning methods for liquid metals for soft electronics.[54,210] Firstly, the adhesion of the liquid metal oxide to a substrate plays a critical role in determining its printability. As such, the adhesion between oxide and silicone surface – the common materials for soft electronics,[211,212] is still not fully understood. There remain opportunities to improve direct-write processes of liquid metals for soft electronics. Next, gallium tends to reactively wet or alloy with other metals, which can lead to embrittlement[188,213] and increased contact resistance.[214] Gallium/gold (Ga/Au) biphasic alloys have also been shown to have increased skin-migration currents that can be problematic for interconnects; this challenge can be mitigated by using alternating currents.[215] This behavior poses a great risk for utilizing liquid metals in soft electronics. A recent work has shown that graphene multilayers and other conductive carbon materials prove to be promising conductive barrier coatings for electrical contacts.[204,216] Finally, the ability to print the oxide alone is an emerging topic because it can be a template for chalcogenides, 2D materials, and high-speed and soft transistors.[189–193]

### 8.6.4 Subtractive

Unlike additive patterning, subtractive techniques are those that selectively remove metal from a film to leave behind a pattern of metal. One of the methods is to laser ablate films of liquid-metal films in an elastomer.[217,218] To date, this approach has realized patterns with 5 μm resolution of line widths.[219] At this length scale, the metal lines are visibly imperceptible; accordingly, pattering liquid metals onto a transparent substrate (i.e., elastomer or gels) renders a composite material that is stretchable, conductive, and fully transparent. It is also possible to selectively remove the metal from complex microchannels or surfaces[145] by electrochemical reduction of the oxide.[141,220] After spreading a metal film on a surface, it is possible to induce localized and temporary capillary withdrawal of the metal by electrochemically reducing the oxide.[220] This technique is termed 'recapillarity' due to the use of reductive potentials to induce capillary behavior. Applying a reducing potential to an electrode (e.g. a wire) with a drop of water in contact with the surface of the film removes the oxide layer and allows the metal to flow. With both laser and electrochemical approaches, the metal recedes into the surrounding regions; thus, the total mass of the metal is maintained when transitioning from a film to patterned feature. Similarly, it may be possible to remove the metal using chemical treatments that etch the oxide, such as acids or bases. Although it is a low resolution (~ mm) technique, it offers a simple 'subtractive' way to induce the metal to flow locally and thereby relocate to other parts of the film without any net loss of the metal in the process. Examples of such subtractive liquid metal patterning processes are shown in Figure 8.12.[217,219,220]

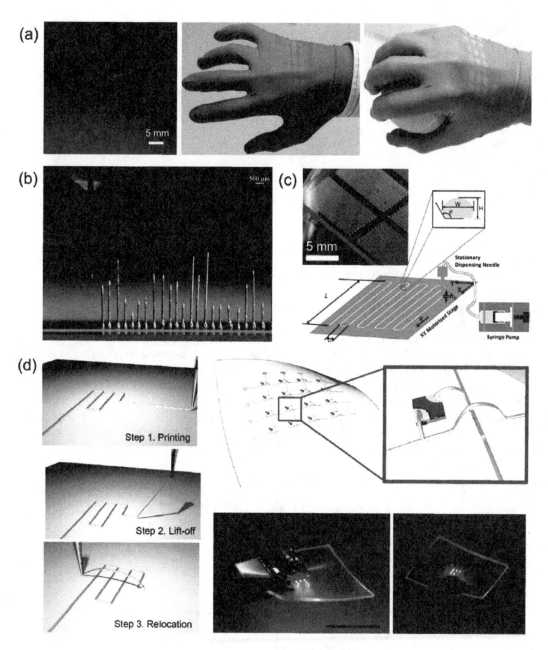

**FIGURE 8.11**

Additive patterning of liquid metal alloys. (a) Photographs of (left) an inkjet system printing EGaIn nanoparticles dispersion; (center) human hand wearing inkjet functionalized nitrile glove with arrays of strain gauges, intricate wiring, and contact pads; and (right) same hand holding a tennis ball, demonstrating stretchability of the electronics.[177] (b) A series of free-standing wires 3D printed by a syringe being withdrawn from a substrate while simultaneously extruding liquid metal. Applying vacuum terminates the wires on demand.[54] (c) Photograph of strain gauge devices on glass and the schematic of a direct-writing system patterning a serpentine pattern of length $L$, center-to-center line spacing $p$ with a writing speed $v$, flow rate $Q$, and a needle standoff distance $h_0$ with a cross-sectional view of a written trace of width $W$, height $H$, and contact angle $\theta$.[180] (d) Schematic illustrations of each step of printing and reconfiguration process (left), microLED array with reconfigured 3D interconnects (top right), and photographs of light emission of the microLED array. Scale bars, 1 cm (bottom right).[199]

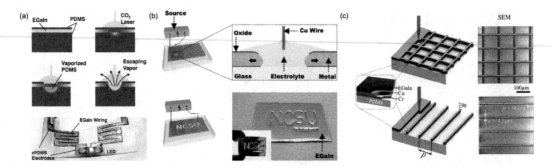

**FIGURE 8.12**
Subtractive patterning of liquid metal alloys. (a) Proposed mechanism for a direct-laser-ablation process which creates traces of liquid metals and other soft conductors in a rapid and inexpensive fashion.[217] (b) Illustration of the recapillarity process where the application of a voltage induces capillary withdrawal of the metal via local-ized reduction of the oxide.[220] (c) The schematic and SEM images of the square grid and parallel line patterns of 4.5-μm-wide EGaIn traces; the liquid metal patterns are patterned by microscale laser ablation of the thin-film architecture (liquid metal/copper/chromium) on a PDMS substrate.[219]

As we have seen in the different liquid metal patterning techniques, although the oxide may be a nuisance in some cases, its properties can be harnessed to enable patterning of liquid metals. For example, the adhesion of the oxide may be problematic for handling the metal; however, selective wetting of the metal oxide can be used to pattern the metal. In other cases, the wetting can stabilize the metal into patterned shapes, including non-equilibrium shapes and free-standing 3D structures. Therefore, one may alternatively categorize patterning methods based on whether the oxide guides the pattern by wet-ting or stabilizes the pattern due to its mechanical properties. This approach is shown in Figure 8.13. For further details on patterning techniques, we refer the reader to several excellent review papers.[221–224]

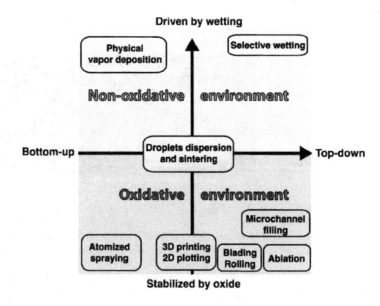

**FIGURE 8.13**
Figure showing the classification of different liquid metal patterning methods signifying the importance of presence or absence of the surface oxide skin.[221]

## 8.7 Applications of Gallium-Based Liquid Metals in Soft Electronics

There are hundreds of papers in the literature that exhibit different applications of liquid metals. This section provides the reader with some of examples under different categories, namely, (i) stretchable interconnects, antennas, and self-healing conductors, (ii) soft sensors, (iii) soft composite devices, and (iv) soft and reconfigurable devices.

### 8.7.1 Stretchable Interconnects, Antennas, and Self-Healing Conductors

Liquid metals are intrinsically stretchable with their mechanical properties determined by the embedding material and, hence, have been applied to fabricate stretchable wires that can be used as interconnects. Most examples in literature use crosslinked PDMS as the embedding polymer due to its biocompatibility and widespread use in the microfluidics community. However, PDMS fails at about 100–200% strain. It is however possible to improve the mechanical properties by microstructuring the PDMS[82] as shown in Figure 8.14a or using other commercial silicone polymers. In addition, we can use thermoplastic elastomers that can be melt-processed such as poly(styrene-*b*- (ethylene-co-butylene)-*b*-styrene) or gels that can allow creation of fibers[83,225,226] and microchannels[227] injected with liquid metals allowing them to maintain their metallic conductivity while stretching devices to extreme strains as high as 600–700% as shown in Figure 8.14b. These fibers may find applications in electronic textiles or stretchable wiring. The devices can be fabricated by first patterning the liquid metal and then using 'pick-and-place' techniques to arrange

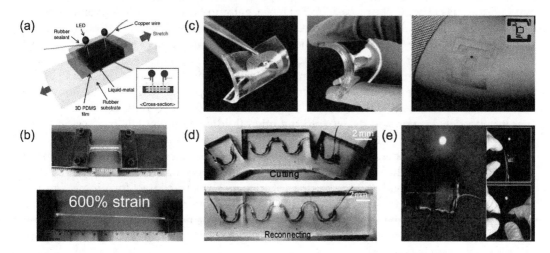

**FIGURE 8.14**
Examples of liquid metal-based stretchable interconnects, antennas, and self-healing electronics. (a) Illustration of a hyper-elastic nanostructured elastomer infiltrated with liquid metal, stretchable up to ~220% strain.[82] (b) Images of stretching a liquid metal injected microchannel molded with a tough thermoplastic elastomer showing elongations at break as high as ~600%.[227] (c) Images of a deformable coil (left) for wireless power transfer,[245] an elastomer patch antenna (center),[248] and a tattoo of an RFID antenna (right) made with liquid metal and a chip in the center placed on an arm.[161] (d) Physical separation of an LED-integrated galinstan circuit (top) with a blade and (bottom) reconnection of the circuit.[271] (e) Top-down image of the final reconnected circuit with a self-healing stretchable wire with the insets showing the (top right) disconnection and the (bottom right) reconnection steps.[272]

the electrical circuit elements at their corresponding location.[161,228] In other cases, we pattern liquid metals using injection,[229] direct-writing,[126] and dipping/plunging[194] to make contact with the circuit elements on the surface or inside a molded microfluidic channel without the need for any soldering as needed on conventional circuit boards. Gallium and its alloys have also been used as conductors for unconventional electronic applications, including, bonding for flip chips,[230] electrodes for fiber LEDs,[231] thermoelectrics,[232–234] and acoustic devices (for coupling).[235] It is also possible to form interconnects for interfacing with external rigid electronic components such as sensors,[236] integrated circuits,[237] and super capacitors.[238] Finally, liquid metals can also be utilized as vias for 'reworkable' electronics[239] and to interface with other stretchable conductors.[35,240,241]

In addition to interconnects and wires, liquid metals have also been applied to make efficient antennas.[242,243] The spectral properties of an antenna are a function of its shape and, hence, liquid metals are a promising candidate here due to them being physically deformable, allowing different shapes to fabricate frequency tunable and reconfigurable antennas.[10] In most examples shown in literature, liquid metals have been injected into prefabricated microchannels to make different types of antennas such as dipoles as shown in Figure 8.2c,[10] loops,[244] coils, and inductors as exhibited in Figure 8.14c (left),[245–247] patch antennas as demonstrated in Figure 8.14c (center),[248,249] radio-frequency antennas as shown in Figure 8.14c (right),[250] spherical caps,[251] phase-shifting coaxial transmission lines,[252] monopoles,[253] reconfigurable antennas,[254–260] and filters.[261] Conventionally, copper is the most popular material for fabricating antennas due to its high electrical conductivity and relatively low cost. For some antenna geometries, such as a dipole, there is minimal degradation of radiation efficiency, between gallium-based liquid metal alloys and copper allowing efficiencies greater than 90%.[10] However, for patch antennas, the efficiency numbers are lower with liquid metals when compared to copper, since liquid metals need to be encapsulated within dielectrics that are lossy.[245,248] Hence, gallium-based liquid metal antennas are only built for additional benefits such as complex metal bodies, flexibility, stretchability, or reconfigurability of the final antenna element. In addition to antennas, liquid metals can also be used for fabricating mirrors,[262] diffraction gratings,[263] frequency selective surfaces,[264] reconfigurable plasmonics,[265,266] tunable metamaterials,[267,268] and radiation shielding.[269]

Liquid metals are also known for their self-healing properties[10,270,271] as shown in Figure 8.14d.[271] The self-healing occurs as the metal oxidizes so rapidly when it is cut that it neither leaks, nor withdraws into the microfluidic channels. The oxide skin holds the metal flush against the disconnected interface, similar to how a scab forms over a cut on the human body. When the physically disconnected surfaces are brought in contact together, the metal self-heals electrically. It is not fully known as to what happens at the interface, but it is proposed that the oxide ruptures upon the re-contact forming a continuous metal bridge across the interface. In addition, if the polymer is capable of self-healing, then it is possible that the liquid metal circuit also self-heals mechanically as shown in Figure 8.14e.[272] The authors guide the readers to additional literature[10,273–276] on this subject for more of such examples.

### 8.7.2 Soft Sensors

When stretched, liquid metal deforms and hence changes its shape. This change in shape can trigger a corresponding change in electrical resistance, inductance, or capacitance, which can be then translated from physical changes in the device such as sensing touch, pressure, and strain.[31,167,277–285] Some of these examples have been shown in Figure 8.15. We will review a few of them in this section.

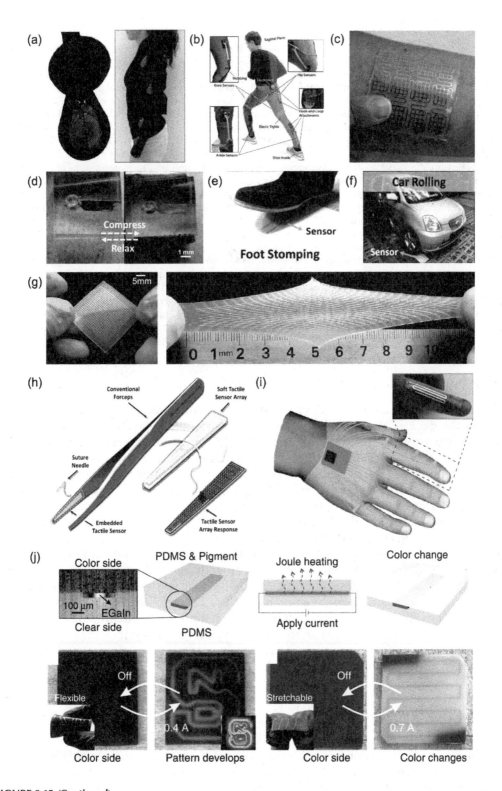

**FIGURE 8.15** *(Continued)*

**FIGURE 8.15** *(Continued)*

Examples of liquid metal-based soft sensors. (a) Images of the Polipo pressure-sensing system. Twelve pressure sensors are integrated to a wearable system such as a space suit to measure the pressure between the space suit and person's arm during dynamic movement.[292] (b) Wearable soft sensing suit for human gait measurement.[15] Soft strain sensors were placed at each lower limb joint to capture motion in the sagittal plane. In the case of the knee and ankle sensors (insets on left), webbing directed motion over the joint. (c) Thin sheets of PDMS silicone rubber embedded with EGaIn microchannels for pressure sensing applications.[290] A 12-key keypad resting on a wrist. The channels are 20 µm in height and 200 µm in width. The entire device thickness is approximately 700 µm. (d) Images of an oxygen plasma-treated PDMS microchannel that buckles under compression creating optical diffraction patterns.[263] (e, f) Images of sensitive yet durable liquid metal-based microfluidic tactile sensors for pressure differentiation and measurement that can survive activities like foot stomping and car rolling over them.[293] (g) Images of a prototype sensor (left) made with a 23 × 23 microchannel array with filament diameters of 200 µm and thickness of 1 mm. The thin sensing array yields high stretchability (right), well over 400%.[294] (h) Conceptual drawing of a soft tactile sensor array mounted on conventional forceps and responding to forces used to grasp a needle.[15] (i) Elastomer-based curvature sensors allow mechanically noninvasive measurement of human body motion and robot kinematics. An illustration of a human hand covered with curvature sensors at every joint with the inset showing an actual sensor mounted on a host finger. The sensor consists of a 300-µm-thick elastomer sheet embedded with a microchannel filled with a conductive fluid.[290] (j) A platform for materials logic via pigmented silicones innervated with liquid metal. Liquid metals patterned within an elastomer can Joule heat to invoke color changes. (top) Schematic of a simple device consisting of a liquid metal circuit between two layers of PDMS. Joule heating causes color change of the pigment in the PDMS. (bottom left) Localized Joule heating changes the color from red to white, resulting in a passive display. (bottom right) A device with blue thermochromic completely changes color to white in response to electrical current.[285]

◄

### 8.7.2.1 Strain Measurement

Strain can be measured when a device is stretched, which inherently changes the resistance of the device. In case of liquid metals, for example, when a microfluidic device or a hollow elastomeric fiber is stretched,[83,286,287] the metal lengthens due to strain and thins out due to the Poisson effect; causing the resistance to increase while the resistivity of the metal remains constant. The change in resistance can be calibrated to the strain incurred.[282,288] Such strain sensors have been applied for sensing human movement such as gait without restricting the natural kinematics of the movement[31] as shown in Figure 8.15b.[15] Such sensors can also be used for detecting joint motion[279,289] as seen in Figure 8.15i[290] and for detecting deformation in soft robots made with origami. These sensors are soft with a modulus of 0.1–10 MPa, yet, stretchable with strains at failure varying from 100 to 1000% up to thousands of cycles. In addition to the variations in electrical resistance, resonant frequency changes can also assist in strain measurement due to physical elongation using liquid metal antennas as described in the earlier application section.[10] The advantage of using liquid metals here is that it maintains a consistent surface resistance over multiple stretch/relax cycles, necessary for efficient radiation. Finally, capacitance changes due to changes in geometry during deformation can also be used to measure strain in a liquid metal device.[247]

### 8.7.2.2 Pressure and Touch Detection

When a liquid metal device embedded inside an elastomer is deformed, the changes in resistance or capacitance due to compression of the liquid metal can be correlated to pressure and touch detection. As seen in Figure 8.10d, an electronic skin made with liquid metal embedded inside a soft polymer can detect touch,[167] other examples include, spacesuits for detecting pressures as low as 5 kPa[291] as seen in Figure 8.15a,[292] tactile keyboards made from 100% soft materials,[277] as exhibited in Figure 8.15c,[290] embedded sensors in micromanipulators such as forceps,[278] as seen in Figure 8.15h.[15] These soft pressure sensors are also fairly robust and durable, as seen in Figure 8.15e–f where such sensors remain intact even when being stomped on by foot and run over by a car wheel.[293] The sensors can be made more or less sensitive using different types of microchannel geometries and elastomer used in the process. In addition to resistance-based sensors, liquid metals can also be utilized as electrodes for soft capacitors in both self and mutual capacitive modes.[294] For examples, as seen in Figure 8.15g, an array of ~200-μm liquid metal wires embedded inside 1-mm-thin PDMS can create a tactile sensor array based on capacitance that can be stretched all the way up to 400% without loss of sensitivity.[294,295] These capacitive sensor arrays can combine with resistance sensors for multimodal sensing in the same device (i.e., strain and touch). Recently, a completely soft, stretchable silicone composite doped with thermochromic pigments and innervated with liquid metal, as seen in Figure 8.15j,[285] has been used for distributed decision-making. The ability to deform the liquid metal couples the geometric changes to Joule heating, thus enabling tunable thermo-mechanochromic sensing of touch and strain.

Though most examples exhibited until now use electrical measurements to sense deformation, it is possible to use optical effects to detect deformation in devices made with liquid metals. When an elastomeric microchannel made with PDMS is deformed, the walls buckle due to compressive forces, creating a soft diffraction grating as seen in Figure 8.15d.[263] The key step in the process is the plasma oxidation that is used to seal the top and bottom layers of a microchannel that creates a rigid oxide later on the PDMS channel wall – resulting into buckling under compression. Due to the optical properties of liquid metal that conforms to the PDMS channel walls, the diffraction effect is visible due to the light reflecting off the surface of the metal. The PDMS buckles diffract light from the liquid metal that can be used as color-changing surfaces or in optical sensor applications.

Finally, soft electrodes made with liquid metal can be used to sense changes in temperature, oxygen, concentration, and humidity by monitoring changes in conductivity or capacitance through ionic liquids.[296,297] Liquid metals inside carbon nanotubes (CNTs) can also be used for making tiny thermometers[170] and interfacing junctions between liquid metals and conventional metals such as copper can be harnessed to create thermocouples.[298]

### 8.7.3 Soft Composite Devices

Since gallium-based liquid metals are fluidic in nature, they need to be embedded inside soft polymers such as PDMS to fabricate the final device. In some cases, we can also mix the liquid metal particles into the elastomer matrix to make composite materials. There is extensive literature available on liquid metal composites to build soft and stretchable electronic devices.[299] The liquid metal particles can be formulated using sonication,[300] microfluidic flow focusing,[301,302] molding,[303] or shear mixing within an elastomer.[304] The advantage of physically mixing liquid metal particles inside polymer, apart from the mechanical stability of final patterns, is the enhancement of thermal conductivity of the polymers and its heat transfer coefficients as seen in Figure 8.16a and b.[305] At sufficiently high metal particle loadings, the composites as shown in Figure 8.16c,[304] can be rendered conductive by pressing onto them to 'mechanically sinter'[177] the particles to create conductive pathways as exhibited in Figure 8.16d and e.[304] Researchers have used this approach to create 'handwritten soft circuit boards' that can be made conductive on-demand by mechanically sintering the liquid metal nanoparticles mixed inside the elastomer.[194]

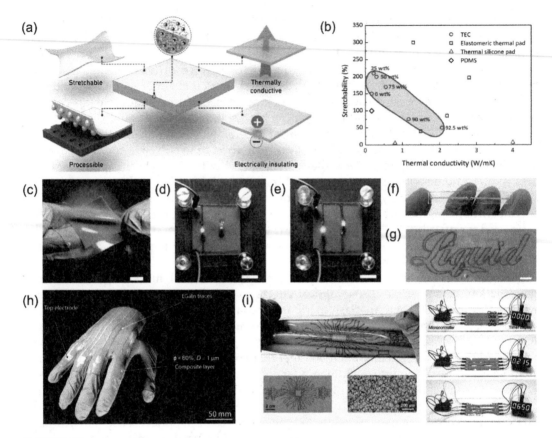

**FIGURE 8.16**
Liquid-metal particles dispersed in elastomer can be used to make soft composites. (a) Schematic illustration of the concept of a liquid alloy droplet embedded elastomer composite for high heat conduction and insulating with a soft, stretchable behavior.[305] (b) A plot showing thermal conductivity vs. maximum stretchability of thermal elastomer composites (TECs) compared to the reported works and commercialized products of electrically insulating polymer-based thermally conductive materials.[305] (c) A square sheet of the stretchable PDMS-galinstan embedded in a thin layer of PDMS. Scale bar is 10 mm. Circuits can be drawn into this plain sheet of material by using a thin tipped tool to selectively apply compression.[304] (d) An LED embedded into the nodes of a drawn circuit lights up, but a newly added LED will not turn on until more selective compression is used to draw connections to the existing circuit.[304] Scale bar is 10 mm. (e) Compression causes the material to darken, allowing the drawn conductive traces to be visible.[304] Scale bar is 10 mm. (f) A microchannel filled with a suspension of liquid-metal particles can be rendered conductive enough to form a tunable dipole antenna by pressing on the elastomer resulting into mechanical sintering.[194] (g) The particles can also be rendered conductive by exposure to a laser sintering process.[194] Scale bar is 3 mm. (h) Wearable sensing glove realized by attaching the sensors at finger joints to detect hand motion, specifically the grasping of objects.[309] (i) (left) Photograph of a liquid metal-elastomer composite being stretched and twisted with an intricate design of electrically conductive traces. The lower-left inset shows the undeformed sample and lower right inset is an optical micrograph showing the liquid metal microdroplets in the elastomer at $\phi = 50\%$.[310] (right) Example of a reconfigurable material ($\phi = 50\%$) transmitting DC power ($V_{cc}$, positive supply voltage; GND, ground) and digital communication signals to operate a counter display. As severe damage is induced, the counter maintains operation, requiring all four traces to constantly maintain electrical conductivity.[310]

Same liquid metal nanoparticles can be injected inside microfluidic channels to build antennas of a desired frequency by regulating the sintered length of the antenna as seen in Figure 8.16f.[194] Apart from mechanical sintering, lasers can also be used to merge/sinter the liquid metal particles into conductive traces. Laser sintering[196,197,306,307] gives the advantage of having locations becoming conductive at precise locations with high resolution as shown in Figure 8.16g.[194] In addition to tuning particle loading, we can also manipulate liquid metal particle sizes to enable programmable liquid metal composites for soft robotics and stretchable electronics where flexibility and tunable functional response are critical. An example of such a composite sensor is seen in Figure 8.16h[308,309] where a wearable sensing glove is physically realized by attaching the sensors at finger joints to detect hand motion, specifically, grasping of objects, which needs finer control over the movement of the fingers. Another advantage of using a liquid metal-elastomer composite is the ability to be reconfigurable and self-healable allows the final electronic device to be electromechanically stable under typical loading conditions. When the device is damaged, the liquid metal-elastomer droplets rupture to make a new connection, spontaneously, with the neighbors to reroute the electrical signals without any discontinuity or the presence of an external trigger like heat. An example of such a device is shown in Figure 8.16i, where a self-healing and reconfigurable liquid metal-elastomer composite is fabricated using an intricate design of electrically conductive liquid metal traces.[310]

Despite several applications, there is significant work that can be carried out to understand the physics behind the process of impact when liquid metal particles – having high surface tension and oxide skins – merge, on the mechanical properties of the composite formed, and the soft electronic device resulting from it.

### 8.7.4 Reconfigurable Electronics

Reconfigurability is the ability to control the shape or position of an object. Liquid metals, being viscoelastic fluids can be molded into various shapes that can serve different functions. This attribute of shape-shifting liquid metals has been applied into several applications, including reconfigurable antennas,[254–256,311,312] switches,[313] metamaterials,[314] and plasmonics.[266]

The most common technique to change the shape of liquid metals is using an electric field or modest voltages (~1 V) at room temperature.[142] It is a preferred way to shape liquid metals as it can be easily controlled and applied. The most common techniques include electrowetting on a dielectric, electrocapillarity, continuous electrowetting, and electrochemically controlled capillarity. Electrowetting on dielectric refers to a process of applying voltage between a drop of liquid metal and a dielectric-coated electrode. This results into an electrostatic force that causes the liquid metal to physically wet the substrate. Due to the presence of oxide skin on the surface, this method is of limited use. Due to the surrounding medium being insulating, it is not possible to use acid or base to remove the oxide. However, recently an acidified yet insulating oil has been proposed to counter this problem.[315] Electrocapillarity uses charges already present at the interface between the liquid metal and electrolyte to reduce the interfacial tension. The change in electrical potential results in a change in interfacial tension, causing the droplets and plugs of metal to move via continuous electrowetting.[316–320] However, electrocapillarity phenomena results in only modest changes in interfacial tension and, hence, only modest movements can be achieved. Electrochemical oxidation on the surface of the metal can significantly lower the interfacial tension resulting in transforming a spherical droplet into a pancake-like structure as seen in Figure 8.17a.[141,382] Despite the oxide skin

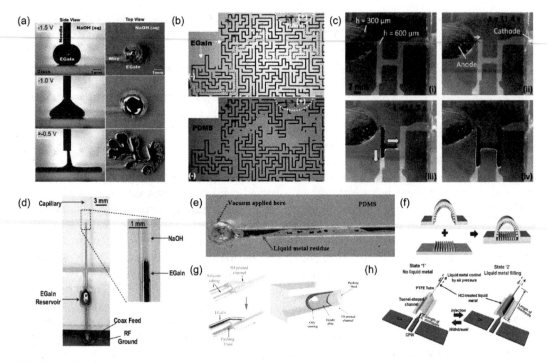

**FIGURE 8.17**
Examples of liquid metal-based reconfigurable electronics. (a) Oxidative spreading of a bead of liquid metal in 1 M NaOH solution.[141] (left) A needle serves as a top electrical contact to the droplet. (right) A wire serves as a bottom electrical contact to the droplet. (top) The drop assumes a spherical shape due to its large surface tension; (center) upon application of an oxidative potential, the metal assumes a new equilibrium shape; (bottom) above a critical potential, the metal flattens and spreads, ultimately forming fingering patterns that further increase its surface area and destabilize the metal. (b) Liquid metal in a ~100 μm resolution microfluidic maze made with soft lithography (top) can be removed selectively by electrochemically removing the oxide (bottom) that sets the metal in a state of high tension and allows it to withdraw along the electrical path.[145] (c) Liquid metal in a 300 μm high reservoir is connected to a 600 μm high target chamber in the shape of an H.[321] The liquid metal is electrically connected via an external conductor to a copper cathode on the opposite side of the chamber and the chamber itself is filled with 1 M NaOH solution. (i) When a few drops of NaOH bridge the chamber to the cathode, galvanic actuation is initiated, and the liquid metal begins flowing into the connecting channel. (ii) Once the liquid metal reaches the end of the channel, (iii–iv) its internal Laplace pressure drives it to rapidly fill the taller target chamber in order to lower its free surface energy. (d) A photograph of the antenna, feed, and reservoir of the tunable monopole antenna.[256] The inset shows a zoomed image of the EGaIn-NaOH interface under no bias (oxide skin removed). (e) Photograph showing how the oxide adhering to the surface of the ~100-μm-thin microchannels, making it difficult to use pneumatics and pump liquid metal in and out of the channels.[145] (f) Fabrication sequence of the super-lyophobic microfluidic channel.[15] PDMS-PDMS bonding to create the microfluidic channel platform. (g) The experimental setup for injecting the EGaIn plugs into the channels. (left) The length of the plugs can be easily controlled by adjusting the time and flow rate of EGaIn infusion. (right) The cross-sectional view of EGaIn plug traversing a 3D printed channel in presence of oleic acid as a pushing fluid.[326] (h) Conceptual image of liquid metal actuation-based monopole antenna.[328] Without liquid metal filling (state 1), the resonant frequency is determined by the dimensions of the monopole antenna fabricated by Cu. After the liquid metal filling (state 2), the liquid metal is injected and connected to the Cu monopole stub resulting in the frequency tuning. States 1 and 2 are reversible by controlling the air pressure to inject and withdraw the liquid metal slug into/from the channel.

acting as a physical barrier to the flow of liquid metal, in presence of a base – like aqueous NaOH, the oxide is dissolved while competing with the electrochemical deposition of the skin. It has also been shown that electrochemical reactions can be used to remove the oxide, resulting into the metal achieving a state of high interfacial tension – thus acquiring a spherical shape. This feature is used to selectively withdraw liquid metal from a maze as seen in Figure 8.17b where the liquid metal follows the path with least electrical resistance.[145] The same technique can be used to withdraw liquid metal from a thin film of the metal spread on the surface.[220] Combining the ability to increase and decrease the interfacial tension with modest electrical voltages in the presence of an electrolyte provides a scheme to pump and manipulate the shape of liquid metal as shown in Figure 8.17c.[321] A similar pump-free control version of the electrochemically controlled capillarity to fabricate a reconfigurable antenna where EGaIn forms the radiating element is exhibited in Figure 8.17d.[256]

Apart from electrically reconfiguring liquid metals, there are several examples using mechanical and pneumatic reconfigurability to move and manipulate liquid metals. Soft liquid metal devices can be reshaped mechanically using stretching to help change the function of the device. As discussed in the earlier application sections (Section 8.7), a stretchable antenna is a common example where the resonant frequency of the antenna can be altered by stretching the antenna.[10,322] Similar technique can be applied to metamaterials where liquid metal is embedded inside an elastomer matrix.[323]

Pneumatic reconfigurability refers to using pumping liquid metals through microfluidic channels and capillaries to change its shape. As mentioned before, the presence of oxide skin on the surface of gallium-based liquid metals limits the applicability here as the metal ends up sticking on the sidewalls of the channels or capillaries as it travels through the channels. Hence, there is always a residue remaining on the inner walls when a metal plug is travelling in or out of a channel as seen in Figure 8.17e. The oxide adhesion can be avoided by several means such as rough surfaces that can help prevent sticking of the oxide on surface[184,187,324] as shown in Figure 8.17f.[15] However, this technique is tough to implement on the interior surfaces of small microcapillaries.[184,188,325] Apart from rough surfaces, microchannels can also be pre-filled with a 'carrier fluid' before injection of the liquid metal as seen in Figure 8.17g.[326] This fluid creates a 'slip layer' between the liquid metal slug and the wall that prevents oxide adhesion.[135,257,326] Although, a modest repeatability of ~1000 cycles is achieved, this technique eventually fails due to mixing between the carrier fluid and liquid metal. Water can also be used as a carrier fluid since it is neither acidic or basic, especially useful for continuous electrowetting to move plugs of liquid metal inside capillaries.[135] Finally, the easiest method of all is to remove the oxide skin using an acid such as HCl in liquid or vapor state, or base, such as NaOH of a suitable strength.[327–332] The liquid metal plugs can then move without leaving any residue across the length of the microfluidic channel or capillary as shown in Figure 8.17h.[328]

One of the major concerns with liquid metal soft electronics is their interaction with solid metal contacts resulting in unwanted changes in the electrical performance and delamination of solid metal contacts due to the atomic diffusion of gallium at the liquid metal/solid interface. Recently, a solution to this problem has been proposed by implementing Laplace barriers[333] that control fluid flow and position via pressure-sensitive thresholds to facilitate physical movement of the fluids within the channels. Although gallium-based liquid metal alloys are nonresponsive to magnetic fields, it is possible to manipulate droplets magnetically by coating them with magnetic powders.[334]

In addition to these broad application categories, there are several other types of electronic components that can be fabricated using liquid metals such as memory devices (memristors),[335,336] soft electrodes,[35,337–345] diodes,[346] capacitors,[297] bio-electrodes,[347–351] energy capture/storage devices,[276,352,353] and microfluidics[166,243,347,352,354–363] that have eventually been integrated in electronic devices and showing electrical performance comparable to conventional metals like copper, commonly used in such applications.

## 8.8 Outlook: Opportunities and Challenges

Although liquid metals have the best of both worlds – high electrical conductivity at maximum strains of any electrical conductors—making it very promising for rapid prototyping of flexible and stretchable electronic devices—there are some well-known challenges and hence opportunities associated with liquid metals in the context of soft electronics.

### 8.8.1 Mechanics

Perhaps, the most important challenge with liquid metals is the mismatch in mechanical properties (or mechanical impedance) between the soft liquid metal and other harder components (elastomers, metal contacts, etc.). This mismatch may lead to an undesired build-up of mechanical strain as well as inconsistent electrical contacts. In addition, due to the softness of the metal, 3D structures of liquid metals require encapsulation to maintain their structural integrity. Finally, the ability of the liquid metal to deform easily may change its geometry and morphology over time or cause it to break apart when subjected to high-energy perturbations (i.e., ultrasonic waves, high G-forces in aerospace applications). Hence, circuits need to be designed to both tolerate these changes and embrace them for tuning or sensing.

### 8.8.2 Electrical Contacts

While clever implementations of liquid metal electronic devices avoid the need to make a rigid metal to liquid metal interface (for example, parasitic RF antennas),[326] most devices require the liquid metal to make intimate contact to the electrical leads and interface with other rigid electrical components. We can take advantage of the wetting properties of liquid metals to make electrical contact to other metals without the need to solder.[364] For example, in some of devices such as the flexible and wearable TEGs,[83,233,234,270] copper wires are inserted into the metal to make electrical leads. This technique has been reliable for simple circuit demonstrations. However, it remains a challenge to interface with integrated circuit chips and other complex microelectronic components, which are more common in the industry. The reason for this is the lack of understanding about the role of the surface oxide skin on the interfacial resistance between the liquid metal and the metal surface it contacts, in addition to what happens when two liquid metal surfaces contact each other. The oxide usually provides minimal resistance in sensitive electrical measurements of SAMs (self-assembled monolayers).[365]

However, in some other cases when interfacing with metals such as platinum, gold, or silver, prewetting of liquid metal without oxide on the surface is required to decrease the contact resistance,[233,234] whereas in other examples, we need to use electrochemical reactions, acids, bases, oxygen free environments, or reducing agents to make metal-metal contacts in absence of the oxide skin.

### 8.8.3 Toxicity

Although several studies have suggested gallium-based liquid metals have low toxicity,[105] extensive studies are needed to understand the implications of gallium and its alloys on human health, especially in bioelectronic devices to be implanted in the human body as energy harvesting devices. Since gallium has zero or negligible vapor pressure with limited solubility in water, the metal cannot enter the human body as vapor. However, since gallium is a liquid metal, it can penetrate the human body more easily than other solid metals if it leaks from the polymer encapsulating it, further signifying the importance of effectively and reliably embedding the liquid metal.

### 8.8.4 Cost

For commercial applications, there remain certain techno-economic considerations when using liquid metals. Firstly, gallium alloys are not as abundant as other bulk metals commonly available, such as copper. Thus, there may be challenges with supply chain if large quantities of liquid metals are required. Similarly, the cost of these liquid metals (~$0.25/g) is much higher than other bulk metals (i.e., Cu). Gallium is abundantly present in the earth, but it cannot be mined directly; it is an impurity often found in alumina, a precursor to aluminum. For most of soft electronic applications, the volume of liquid metal used is small enough that cost may not be an issue but if the devices go into manufacturing, the procurement costs must be accounted for. Thus, developing new techniques to synthesize liquid metals (i.e., using electrochemical approaches) may be useful.

### 8.8.5 Electrical Conductivity

Although the electrical conductivity of gallium is orders of magnitude higher than other fluids, such as saltwater, it is still more than an order of magnitude more resistive than copper – the state-of-the-art metal used for antennas and interconnects. In some applications involving electromagnetic, millimeter, or microwave devices where electrical conductivities affect the final performance of the device, gallium or its alloys may not be best candidate to be used as the soft conductors.[366] The addition of other metals to gallium changes its conductivity modestly. Hence, there is a room to explore and develop metallic alloys and intermetallic systems that have high electrical conductivities and can be processed with the patterning methods outlined in this book chapter, without losing the fluidic properties.

In addition, liquid metals as well as other flexible or stretchable conductors have non-zero temperature coefficient of resistance; that is, the electrical resistance of the system varies with temperature. For systems or applications where temperature varies or drifts, this property may cause an invariance in system performance. In particular, this property or behavior can lead to inaccurate sensors.[367]

### 8.8.6 Scalability, Stretchability, and Reconfigurability

As of today, there are no commercially available liquid metal electronic devices in the market. The main reason for absence of these devices is the lack of understanding on how to safely integrate the current processing and patterning techniques of liquid metals into existing manufacturing processes with minimum modification.

In addition, to date, liquid metal injected wires have been stretched to nearly 800–1000% strain.[83,227] However, there is limited knowledge available if the limit is defined only by the encasing material or are other factors important, such as channel collapse or capillary instabilities. Any stretchable or flexible device will need to withstand many cycles of deformation. In one study, liquid-metal wires showed minimal changes in resistance after hundreds to a thousand of strain cycles[233,234,368] and soft sensors have been shown to withstand more than a thousand cycles.[31] Additional studies need to be performed to understand the longevity of devices.

There are opportunities to create shape reconfigurable metallic structures, but the tendency of the oxide to adhere to surfaces and disrupt fluidity remains a challenge.

### 8.8.7 Resolution

Current liquid metal-patterning techniques suffer from minimum feature size being about 4–5 orders of magnitude wider than photolithography.[369] It is important to know how small liquid metals can be patterned and at what length scales will capillary forces induce fluid instabilities.[137] Liquid metal has been injected into ~150-nm capillaries using large pressures[370] and liquid metal has been induced inside carbon nanotubes.[170,371] However, most methods for patterning the metal have yet to achieve these small length scales.

In addition to low resolution of liquid metal patterning, it has been proven difficult to deposit thin, uniform films of liquid metals with reasons not well understood.

### 8.8.8 Oxide Wetting Properties

The wetting behavior of the oxide skin of liquid metals on surfaces is important for patterning all soft electronic devices. Gallium is corrosive to most metals; for example, gallium will 'reactively wet' and diffuse into aluminum. As a result, the aluminum structure will embrittle quickly (within minutes). This behavior poses a great risk for structural materials, particularly in aerospace applications, in addition to degrading ohmic contacts and increase in resistivity over time. For example, gallium penetrates into silver[372] or aluminum,[373] resulting in embrittlement in the latter case. Barrier metals (e.g. W, Mo)[374] or conductive barrier materials (e.g., W, Mo, graphite, and graphene)[375,376] may help address this issue but long-term device reliability studies need to be undertaken. Hence, the liquid-metal structures must be embedded in a way that they do not leak externally or internally.

The surface oxide creates two complications when performing conventional wetting experiments. Firstly, the oxide provides a mechanical barrier that prevents the metal from adopting shapes that minimize interfacial energy – which forms the basis for many wetting measurements such as the ones based on contact angle measurement. Hence, contact angle studies cannot provide much useful information for understanding liquid metal adhesion on different surfaces.[171] For example, on many surfaces, the metal can be physically manipulated to assume nearly any contact angle due to the stabilizing effects of the oxide (i.e., it gets pinned). Secondly, it forms a physical barrier between the substrate and the metal. Contacting the liquid metal to a substrate likely ruptures the oxide, allowing fresh oxide to form, that affects the wetting.[377] The role of surface roughness and chemical composition on wetting are hence important,[171,188] but have yet to be generalized. In summary, there are many opportunities to better understand wetting of these oxide-coated metals,[377] its interaction with nonmetals,[378] especially in presence of water.[135]

## 8.9 Summary

Gallium-based liquid metals currently have the best combination of conductivity and stretchability of any conventional conductor and, hence, is well suited for flexible and stretchable electronic applications. Unlike mercury, gallium-based liquid metal alloys have shown low toxicity in addition to forming surface oxide skin which allows them to be patterned into useful electronic components such as ultrastretchable wires, antennas, interconnects, self-healing conductors, soft sensors (to detect strain, touch, curvature, and bending), 100% soft composites, and reconfigurable electronics. In addition, liquid metals are also used as active components in soft memory devices (memristors), soft electrodes, diodes, capacitors, bio-electrodes, energy capture/storage devices, and microfluidics.

This chapter summarized the need, properties, patterning techniques, and application/ device integration of these soft-flexible and stretchable liquid metals. We hope that this book chapter will inspire discovery of novel applications that take advantage of the unique properties of liquid metals while solving the fundamental challenges associated with patterning these liquid metals in the future.

## Acknowledgments

We would like to thank Dr. Katsuyuki Sakuma-san of IBM Research for his constant support and feedback during the writing process. I.D.J. also acknowledges support from the Department of Energy. This work was performed under the auspices of the U.S. Department of Energy by Lawrence Livermore National Laboratory under Contract DE-AC52-07NA27344.IM Release # LLNL-BOOK-805259.

## References

[1] E. Rivin, *Stiffness and Damping in Mechanical Design*, CRC Press, **1999**.

[2] S. Wang, J. Song, D. H. Kim, Y. Huang, J. A. Rogers, *Appl. Phys. Lett.* **2008**, *93*, 023126.

[3] H. C. Ko, G. Shin, S. Wang, M. P. Stoykovich, J. W. Lee, D. H. Kim, J. S. Ha, Y. Huang, K.-C. Hwang, J. A. Rogers, Small Weinh. *Bergstr. Ger.* **2009**, *5*, 2703.

[4] X. Zang, C. Shen, Y. Chu, B. Li, M. Wei, J. Zhong, M. Sanghadasa, L. Lin, *Adv. Mater.* **2018**, *30*, 1800062.

[5]  Z. Liu, H. Du, J. Li, L. Lu, Z. Y. Li, N. X. Fang, *Sci. Adv.* **2018**, *4*, eaat4436.

[6]  J. Park, M. Kim, Y. Lee, H. S. Lee, H. Ko, *Sci. Adv.* **2015**, *1*, e1500661.

[7]  S. Lin, H. Yuk, T. Zhang, G. A. Parada, H. Koo, C. Yu, X. Zhao, *Adv. Mater.* **2016**, *28*, 4497.

[8]  J. Xu, S. Wang, G. J. N. Wang, C. Zhu, S. Luo, L. Jin, X. Gu, S. Chen, V. R. Feig, J. W. F. To, S. Rondeau-Gagné, J. Park, B. C. Schroeder, C. Lu, J. Y. Oh, Y. Wang, Y. H. Kim, H. Yan, R. Sinclair, D. Zhou, G. Xue, B. Murmann, C. Linder, W. Cai, J. B. H. Tok, J. W. Chung, Z. Bao, *Science* **2017**, *355*, 59.

[9]  L. Cai, J. Li, P. Luan, H. Dong, D. Zhao, Q. Zhang, X. Zhang, M. Tu, Q. Zeng, W. Zhou, S. Xie, *Adv. Funct. Mater.* **2012**, *22*, 5238.

[10]  J. H. So, J. Thelen, A. Qusba, G. J. Hayes, G. Lazzi, M. D. Dickey, *Adv. Funct. Mater.* **2009**, *19*, 3632.

[11]  D. Son, Z. Bao, *ACS Nano* **2018**, *12*, 11731.

[12]  S. Wang, J. Y. Oh, J. Xu, H. Tran, Z. Bao, *Acc. Chem. Res.* **2018**, *51*, 1033.

[13]  M. D. Dickey, *ACS Appl. Mater. Interfaces* **2014**, *6*, 18369.

[14]  M. D. Dickey, in Stretchable Bioelectron. *Med. Devices Syst.*, Springer, Cham, **2016**, pp. 3–30.

[15]  M. D. Dickey, *Adv. Mater.* **2017**, *29*, 1606425.

[16]  H. J. Koo, O. D. Velev, *Biomicrofluidics* **2013**, *7*, 031501.

[17]  H. Chun, T. D. Chung, *Annu. Rev. Anal. Chem. Vol 8* **2015**, *8*, 441.

[18]  E. M. Ahmed, *J. Adv. Res.* **2015**, *6*, 105.

[19]  O. J. Cayre, S. T. Chang, O. D. Velev, *J. Am. Chem. Soc.* **2007**, *129*, 10801.

[20]  J. H. Han, K. B. Kim, H. C. Kim, T. D. Chung, *Angew. Chem.-Int. Ed.* **2009**, *48*, 3830.

[21]  H. J. Koo, S. T. Chang, J. M. Slocik, R. R. Naik, O. D. Velev, *J. Mater. Chem.* **2011**, *21*, 72.

[22]  C. Keplinger, J. Y. Sun, C. C. Foo, P. Rothemund, G. M. Whitesides, Z. Suo, *Science* **2013**, *341*, 984.

[23]  J. Y. Sun, C. Keplinger, G. M. Whitesides, Z. Suo, *Adv. Mater.* **2014**, *26*, 7608.

[24]  J. Le Bideau, L. Viau, A. Vioux, *Chem. Soc. Rev.* **2011**, *40*, 907.

[25]  B. Chen, J. J. Lu, C. H. Yang, J. H. Yang, J. Zhou, Y. M. Chen, Z. Suo, Acs Appl. *Mater. Interfaces* **2014**, *6*, 7840.

[26]  Saricilar, D. Antiohos, K. Shu, P. G. Whitten, K. Wagner, C. Wang, G. G. Wallace, *Electrochem. Commun.* **2013**, *32*, 47.

[27]  M. Kaltenbrunner, G. Kettlgruber, C. Siket, R. Schwoediauer, S. Bauer, *Adv. Mater.* **2010**, *22*, 2065.

[28]  G. Kettlgruber, M. Kaltenbrunner, C. M. Siket, R. Moser, I. M. Graz, R. Schwödiauer, S. Bauer, *J. Mater. Chem. A* **2013**, *1*, 5505.

[29]  K. Suganuma, *Introduction to Printed Electronics*, Springer, New York, **2014**.

[30]  E. Cantatore, *Applications of Organic and Printed Electronics: A Technology-Enabled Revolution*, Springer, **2012**.

[31]  Y. Mengüç, Y. L. Park, H. Pei, D. Vogt, P. M. Aubin, E. Winchell, L. Fluke, L. Stirling, R. J. Wood, C. J. Walsh, *Int. J. Robot. Res.* **2014**, *33*, 1748.

[32]  T. Sekitani, Y. Noguchi, K. Hata, T. Fukushima, T. Aida, T. Someya, *Science* **2008**, *321*, 1468.

[33]  T. Sekitani, H. Nakajima, H. Maeda, T. Fukushima, T. Aida, K. Hata, T. Someya, *Nat. Mater.* **2009**, *8*, 494.

[34]  N. Matsuhisa, M. Kaltenbrunner, T. Yokota, H. Jinno, K. Kuribara, T. Sekitani, T. Someya, *Nat. Commun.* **2015**, *6*, 7461.

[35]  D. J. Lipomi, M. Vosgueritchian, B. C. K. Tee, S. L. Hellstrom, J. A. Lee, C. H. Fox, Z. Bao, *Nat. Nanotechnol.* **2011**, *6*, 788.

[36]  S. Rosset, H. R. Shea, *Appl. Phys. Mater. Sci. Process.* **2013**, *110*, 281.

[37]  K. Jain, M. Klosner, M. Zemel, S. Raghunandan, *Proc. IEEE* **2005**, *93*, 1500.

[38]  J. A. Rogers, Z. Bao, K. Baldwin, A. Dodabalapur, B. Crone, V. R. Raju, V. Kuck, H. Katz, K. Amundson, J. Ewing, P. Drzaic, *Proc. Natl. Acad. Sci.* **2001**, *98*, 4835.

[39]  G. H. Gelinck, H. E. A. Huitema, E. van Veenendaal, E. Cantatore, L. Schrijnemakers, J. B. P. H. van der Putten, T. C. T. Geuns, M. Beenhakkers, J. B. Giesbers, B.-H. Huisman, E. J. Meijer, E. M. Benito, F. J. Touwslager, A. W. Marsman, B. J. E. van Rens, D. M. de Leeuw, *Nat. Mater.* **2004**, *3*, 106.

[40] Y. Chen, J. Au, P. Kazlas, A. Ritenour, H. Gates, M. McCreary, *Nature* **2003**, *423*, 136.

[41] J. Heikenfeld, P. Drzaic, J. S. Yeo, T. Koch, *J. Soc. Inf. Disp.* **2011**, *19*, 129.

[42] W. Gao, S. Emaminejad, H. Y. Y. Nyein, S. Challa, K. Chen, A. Peck, H. M. Fahad, H. Ota, H. Shiraki, D. Kiriya, D. H. Lien, G. A. Brooks, R. W. Davis, A. Javey, *Nature* **2016**, *529*, 509.

[43] Y. Liu, M. Pharr, G. A. Salvatore, *ACS Nano* **2017**, DOI 10.1021/acsnano.7b04898.

[44] C. Pang, C. Lee, K. Y. Suh, *J. Appl. Polym. Sci.* **2013**, *130*, 1429.

[45] Y. Miyoshi, K. Miyajima, H. Saito, H. Kudo, T. Takeuchi, I. Karube, K. Mitsubayashi, *Sens. Actuators B Chem.* **2009**, *142*, 28.

[46] J. C. G. Matthews, G. Pettitt, in *2009 3rd Eur. Conf. Antennas Propag.*, **2009**, pp. 273–277.

[47] C. Cibin, P. Leuchtmann, M. Gimersky, R. Vahldieck, S. Moscibroda, in *IEEE Antennas Propag. Soc. Symp. 2004*, **2004**, *4*, pp. 3589–3592.

[48] L. Hu, H. Wu, F. La Mantia, Y. Yang, Y. Cui, *ACS Nano* **2010**, *4*, 5843.

[49] M. Koo, K. I. Park, S. H. Lee, M. Suh, D. Y. Jeon, J. W. Choi, K. Kang, K. J. Lee, *Nano Lett.* **2012**, *12*, 4810.

[50] G. A. Salvatore, N. Münzenrieder, T. Kinkeldei, L. Petti, C. Zysset, I. Strebel, L. Büthe, G. Tröster, *Nat. Commun.* **2014**, *5*, DOI 10.1038/ncomms3982.

[51] W. S. Wong, A. Salleo, *Flexible Electronics: Materials and Applications*, Springer Science & Business Media, **2009**.

[52] G. Schwartz, B. C. K. Tee, J. Mei, A. L. Appleton, D. H. Kim, H. Wang, Z. Bao, *Nat. Commun.* **2013**, *4*, 1859.

[53] B. P. Timko, T. Cohen-Karni, G. Yu, Q. Qing, B. Tian, C. M. Lieber, *Nano Lett.* **2009**, *9*, 914.

[54] D. P. Parekh, Additive Patterning of Gallium-Based Liquid Metal Alloys at Room Temperature for Rapid Prototyping of Multilayered Microfluidics and 3D Printed Soft Electronics, PhD Thesis, North Carolina State University, **2018**.

[55] T. Yokota, P. Zalar, M. Kaltenbrunner, H. Jinno, N. Matsuhisa, H. Kitanosako, Y. Tachibana, W. Yukita, M. Koizumi, T. Someya, *Sci. Adv.* **2016**, *2*, e1501856.

[56] C. Larson, B. Peele, S. Li, S. Robinson, M. Totaro, L. Beccai, B. Mazzolai, R. Shepherd, *Science* **2016**, *351*, 1071.

[57] A. Miyamoto, S. Lee, N. F. Cooray, S. Lee, M. Mori, N. Matsuhisa, H. Jin, L. Yoda, T. Yokota, A. Itoh, M. Sekino, H. Kawasaki, T. Ebihara, M. Amagai, T. Someya, *Nat. Nanotechnol.* **2017**, *12*, 907.

[58] I. R. Minev, P. Musienko, A. Hirsch, Q. Barraud, N. Wenger, E. M. Moraud, J. Gandar, M. Capogrosso, T. Milekovic, L. Asboth, R. F. Torres, N. Vachicouras, Q. Liu, N. Pavlova, S. Duis, A. Larmagnac, J. Vörös, S. Micera, Z. Suo, G. Courtine, S. P. Lacour, *Science* **2015**, *347*, 159.

[59] S. P. Lacour, G. Courtine, J. Guck, *Nat. Rev. Mater.* **2016**, *1*, 16063.

[60] Z. Fekete, A. Pongrácz, *Sens. Actuators B Chem.* **2017**, *243*, 1214.

[61] S. Xu, Y. Zhang, J. Cho, J. Lee, X. Huang, L. Jia, J. A. Fan, Y. Su, J. Su, H. Zhang, H. Cheng, B. Lu, C. Yu, C. Chuang, T. Kim, T. Song, K. Shigeta, S. Kang, C. Dagdeviren, I. Petrov, P. V. Braun, Y. Huang, U. Paik, J. A. Rogers, *Nat. Commun.* **2013**, *4*, 1543.

[62] A. M. Zamarayeva, A. E. Ostfeld, M. Wang, J. K. Duey, I. Deckman, B. P. Lechêne, G. Davies, D. A. Steingart, A. C. Arias, *Sci. Adv.* **2017**, *3*, e1602051.

[63] Y. Zhang, Y. Huang, J. A. Rogers, *Curr. Opin. Solid State Mater. Sci.* **2015**, *19*, 190.

[64] M. S. White, M. Kaltenbrunner, E. D. Głowacki, K. Gutnichenko, G. Kettlgruber, I. Graz, S. Aazou, C. Ulbricht, D. A. M. Egbe, M. C. Miron, Z. Major, M. C. Scharber, T. Sekitani, T. Someya, S. Bauer, N. S. Sariftci, *Nat. Photonics* **2013**, *7*, 811.

[65] T. Sekitani, H. Nakajima, H. Maeda, T. Fukushima, T. Aida, K. Hata, T. Someya, *Nat. Mater.* **2009**, *8*, 494.

[66] J. Byun, B. Lee, E. Oh, H. Kim, S. Kim, S. Lee, Y. Hong, *Sci. Rep.* **2017**, *7*, 45328.

[67] D. H. Kim, J. H. Ahn, W. M. Choi, H. S. Kim, T. H. Kim, J. Song, Y. Y. Huang, Z. Liu, C. Lu, J. A. Rogers, *Science* **2008**, *320*, 507.

[68] D. H. Kim, J. Song, W. M. Choi, H. S. Kim, R. H. Kim, Z. Liu, Y. Y. Huang, K. C. Hwang, Y. Zhang, J. A. Rogers, *Proc. Natl. Acad. Sci.* **2008**, *105*, 18675.

[69] H. C. Ko, M. P. Stoykovich, J. Song, V. Malyarchuk, W. M. Choi, C. J. Yu, J. B. G. Iii, J. Xiao, S. Wang, Y. Huang, J. A. Rogers, *Nature* **2008**, *454*, 748.

[70] J. A. Rogers, T. Someya, Y. Huang, *Science* **2010**, *327*, 1603.

[71] D. H. Kim, N. Lu, R. Ma, Y. S. Kim, R. H. Kim, S. Wang, J. Wu, S. M. Won, H. Tao, A. Islam, K. J. Yu, T. Kim, R. Chowdhury, M. Ying, L. Xu, M. Li, H. J. Chung, H. Keum, M. McCormick, P. Liu, Y. W. Zhang, F. G. Omenetto, Y. Huang, T. Coleman, J. A. Rogers, *Science* **2011**, *333*, 838.

[72] J. W. Jeong, W. H. Yeo, A. Akhtar, J. J. S. Norton, Y. J. Kwack, S. Li, S. Y. Jung, Y. Su, W. Lee, J. Xia, H. Cheng, Y. Huang, W. S. Choi, T. Bretl, J. A. Rogers, *Adv. Mater.* **2013**, *25*, 6839.

[73] W. H. Yeo, Y. S. Kim, J. Lee, A. Ameen, L. Shi, M. Li, S. Wang, R. Ma, S. H. Jin, Z. Kang, Y. Huang, J. A. Rogers, *Adv. Mater.* **2013**, *25*, 2773.

[74] S. Wagner, S. Bauer, *MRS Bull.* **2012**, *37*, 207.

[75] D. Rus, M. T. Tolley, *Nature* **2015**, *521*, 467.

[76] F. Xu, Y. Zhu, *Adv. Mater.* **2012**, *24*, 5117.

[77] M. Park, J. Im, M. Shin, Y. Min, J. Park, H. Cho, S. Park, M. B. Shim, S. Jeon, D. Y. Chung, J. Bae, J. Park, U. Jeong, K. Kim, *Nat. Nanotechnol.* **2012**, *7*, 803.

[78] Y. Kim, J. Zhu, B. Yeom, M. Di Prima, X. Su, J. G. Kim, S. J. Yoo, C. Uher, N. A. Kotov, *Nature* **2013**, *500*, 59.

[79] K. Y. Chun, Y. Oh, J. Rho, J. H. Ahn, Y. J. Kim, H. R. Choi, S. Baik, *Nat Nano* **2010**, *5*, 853.

[80] H. Stoyanov, M. Kollosche, S. Risse, R. Waché, G. Kofod, *Adv. Mater.* **2013**, *25*, 578.

[81] B. Y. Ahn, E. B. Duoss, M. J. Motala, X. Guo, S. I. Park, Y. Xiong, J. Yoon, R. G. Nuzzo, J. A. Rogers, J. A. Lewis, *Science* **2009**, *323*, 1590.

[82] J. Park, S. Wang, M. Li, C. Ahn, J. K. Hyun, D. S. Kim, D. K. Kim, J. A. Rogers, Y. Huang, S. Jeon, *Nat. Commun.* **2012**, *3*, 916.

[83] S. Zhu, J. H. So, R. Mays, S. Desai, W. R. Barnes, B. Pourdeyhimi, M. D. Dickey, *Adv. Funct. Mater.* **2013**, *23*, 2308.

[84] B. Fahlman, *Materials Chemistry*, Springer, Netherlands, **2011**.

[85] D. J. Steele, *The Chemistry of the Metallic Elements: The Commonwealth and International Library: Intermediate Chemistry Division*, Pergamon Press Ltd., Oxford, London, **2017**.

[86] C. S. Ivanoff, A. E. Ivanoff, T. L. Hottel, *Food Chem. Toxicol.* **2012**, *50*, 212.

[87] R. K. Zalups, *Pharmacol. Rev.* **2000**, *52*, 113.

[88] T. W. Clarkson, L. Magos, *Crit. Rev. Toxicol.* **2006**, *36*, 609.

[89] G. D. Sprouse, L. A. Orozco, J. E. Simsarian, W. Shi, W. Z. Zhao, *Nucl. Instrum. Methods Phys. Res. Sect. B Beam Interact. Mater. At.* **1997**, *126*, 370.

[90] J. Rumble, Ed., *CRC Handbook of Chemistry and Physics, 98th Edition*, CRC Press, Boca Raton London New York, **2017**.

[91] T. W. Clarkson, L. Magos, G. J. Myers, *N. Engl. J. Med.* **2003**, *349*, 1731.

[92] L. J. Norrby, *J. Chem. Educ.* **1991**, *68*, 110.

[93] J. F. Bates, A. G. Knapton, *Int. Mater. Rev.* **1977**, *22*, 39.

[94] R. Bharti, K. K. Wadhwani, A. P. Tikku, A. Chandra, *J. Conserv. Dent. JCD* **2010**, *13*, 204.

[95] G. Guzzi, P. D. Pigatto, in *Met. Allergy*, Springer, Cham, **2018**, pp. 397–421.

[96] H. Herø, C. J. Simensen, R. B. Jørgensen, *Biomaterials* **1996**, *17*, 1321.

[97] N. E. Selin, *J. Environ. Monit.* **2011**, *13*, 2389.

[98] U. Park, K. Yoo, J. Kim, *Sens. Actuators Phys.* **2010**, *159*, 51.

[99] J. C. Hafele, R. E. Keating, *Science* **1972**, *177*, 166.

[100] A. Saasen, O. H. Jordal, D. Burkhead, P. C. Berg, G. Løklingholm, E. S. Pedersen, J. Turner, M. J. Harris, Society of Petroleum Engineers, **2002**.

[101] H. A. Storms, K. F. Brown, J. D. Stein, *Anal. Chem.* **1977**, *49*, 2023.

[102] T. J. Yasunari, A. Stohl, R. S. Hayano, J. F. Burkhart, S. Eckhardt, T. Yasunari, *Proc. Natl. Acad. Sci. USA* **2011**, *108*, 19530.

[103] N. B. Morley, J. Burris, L. C. Cadwallader, M. D. Nornberg, *Rev. Sci. Instrum.* **2008**, *79*, 056107.

[104] L. R. Bernstein, *Pharmacol. Rev.* **1998**, *50*, 665.

[105] Y. Lu, Q. Hu, Y. Lin, D. B. Pacardo, C. Wang, W. Sun, F. S. Ligler, M. D. Dickey, Z. Gu, *Nat. Commun.* **2015**, *6*, 10066.

[106] J. L. Domingo, J. Corbella, *Trace Elem. Med.* **1991**, *8*, 56.
[107] F. Gray, D. A. Kramer, J. D. Bliss, in *Kirk-Othmer Encycl. Chem. Technol.*, American Cancer Society, **2013**, pp. 1–26.
[108] G. Marinenko, *J. Res. Natl. Bur. Stand. Sect. Phys. Chem.* **1977**, *81A*, 1.
[109] R. R. Moskalyk, *Miner. Eng.* **2003**, *16*, 921.
[110] J. S. Blakemore, *J. Appl. Phys.* **1982**, *53*, R123.
[111] J. I. Pankove, *MRS Online Proc. Libr. Arch.* **1987**, *97*, DOI 10.1557/PROC-97-409.
[112] H. Okamoto, *Desk Handbook: Phase Diagrams for Binary Alloys*, ASM International, **2000**.
[113] M. D. Dickey, R. C. Chiechi, R. J. Larsen, E. A. Weiss, D. A. Weitz, G. M. Whitesides, *Adv. Funct. Mater.* **2008**, *18*, 1097.
[114] D. Zrnic, D. S. Swatik, *J. Common Met.* **1969**, *18*, 67.
[115] K. E. Spells, *Proc. Phys. Soc.* **1936**, *48*, 299.
[116] H. E. Sostman, *Rev. Sci. Instrum.* **1977**, *48*, 127.
[117] C. Dodd, *Proc. Phys. Soc. Sect. B* **1950**, *63*, 662.
[118] G. J. Abbaschian, *J. Common Met.* **1975**, *40*, 329.
[119] M. J. Duggin, *Phys. Lett. A* **1969**, *29*, 470.
[120] G. N. van Ingen, J. Kapteijn, J. L. Meijering, *Scr. Metall.* **1970**, *4*, 733.
[121] Y. Plevachuk, V. Sklyarchuk, S. Eckert, G. Gerbeth, R. Novakovic, *J. Chem. Eng. Data* **2014**, *59*, 757.
[122] S. Yu, M. Kaviany, *J. Chem. Phys.* **2014**, *140*, 064303.
[123] W. H. Hoather, *Proc. Phys. Soc.* **1936**, *48*, 699.
[124] A. J. Downs, *Chemistry of Aluminium, Gallium, Indium and Thallium*, Springer Science & Business Media, **1993**.
[125] T. Liu, P. Sen, C. J. Kim, *J. Microelectromechanical Syst.* **2012**, *21*, 443.
[126] C. Ladd, J. H. So, J. Muth, M. D. Dickey, *Adv. Mater.* **2013**, *25*, 5081.
[127] R. C. Chiechi, E. A. Weiss, M. D. Dickey, G. M. Whitesides, *Angew. Chem. Int. Ed.* **2008**, *47*, 142.
[128] I. D. Joshipura, H. R. Ayers, C. Majidi, M. D. Dickey, *J. Mater. Chem. C* **2015**, *3*, 3834.
[129] M. J. Regan, H. Tostmann, P. S. Pershan, O. M. Magnussen, E. DiMasi, B. M. Ocko, M. Deutsch, *Phys. Rev. B* **1997**, *55*, 10786.
[130] F. Scharmann, G. Cherkashinin, V. Breternitz, C. Knedlik, G. Hartung, T. Weber, J. A. Schaefer, *Surf. Interface Anal.* **2004**, *36*, 981.
[131] A. Plech, U. Klemradt, H. Metzger, J. Peisl, *J. Phys. Condens. Matter* **1998**, *10*, 971.
[132] J. M. Chabala, *Phys. Rev. B* **1992**, *46*, 11346.
[133] H. Tostmann, E. DiMasi, P. S. Pershan, B. M. Ocko, O. G. Shpyrko, M. Deutsch, *Phys. Rev. B* **1999**, *59*, 783.
[134] M. Passlack, E. F. Schubert, W. S. Hobson, M. Hong, N. Moriya, S. N. G. Chu, K. Konstadinidis, J. P. Mannaerts, M. L. Schnoes, G. J. Zydzik, *J. Appl. Phys.* **1995**, *77*, 686.
[135] M. R. Khan, C. Trlica, J. H. So, M. Valeri, M. D. Dickey, *ACS Appl. Mater. Interfaces* **2014**, *6*, 22467.
[136] A. Hirsch, H. O. Michaud, A. P. Gerratt, S. de Mulatier, S. P. Lacour, *Adv. Mater.* **2016**, *28*, 4507.
[137] Y. Lin, C. Ladd, S. Wang, A. Martin, J. Genzer, S. A. Khan, M. D. Dickey, *Extreme Mech. Lett.* **2016**, *7*, 55.
[138] L. J. Briggs, *J. Chem. Phys.* **1957**, *26*, 784.
[139] M. D. Dickey, *ACS Appl. Mater. Interfaces* **2014**, *6*, 18369.
[140] M. Pourbaix, *Atlas of Electrochemical Equilibria in Aqueous Solutions*, National Association of Corrosion Engineers, TX, USA, **1974**.
[141] M. R. Khan, C. B. Eaker, E. F. Bowden, M. D. Dickey, *Proc. Natl. Acad. Sci.* **2014**, *111*, 14047.
[142] C. B. Eaker, M. D. Dickey, *Appl. Phys. Rev.* **2016**, *3*, 031103.
[143] R. J. Larsen, M. D. Dickey, G. M. Whitesides, D. A. Weitz, *J. Rheol.* **2009**, *53*, 1305.
[144] A. R. Jacob, D. P. Parekh, M. D. Dickey, L. C. Hsiao, *Langmuir* **2019**, *35*, 11774.
[145] M. R. Khan, C. Trlica, M. D. Dickey, *Adv. Funct. Mater.* **2015**, *25*, 671.
[146] B. A. Gozen, A. Tabatabai, O. B. Ozdoganlar, C. Majidi, *Adv. Mater.* **2014**, *26*, 5211.
[147] S. H. Jeong, K. Hjort, Z. Wu, *Sensors* **2014**, *14*, 16311.
[148] J. Wissman, T. Lu, C. Majidi, in *2013 IEEE Sens.*, **2013**, pp. 1–4.

[149] A. Tabatabai, A. Fassler, C. Usiak, C. Majidi, *Langmuir* **2013**, *29*, 6194.

[150] R. K. Kramer, C. Majidi, R. J. Wood, *Adv. Funct. Mater.* **2013**, *23*, 5292.

[151] C. W. Park, Y. G. Moon, H. Seong, S. W. Jung, J. Y. Oh, B. S. Na, N. M. Park, S. S. Lee, S. G. Im, J. B. Koo, *ACS Appl. Mater. Interfaces* **2016**, *8*, 15459.

[152] R. David, N. Miki, *Nanoscale* **2019**, *11*, 21419.

[153] Y. Wu, Z. Deng, Z. Peng, R. Zheng, S. Liu, S. Xing, J. Li, D. Huang, L. Liu, *Adv. Funct. Mater.* **2019**, *29*, 1903840.

[154] R. David, N. Miki, *Langmuir* **2018**, *34*, 10550.

[155] F. Hoshyargar, J. Crawford, A. P. O'Mullane, *J. Am. Chem. Soc.* **2017**, *139*, 1464.

[156] S. Y. Tang, J. Zhu, V. Sivan, B. Gol, R. Soffe, W. Zhang, A. Mitchell, K. Khoshmanesh, *Adv. Funct. Mater.* **2015**, *25*, 4445.

[157] M. Kim, C. Kim, H. Alrowais, O. Brand, *Adv. Mater. Technol.* **2018**, *3*, 1800061.

[158] M. Kim, H. Alrowais, S. Pavlidis, O. Brand, *Adv. Funct. Mater.* **2017**, *27*, 1604466.

[159] J. L. Melcher, K. S. Elassy, R. C. Ordonez, C. Hayashi, A. T. Ohta, D. Garmire, *Micromachines* **2019**, *10*, 54.

[160] S. H. Jeong, K. Hjort, Z. Wu, *Sci. Rep.* **2015**, *5*, 8419.

[161] S. H. Jeong, A. Hagman, K. Hjort, M. Jobs, J. Sundqvist, Z. Wu, *Lab. Chip* **2012**, *12*, 4657.

[162] G. Li, X. Wu, D. W. Lee, *Sens. Actuators B Chem.* **2015**, *221*, 1114.

[163] G. Li, D. W. Lee, *Lab. Chip* **2017**, *17*, 3415.

[164] N. Lazarus, S. S. Bedair, I. M. Kierzewski, *ACS Appl. Mater. Interfaces* **2017**, *9*, 1178.

[165] Y. G. Moon, J. B. Koo, N. M. Park, J. Y. Oh, B. S. Na, S. S. Lee, S. D. Ahn, C. W. Park, *IEEE Trans. Electron Devices* **2017**, 1.

[166] J. H. So, M. D. Dickey, *Lab. Chip* **2011**, *11*, 905.

[167] Y. L. Park, B.-R. Chen, R. J. Wood, *IEEE Sens. J.* **2012**, *12*, 2711.

[168] W. Zhao, J. L. Bischof, J. Hutasoit, X. Liu, T. C. Fitzgibbons, J. R. Hayes, P. J. A. Sazio, C. Liu, J. K. Jain, J. V. Badding, M. H. W. Chan, *Nano Lett.* **2015**, *15*, 153.

[169] Y. B. Li, Y. Bando, D. Golberg, Z. W. Liu, *Appl. Phys. Lett.* **2003**, *83*, 999.

[170] Y. Gao, Y. Bando, *Nature* **2002**, *415*, 599.

[171] I. D. Joshipura, H. R. Ayers, G. A. Castillo, C. Ladd, C. E. Tabor, J. J. Adams, M. D. Dickey, *ACS Appl. Mater. Interfaces* **2018**, *10*, 44686.

[172] S. Zhang, B. Wang, J. Jiang, K. Wu, C. F. Guo, Z. Wu, *ACS Appl. Mater. Interfaces* **2019**, *11*, 7148.

[173] Q. Wang, Y. Yu, J. Yang, J. Liu, *Adv. Mater.* **2015**, *27*, 7109.

[174] R. Guo, J. Tang, S. Dong, J. Lin, H. Wang, J. Liu, W. Rao, *Adv. Mater. Technol.* **2018**, *3*, 1870045.

[175] Y. Lin, O. Gordon, M. R. Khan, N. Vasquez, J. Genzer, M. D. Dickey, *Lab. Chip* **2017**, *17*, 3043.

[176] V. Bharambe, D. P. Parekh, C. Ladd, K. Moussa, M. D. Dickey, J. J. Adams, *Addit. Manuf.* **2017**, *18*, 221.

[177] J. W. Boley, E. L. White, R. K. Kramer, *Adv. Mater.* **2015**, *27*, 2355.

[178] G. Li, X. Wu, D.-W. Lee, *Lab. Chip* **2016**, DOI 10.1039/C6LC00046K.

[179] Lord Rayleigh, *Proc. Lond. Math. Soc.* **1878**, *s1-10*, 4.

[180] J. W. Boley, E. L. White, G. T. C. Chiu, R. K. Kramer, *Adv. Funct. Mater.* **2014**, *24*, 3501.

[181] C. Trlica, D. Parekh, L. Panich, C. Ladd, M. Dickey, Micro- and Nanotechnology Sensors, Systems, and Applications VI **2014**, *Proc. SPIE 9083*, DOI 10.1117/12.2050212.

[182] Y. Zheng, Q. Zhang, J. Liu, *AIP Adv.* **2013**, *3*, 112117.

[183] C. B. Eaker, I. D. Joshipura, L. R. Maxwell, J. Heikenfeld, M. D. Dickey, *Lab. Chip* **2017**, *17*, 1069.

[184] D. Kim, D. Jung, J. H. Yoo, Y. Lee, W. Choi, G. S. Lee, K. Yoo, J. B. Lee, *J. Micromechanics Microengineering* **2014**, *24*, 055018.

[185] S. S. Kadlaskar, J. H. Yoo, Abhijeet, J. B. Lee, W. Choi, *J. Colloid Interface Sci.* **2017**, *492*, 33.

[186] Y. Jiang, S. Su, H. Peng, H. S. Kwok, X. Zhou, S. Chen, *J. Mater. Chem. C* **2017**, *5*, 12378.

[187] D. Kim, D. W. Lee, W. Choi, J. B. Lee, *J. Microelectromechanical Syst.* **2013**, *22*, 1267.

[188] R. K. Kramer, J. W. Boley, H. A. Stone, J. C. Weaver, R. J. Wood, *Langmuir* **2014**, *30*, 533.

[189] N. Syed, A. Zavabeti, K. A. Messalea, E. Della Gaspera, A. Elbourne, A. Jannat, M. Mohiuddin, B. Y. Zhang, G. Zheng, L. Wang, S. P. Russo, D. Esrafilzadeh, C. F. McConville, K. Kalantar-Zadeh, T. Daeneke, *J. Am. Chem. Soc.* **2019**, *141*, 104.

[190] N. Syed, A. Zavabeti, J. Z. Ou, M. Mohiuddin, N. Pillai, B. J. Carey, B. Y. Zhang, R. S. Datta, A. Jannat, F. Haque, K. A. Messalea, C. Xu, S. P. Russo, C. F. McConville, T. Daeneke, K. Kalantar-Zadeh, *Nat. Commun.* **2018**, *9*, 3618.

[191] B. J. Carey, J. Z. Ou, R. M. Clark, K. J. Berean, A. Zavabeti, A. S. R. Chesman, S. P. Russo, D. W. M. Lau, Z. Q. Xu, Q. Bao, O. Kavehei, B. C. Gibson, M. D. Dickey, R. B. Kaner, T. Daeneke, K. Kalantar-Zadeh, *Nat. Commun.* **2017**, *8*, 14482.

[192] M. M. Y. A. Alsaif, S. Kuriakose, S. Walia, N. Syed, A. Jannat, B. Y. Zhang, F. Haque, M. Mohiuddin, T. Alkathiri, N. Pillai, T. Daeneke, J. Z. Ou, A. Zavabeti, *Adv. Mater. Interfaces* **2019**, *6*, 1900007.

[193] M. M. Y. A. Alsaif, N. Pillai, S. Kuriakose, S. Walia, A. Jannat, K. Xu, T. Alkathiri, M. Mohiuddin, T. Daeneke, K. Kalantar-Zadeh, J. Z. Ou, A. Zavabeti, *ACS Appl. Nano Mater.* **2019**, DOI 10.1021/acsanm.9b01133.

[194] Y. Lin, C. Cooper, M. Wang, J. J. Adams, J. Genzer, M. D. Dickey, *Small* **2015**, *11*, 6397.

[195] T. R. Lear, S. H. Hyun, J. W. Boley, E. L. White, D. H. Thompson, R. K. Kramer, *Extreme Mech. Lett.* **2017**, *13*, 126.

[196] S. Liu, M. C. Yuen, E. L. White, J. W. Boley, B. Deng, G. J. Cheng, R. Kramer-Bottiglio, *ACS Appl. Mater. Interfaces* **2018**, *10*, 28232.

[197] S. Liu, S. N. Reed, M. J. Higgins, M. S. Titus, R. Kramer-Bottiglio, *Nanoscale* **2019**, *11*, 17615.

[198] E. L. White, J. C. Case, R. K. Kramer, *Sens. Actuators Phys.* **2017**, *253*, 188.

[199] Y. G. Park, H. S. An, J. Y. Kim, J. U. Park, *Sci. Adv.* **2019**, *5*, eaaw2844.

[200] Y. G. Park, H. Min, H. Kim, A. Zhexembekova, C. Y. Lee, J. U. Park, *Nano Lett.* **2019**, *19*, 4866.

[201] J. B. Andrews, K. Mondal, T. V. Neumann, J. A. Cardenas, J. Wang, D. P. Parekh, Y. Lin, P. Ballentine, M. D. Dickey, A. D. Franklin, *ACS Nano* **2018**, *12*, 5482.

[202] A. Cook, D. P. Parekh, C. Ladd, G. Kotwal, L. Panich, M. Durstock, M. D. Dickey, C. E. Tabor, *Adv. Eng. Mater.* **2019**, *21*, 1900400.

[203] A. Gannarapu, B. A. Gozen, *Extreme Mech. Lett.* **2019**, *33*, 100554.

[204] S. Kim, J. Oh, D. Jeong, J. Bae, *ACS Appl. Mater. Interfaces* **2019**, *11*, 20557.

[205] M. G. Mohammed, R. Kramer, *Adv. Mater.* **2017**, *29*, 1604965.

[206] Y. He, L. Zhou, J. Zhan, Q. Gao, J. Fu, C. Xie, H. Zhao, Y. Liu, *3D Print. Addit. Manuf.* **2018**, *5*, 195.

[207] M. A. H. Khondoker, A. Ostashek, D. Sameoto, *Adv. Eng. Mater.* **2019**, *21*, 1900060.

[208] J. W. Boley, W. M. van Rees, C. Lissandrello, M. N. Horenstein, R. L. Truby, A. Kotikian, J. A. Lewis, L. Mahadevan, *Proc. Natl. Acad. Sci.* **2019**, *116*, 20856.

[209] E. P. Yalcintas, K. B. Ozutemiz, T. Cetinkaya, L. Dalloro, C. Majidi, O. B. Ozdoganlar, *Adv. Funct. Mater.* **2019**, *29*, 1906551.

[210] D. Parekh, D. Cormier, M. Dickey, in *Addit. Manuf.*, CRC Press, **2015**, pp. 215–258.

[211] M. Rutkevicius, M. Geiger, D. Parekh, T. Neumann, M. D. Dickey, S. A. Khan, AIChE, **2017**.

[212] S. Roh, D. P. Parekh, B. Bharti, S. D. Stoyanov, O. D. Velev, *Adv. Mater.* **2017**, *29*, 1701554.

[213] M. Rajagopalan, M. A. Bhatia, M. A. Tschopp, D. J. Srolovitz, K. N. Solanki, *Acta Mater.* **2014**, *73*, 312.

[214] H. O. Michaud, J. Teixidor, S. P. Lacour, *Smart Mater. Struct.* **2015**, *24*, 035020.

[215] H. O. Michaud, S. P. Lacour, *APL Mater.* **2019**, *7*, 031504.

[216] E. B. Secor, A. B. Cook, C. E. Tabor, M. C. Hersam, *Adv. Electron. Mater.* **2018**, *4*, 1.

[217] T. Lu, L. Finkenauer, J. Wissman, C. Majidi, *Adv. Funct. Mater.* **2014**, *24*, 3351.

[218] T. Lu, E. J. Markvicka, Y. Jin, C. Majidi, *ACS Appl. Mater. Interfaces* **2017**, *9*, 22055.

[219] C. Pan, K. Kumar, J. Li, E. J. Markvicka, P. R. Herman, C. Majidi, *Adv. Mater.* **2018**, *30*, 1706937.

[220] M. R. Khan, J. Bell, M. D. Dickey, *Adv. Mater. Interfaces* **2016**, *3*, 1600546.

[221] A. Hirsch, L. Dejace, H. O. Michaud, S. P. Lacour, *Acc. Chem. Res.* **2019**, *52*, 534.

[222] I. D. Joshipura, H. R. Ayers, C. Majidi, M. D. Dickey, *J. Mater. Chem. C* **2015**, *3*, 3834.

[223] M. A. H. Khondoker, D. Sameoto, *Smart Mater. Struct.* **2016**, *25*, 093001.

[224] L. Zhu, B. Wang, S.Handschuh-Wang, X. Zhou, *Small* **2020**, *16*, e1903841.

[225] S. Park, N. Baugh, H. K. Shah, D. P. Parekh, I. D. Joshipura, M. D. Dickey, *Adv. Sci.* **2019**, *6*, 1901579.

[226] C. B. Cooper, I. D. Joshipura, D. P. Parekh, J. Norkett, R. Mailen, V. M. Miller, J. Genzer, M. D. Dickey, *Sci. Adv.* **2019**, *5*, eaat4600.

[227] K. P. Mineart, Y. Lin, S. C. Desai, A. S. Krishnan, R. J. Spontak, M. D. Dickey, *Soft Matter* **2013**, *9*, 7695.

[228] H. J. Kim, C. Son, B. Ziaie, *Appl. Phys. Lett.* **2008**, *92*, 011904.

[229] N. Lazarus, C. D. Meyer, W. J. Turner, *RSC Adv.* **2015**, *5*, 78695.

[230] D. F. Baldwin, R. D. Deshmukh, C. S. Hau, *IEEE Trans. Compon. Packag. Technol.* **2000**, *23*, 360.

[231] H. Yang, C. R. Lightner, L. Dong, *ACS Nano* **2012**, *6*, 622.

[232] C. C. Bradley, *Philos. Mag.* **1963**, *8*, 1535.

[233] F. Suarez, D. P. Parekh, C. Ladd, D. Vashaee, M. D. Dickey, M. C. Öztürk, *Appl. Energy* **2017**, *202*, 736.

[234] M. Ozturk, M. D. Dickey, C. Ladd, D. P. Parekh, V. P. Ramesh, F. Suarez, *Flexible Thermoelectric Generator and Methods of Manufacturing*, **2019**, US10431726B2.

[235] M. O. Culjat, R. S. Singh, S. N. White, R. R. Neurgaonkar, E. R. Brown, *Acoust. Res. Lett. Online-Arlo* **2005**, *6*, 125.

[236] H. Hu, K. Shaikh, C. Liu, in *2007 IEEE Sens.*, **2007**, pp. 815–817.

[237] B. Zhang, Q. Dong, C. E. Korman, Z. Li, M. E. Zaghloul, *Sci. Rep.* **2013**, *3*.

[238] Y. Lim, J. Yoon, J. Yun, D. Kim, S. Y. Hong, S. J. Lee, G. Zi, J. S. Ha, *ACS Nano* **2014**, *8*, 11639.

[239] G. A. Hernandez, D. Martinez, C. Ellis, M. Palmer, M. C. Hamilton, in *Electron. Compon. Technol. Conf. ECTC 2013 IEEE 63rd*, **2013**, pp. 1401–1406.

[240] B. Kim, J. Jang, I. You, J. Park, S. Shin, G. Jeon, J. K. Kim, U. Jeong, *ACS Appl. Mater. Interfaces* **2015**, *7*, 7920.

[241] Y. Jiao, C. W. Young, S. Yang, S. Oren, H. Ceylan, S. Kim, K. Gopalakrishnan, P. C. Taylor, L. Dong, *IEEE Sens. J.* **2016**, *16*, 7870.

[242] K. Entesari, A. P. Saghati, *IEEE Microw. Mag.* **2016**, *17*, 50.

[243] S. Cheng, Z. Wu, *Lab. Chip* **2012**, *12*, 2782.

[244] S. Cheng, A. Rydberg, K. Hjort, Z. Wu, *Appl. Phys. Lett.* **2009**, *94*, 144103.

[245] A. Qusba, A. K. RamRakhyani, J. H. So, G. J. Hayes, M. D. Dickey, G. Lazzi, *IEEE Sens. J.* **2014**, *14*, 1074.

[246] N. Lazarus, C. D. Meyer, S. S. Bedair, H. Nochetto, I. M. Kierzewski, *Smart Mater. Struct.* **2014**, *23*, 085036.

[247] A. Fassler, C. Majidi, *Smart Mater. Struct.* **2013**, *22*, 055023.

[248] G. J. Hayes, J. H. So, A. Qusba, M. D. Dickey, G. Lazzi, *IEEE Trans. Antennas Propag.* **2012**, *60*, 2151.

[249] B. Aïssa, M. Nedil, M. A. Habib, E. Haddad, W. Jamroz, D. Therriault, Y. Coulibaly, F. Rosei, *Appl. Phys. Lett.* **2013**, *103*, 063101.

[250] S. Cheng, Z. Wu, *Lab. Chip* **2010**, *10*, 3227.

[251] M. Jobs, K. Hjort, A. Rydberg, Z. Wu, *Small* **2013**, *9*, 3230.

[252] G. J. Hayes, S. C. Desai, Y. Liu, P. Annamaa, G. Lazzi, M. D. Dickey, *Microw. Opt. Technol. Lett.* **2014**, *56*, 1459.

[253] A. M. Morishita, C. K. Y. Kitamura, A. T. Ohta, W. A. Shiroma, *Electron. Lett.* **2014**, *50*, 19.

[254] A. Pourghorban Saghati, J. Singh Batra, J. Kameoka, K. Entesari, *IEEE Trans. Antennas Propag.* **2015**, *63*, 3798.

[255] A. J. King, J. F. Patrick, N. R. Sottos, S. R. White, G. H. Huff, J. T. Bernhard, *IEEE Antennas Wirel. Propag. Lett.* **2013**, *12*, 828.

[256] M. Wang, C. Trlica, M. R. Khan, M. D. Dickey, J. J. Adams, *J. Appl. Phys.* **2015**, *117*, 194901.

[257] C. Koo, B. E. LeBlanc, M. Kelley, H. E. Fitzgerald, G. H. Huff, A. Han, *J. Microelectromechanical Syst.* **2015**, *24*, 1069.

[258] A. Pourghorban Saghati, J. S. Batra, J. Kameoka, K. Entesari, *IEEE Trans. Microw. Theory Tech.* **2015**, *63*, 1.

[259] M. R. Khan, G. J. Hayes, J. H. So, G. Lazzi, M. D. Dickey, *Appl. Phys. Lett.* **2011**, *99*, 013501.

[260] S. J. Mazlouman, X. J. Jiang, A. N. Mahanfar, C. Menon, R. G. Vaughan, *IEEE Trans. Antennas Propag.* **2011**, *59*, 4406.
[261] M. R. Khan, G. J. Hayes, S. Zhang, M. D. Dickey, G. Lazzi, *IEEE Microw. Wirel. Compon. Lett.* **2012**, *22*, 577.
[262] E. F. Borra, G. Tremblay, Y. Huot, J. Gauvin, *PASP* **1997**, *109*, 319.
[263] M. G. Mohammed, M. D. Dickey, *Sens. Actuators Phys.* **2013**, *193*, 246.
[264] M. Li, B. Yu, N. Behdad, *IEEE Microw. Wirel. Compon. Lett.* **2010**, *20*, 423.
[265] J. Wang, S. Liu, Z. V. Vardeny, A. Nahata, *Opt. Express* **2012**, *20*, 2346.
[266] J. Wang, S. Liu, A. Nahata, *Opt. Express* **2012**, *20*, 12119.
[267] K. Ling, K. Kim, S. Lim, *Opt. Express* **2015**, *23*, 21375.
[268] P. Liu, S. Yang, A. Jain, Q. Wang, H. Jiang, J. Song, T. Koschny, C. M. Soukoulis, L. Dong, *J. Appl. Phys.* **2015**, *118*, 014504.
[269] Y. Deng, J. Liu, *J. Med. Devices* **2015**, *9*, 014502.
[270] E. Palleau, S. Reece, S. C. Desai, M. E. Smith, M. D. Dickey, *Adv. Mater.* **2013**, *25*, 1589.
[271] G. Li, X. Wu, D.-W. Lee, *Lab. Chip* **2016**, *16*, 1366.
[272] E. Palleau, S. Reece, S. C. Desai, M. E. Smith, M. D. Dickey, *Adv. Mater.* **2013**, *25*, 1589.
[273] S. J. Benight, C. Wang, J. B. H. Tok, Z. Bao, *Prog. Polym. Sci.* **2013**, *38*, 1961.
[274] B. J. Blaiszik, S. L. B. Kramer, M. E. Grady, D. A. McIlroy, J. S. Moore, N. R. Sottos, S. R. White, *Adv. Mater.* **2012**, *24*, 398.
[275] P. Cordier, F. Tournilhac, C. Soulié-Ziakovic, L. Leibler, *Nature* **2008**, *451*, 977.
[276] R. D. Deshpande, J. Li, Y. T. Cheng, M. W. Verbrugge, *J. Electrochem. Soc.* **2011**, *158*, A845.
[277] R. K. Kramer, C. Majidi, R. J. Wood, in *2011 IEEE Int. Conf. Robot. Autom. ICRA*, **2011**, pp. 1103–1107.
[278] F. L. Hammond, R. K. Kramer, Q. Wan, R. D. Howe, R. J. Wood, *IEEE Sens. J.* **2014**, *14*, 1443.
[279] R. K. Kramer, C. Majidi, R. Sahai, R. J. Wood, in *2011 IEEERSJ Int. Conf. Intell. Robots Syst. IROS*, **2011**, pp. 1919–1926.
[280] R. Matsuzaki, K. Tabayashi, *Adv. Funct. Mater.* **2015**, *25*, 3806.
[281] K. Noda, E. Iwase, K. Matsumoto, I. Shimoyama, in *2010 IEEE Int. Conf. Robot. Autom.* **2010**, pp. 4212–4217.
[282] J. T. B. Overvelde, Y. Mengüç, P. Polygerinos, Y. Wang, Z. Wang, C. J. Walsh, R. J. Wood, K. Bertoldi, *Extreme Mech. Lett.* **2014**, *1*, 42.
[283] J. Park, I. You, S. Shin, U. Jeong, *Chem. Phys. Chem.* **2015**, *16*, 1155.
[284] R. D. Ponce Wong, J. D. Posner, V. J. Santos, *Sens. Actuators Phys.* **2012**, *179*, 62.
[285] Y. Jin, Y. Lin, A. Kiani, I. D. Joshipura, M. Ge, M. D. Dickey, *Nat. Commun.* **2019**, *10*, 4187.
[286] H. Yan, Y. Chen, Y. Deng, L. Zhang, X. Hong, W. Lau, J. Mei, D. Hui, H. Yan, Y. Liu, *Appl. Phys. Lett.* **2016**, *109*, 083502.
[287] J. Choi, S. Kim, J. Lee, B. Choi, *IEEE Sens. J.* **2015**, *15*, 4180.
[288] Y. L. Park, C. Majidi, R. Kramer, P. Bérard, R. J. Wood, *J. Micromechanics Microengineering* **2010**, *20*, 125029.
[289] C. Majidi, R. Kramer, R. J. Wood, *Smart Mater. Struct.* **2011**, *20*, 105017.
[290] R. K. Kramer, Soft Active Materials for Actuation, Sensing, and Electronics, **2012**.
[291] A. Anderson, Y. Mengüç, R. J. Wood, D. Newman, *IEEE Sens. J.* **2015**, *15*, 6229.
[292] A. P. Anderson, Understanding Human-Space Suit Interaction to Prevent Injury during Extravehicular Activity, Thesis, Massachusetts Institute of Technology, **2014**.
[293] J. C. Yeo, Kenry, J. Yu, K. P. Loh, Z. Wang, C. T. Lim, *ACS Sens.* **2016**, *1*, 543.
[294] B. Li, A. K. Fontecchio, Y. Visell, *Appl. Phys. Lett.* **2016**, *108*, 013502.
[295] B. Li, Y. Gao, A. Fontecchio, Y. Visell, *Smart Mater. Struct.* **2016**, *25*, 075009.
[296] H. Ota, K. Chen, Y. Lin, D. Kiriya, H. Shiraki, Z. Yu, T. J. Ha, A. Javey, *Nat. Commun.* **2014**, *5*, 1.
[297] S. Liu, X. Sun, O. J. Hildreth, K. Rykaczewski, *Lab. Chip* **2015**, *15*, 1376.
[298] H. Li, Y. Yang, J. Liu, *Appl. Phys. Lett.* **2012**, *101*, 073511.
[299] N. Kazem, T. Hellebrekers, C. Majidi, *Adv. Mater.* **2017**, *29*, 1605985.
[300] J. N. Hohman, M. Kim, G. A. Wadsworth, H. R. Bednar, J. Jiang, M. A. LeThai, P. S. Weiss, *Nano Lett.* **2011**, *11*, 5104.

[301] J. Thelen, M. D. Dickey, T. Ward, *Lab. Chip* **2012**, *12*, 3961.

[302] T. Hutter, W. A. C. Bauer, S. R. Elliott, W. T. S. Huck, *Adv. Funct. Mater.* **2012**, *22*, 2624.

[303] M. G. Mohammed, A. Xenakis, M. D. Dickey, *Metals* **2014**, *4*, 465.

[304] A. Fassler, C. Majidi, *Adv. Mater.* **2015**, *27*, 1928.

[305] S. H. Jeong, S. Chen, J. Huo, E. K. Gamstedt, J. Liu, S. L. Zhang, Z. B. Zhang, K. Hjort, Z. Wu, *Sci. Rep.* **2015**, *5*, 1.

[306] S. Liu, M. C. Yuen, R. Kramer-Bottiglio, *Flex. Print. Electron.* **2019**, *4*, 015004.

[307] E. L. White, J. C. Case, R. K. Kramer, in *2017 IEEE Sens.*, **2017**, pp. 1–3.

[308] R. Tutika, S. H. Zhou, R. E. Napolitano, M. D. Bartlett, *Adv. Funct. Mater.* **2018**, *28*, 1804336.

[309] R. Tutika, S. Kmiec, A. B. M. T. Haque, S. W. Martin, M. D. Bartlett, *ACS Appl. Mater. Interfaces* **2019**, *11*, 17873.

[310] E. J. Markvicka, M. D. Bartlett, X. Huang, C. Majidi, *Nat. Mater.* **2018**, *17*, 618.

[311] V. Bharambe, D. P. Parekh, C. Ladd, K. Moussa, M. D. Dickey, J. J. Adams, *IEEE Antennas Wirel. Propag. Lett.* **2018**, *17*, 739.

[312] J. Shen, D. P. Parekh, M. D. Dickey, D. S. Ricketts, in *2018 IEEEMTT- Int. Microw. Symp. - IMS*, **2018**, pp. 59–62.

[313] P. Sen, C. J. Kim, *IEEE Trans. Ind. Electron.* **2009**, *56*, 1314.

[314] T. S. Kasirga, Y. N. Ertas, M. Bayindir, *Appl. Phys. Lett.* **2009**, *95*, 214102.

[315] S. Holcomb, M. Brothers, A. Diebold, W. Thatcher, D. Mast, C. Tabor, J. Heikenfeld, *Langmuir* **2016**, *32*, 12656.

[316] J. Lee, C. J. Kim, *J. Microelectromechanical Syst.* **2000**, *9*, 171.

[317] G. Beni, S. Hackwood, J. L. Jackel, *Appl. Phys. Lett.* **1982**, *40*, 912.

[318] R. C. Gough, A. M. Morishita, J. H. Dang, W. Hu, W. A. Shiroma, A. T. Ohta, *IEEE Access* **2014**, *2*, 874.

[319] H. Zeng, A. D. Feinerman, Z. Wan, P. R. Patel, *J. Microelectromechanical Syst.* **2005**, *14*, 285.

[320] Z. Wan, H. Zeng, A. Feinerman, *Appl. Phys. Lett.* **2006**, *89*, 201107.

[321] R. C. Gough, J. H. Dang, M. R. Moorefield, G. B. Zhang, L. H. Hihara, W. A. Shiroma, A. T. Ohta, *ACS Appl. Mater. Interfaces* **2016**, *8*, 6.

[322] M. Kubo, X. Li, C. Kim, M. Hashimoto, B. J. Wiley, D. Ham, G. M. Whitesides, *Adv. Mater.* **2010**, *22*, 2749.

[323] K. Kim, D. Lee, S. Eom, S. Lim, *Sensors* **2016**, *16*, 521.

[324] D. Kim, D. W. Lee, W. Choi, J. B. Lee, in *2012 IEEE 25th Int. Conf. Micro Electro Mech. Syst. MEMS*, **2012**, pp. 1005–1008.

[325] G. Li, M. Parmar, D. W. Lee, *Lab. Chip* **2015**, *15*, 766.

[326] V. T. Bharambe, J. Ma, M. D. Dickey, J. J. Adams, *IEEE Access* **2019**, *7*, 134245.

[327] B. L. Cumby, D. B. Mast, C. E. Tabor, M. D. Dickey, J. Heikenfeld, *IEEE Trans. Microw. Theory Tech.* **2015**, *63*, 3122.

[328] D. Kim, R. G. Pierce, R. Henderson, S. J. Doo, K. Yoo, J.-B. Lee, *Appl. Phys. Lett.* **2014**, *105*, 234104.

[329] N. Ilyas, D. P. Butcher, M. F. Durstock, C. E. Tabor, *Adv. Mater. Interfaces* **2016**, *3*, 1500665.

[330] D. Kim, P. Thissen, G. Viner, D. W. Lee, W. Choi, Y. J. Chabal, J. B. (J. B.) Lee, *ACS Appl. Mater. Interfaces* **2013**, *5*, 179.

[331] G. Li, M. Parmar, D. Kim, J. B. (JB) Lee, D. W. Lee, *Lab. Chip* **2014**, *14*, 200.

[332] B. L. Cumby, G. J. Hayes, M. D. Dickey, R. S. Justice, C. E. Tabor, J. C. Heikenfeld, *Appl. Phys. Lett.* **2012**, *101*, 174102.

[333] A. M. Watson, K. Elassy, T. Leary, M. A. Rahman, A. Ohta, W. Shiroma, C. E. Tabor, in *2019 IEEE MTT- Int. Microw. Symp. IMS*, **2019**, pp. 188–191.

[334] D. Kim, J. B. Lee, *J. Korean Phys. Soc.* **2015**, *66*, 282.

[335] D. B. Strukov, G. S. Snider, D. R. Stewart, R. S. Williams, *Nature* **2008**, *453*, 80.

[336] H. J. Koo, J. H. So, M. D. Dickey, O. D. Velev, *Adv. Mater.* **2011**, *23*, 3559.

[337] Y. Zhang, Z. Zhao, D. Fracasso, R. C. Chiechi, *Isr. J. Chem.* **2014**, *54*, 513.

[338] K. Du, E. Glogowski, M. T. Tuominen, T. Emrick, T. P. Russell, A. D. Dinsmore, *Langmuir* **2013**, *29*, 13640.

[339] E. A. Weiss, R. C. Chiechi, S. M. Geyer, V. J. Porter, D. C. Bell, M. G. Bawendi, G. M. Whitesides, *J. Am. Chem. Soc.* **2008**, *130*, 74.

[340] M. M. Yazdanpanah, S. Chakraborty, S. A. Harfenist, R. W. Cohn, B. W. Alphenaar, *Appl. Phys. Lett.* **2004**, *85*, 3564.

[341] A. Du Pasquier, S. Miller, M. Chhowalla, *Sol. Energy Mater. Sol. Cells* **2006**, *90*, 1828.

[342] F. Ongul, S. A. Yuksel, S. Bozar, G. Cakmak, H. Y. Guney, D. A. M. Egbe, S. Gunes, *J. Phys. Appl. Phys.* **2015**, *48*, 175102.

[343] D. J. Lipomi, B. C. K. Tee, M. Vosgueritchian, Z. Bao, *Adv. Mater.* **2011**, *23*, 1771.

[344] Q. Wang, X. Niu, Q. Pei, M. D. Dickey, X. Zhao, *Appl. Phys. Lett.* **2012**, *101*, 141911.

[345] Y. Liu, M. Gao, S. Mei, Y. Han, J. Liu, *Appl. Phys. Lett.* **2013**, *103*, 064101.

[346] J. So, H. Koo, M. D. Dickey, O. D. Velev, *Adv. Funct. Mater.* **2012**, *22*, 625.

[347] N. Hallfors, A. Khan, M. D. Dickey, A. M. Taylor, *Lab. Chip* **2013**, *13*, 522.

[348] C. Jin, J. Zhang, X. Li, X. Yang, J. Li, J. Liu, *Sci. Rep.* **2013**, *3*, 1.

[349] Y. Yu, J. Zhang, J. Liu, *PLOS ONE* **2013**, *8*, e58771.

[350] H. J. Meiselman, G. R. Cokelet, *Microvasc. Res.* **1975**, *9*, 182.

[351] M. Bradley, A. H. Sacks, *Microvasc. Res.* **1981**, *22*, 210.

[352] T. Krupenkin, J. A. Taylor, *Nat. Commun.* **2011**, *2*, 448.

[353] D. J. Lipomi, Z. Bao, *Energy Environ. Sci.* **2011**, *4*, 3314.

[354] A. C. Siegel, S. K. Y. Tang, C. A. Nijhuis, M. Hashimoto, S. T. Phillips, M. D. Dickey, G. M. Whitesides, *Acc. Chem. Res.* **2010**, *43*, 518.

[355] K. S. Yun, I. J. Cho, J. U. Bu, C. J. Kim, E. Yoon, *J. Microelectromechanical Syst.* **2002**, *11*, 454.

[356] S. Y. Tang, K. Khoshmanesh, V. Sivan, P. Petersen, A. P. O'Mullane, D. Abbott, A. Mitchell, K. Kalantar-zadeh, *Proc. Natl. Acad. Sci.* **2014**, *111*, 3304.

[357] M. Gao, L. Gui, *Lab. Chip* **2014**, *14*, 1866.

[358] M. Knoblauch, J. M. Hibberd, J. C. Gray, A. J. van Bel, *Nat. Biotechnol.* **1999**, *17*, 906.

[359] M. Hodes, R. Zhang, L. S. Lam, R. Wilcoxon, N. Lower, *IEEE Trans. Compon. Packag. Manuf. Technol.* **2014**, *4*, 46.

[360] J. Je, J. Lee, *J. Microelectromechanical Syst.* **2014**, *23*, 1156.

[361] N. Pekas, Q. Zhang, D. Juncker, *J. Micromechanics Microengineering* **2012**, *22*, 097001.

[362] S. Y. Tang, V. Sivan, P. Petersen, W. Zhang, P. D. Morrison, K. Kalantar-zadeh, A. Mitchell, K. Khoshmanesh, *Adv. Funct. Mater.* **2014**, *24*, 5851.

[363] D. P. Parekh, C. Ladd, L. Panich, K. Moussa, M. D. Dickey, *Lab. Chip* **2016**, *16*, 1812.

[364] V. J. King, *Rev. Sci. Instrum.* **1961**, *32*, 1407.

[365] W. F. Reus, M. M. Thuo, N. D. Shapiro, C. A. Nijhuis, G. M. Whitesides, *ACS Nano* **2012**, *6*, 4806.

[366] J. Shen, M. W. Aiken, M. Abbasi, D. P. Parekh, X. Zhao, M. D. Dickey, D. S. Ricketts, in *2017 IEEE MTT- Int. Microw. Symp. IMS*, **2017**, pp. 41–44.

[367] B. C. Marin, S. E. Root, A. D. Urbina, E. Aklile, R. Miller, A. V. Zaretski, D. J. Lipomi, *ACS Omega* **2017**, *2*, 626.

[368] R. Surapaneni, Y. Xie, K. Park, C. Mastrangelo, *Procedia Eng.* **2011**, *25*, 124.

[369] J. Chang, T. Ge, E. Sanchez-Sinencio, *2012 Ieee 55th Int. Midwest Symp. Circuits Syst. Mwscas* **2012**, 582.

[370] W. Zhao, J. L. Bischof, J. Hutasoit, X. Liu, T. C. Fitzgibbons, J. R. Hayes, P. J. A. Sazio, C. Liu, J. Jain, J. Badding, M. H. W. Chan, *Nano Lett.* **2014**, DOI 10.1021/nl503283e.

[371] J. Y. Chen, A. Kutana, C. P. Collier, K. P. Giapis, *Science* **2005**, *310*, 1480.

[372] E. Glickman, M. Levenshtein, L. Budic, N. Eliaz, *Acta Mater.* **2011**, *59*, 914.

[373] E. Pereiro-Lopez, W. Ludwig, D. Bellet, *Acta Mater.* **2004**, *52*, 321.

[374] S. P. Yatsenko, N. A. Sabirzyanov, A. S. Yatsenko, *J. Phys. Conf. Ser.* **2008**, *98*, 062032.

[375] P. Ahlberg, S. H. Jeong, M. Jiao, Z. Wu, U. Jansson, S. L. Zhang, Z. B. Zhang, *IEEE Trans. Electron Devices* **2014**, *61*, 2996.

[376] D. Prasai, J. C. Tuberquia, R. R. Harl, G. K. Jennings, K. I. Bolotin, *ACS Nano* **2012**, *6*, 1102.

[377] K. Doudrick, S. Liu, E. M. Mutunga, K. L. Klein, V. Damle, K. K. Varanasi, K. Rykaczewski, *Langmuir* **2014**, *30*, 6867.

[378] J. V. Naidich, J. N. Chuvashov, *J. Mater. Sci.* **1983**, *18*, 2071.

[379] E. S. Elton, T. C. Reeve, L. E. Thornley, I. D. Joshipura, P. H. Paul, A. J. Pascall, J. R. Jeffries, *J. Rheology* **2019**, *64*, 119.

[380] T. V. Neumann, M. D. Dickey, *Adv. Mater. Tech.* **2020**.

[381] T. V. Neumann, E. G. Facchine, B. Leonardo, S. Khana, M. D. Dickey, *Soft Matter* **2020**, *16*, 6608-6618.

[382] M. Song, K. Kartawira, K. D. Hillaire, C. Li, C. B. Eaker, A. Kiani, K. E. Daniels, M. D. Dickey, PNAS **2020**, *117*, 19026-19032.

# 9

## Advanced Flexible Hybrid Electronics (FHE)

**Takafumi Fukushima**
*Tohoku University*

**Subramanian S. Iyer**
*University of California*

**CONTENTS**

## 9.1 Flexible vs. Rigid Electronics

Table 9.1 shows a simple comparison between rigid inorganic electronics and flexible organic electronics. The former materials of single crystalline semiconductors are generally not flexible. In order to utilize the rigid semiconductors to flexible electronics applications, the encapsulated device chips made of the rigid single crystalline semiconductors are mounted on rigid printed circuit boards (PCBs), and then the rigid PCBs are finally interconnected through a thin flexible printed circuit board (FPC). This configuration is called rigid/flex PCBs and although they have limited flexibility, they are somewhat bendable. Therefore, the single crystal semiconductors are widely used in flexible electronics industry. These rigid/flex PCBs are employed in current smartphones and small wearable devices such as AirPods [1]. The transistor density in Si complementary metal-oxide semiconductor (CMOS) have been dramatically increasing according to Moore's law since 1965. Impressive contributions to large-scale integration (LSI) performance have been achieved by the scaling of technology nodes with gate length, although, nowadays, the half-pitch of Metal 1 (M1) layer is used, as a standard index. The wafer-level microfabrication strategy based on photolithography processes makes it possible to achieve the manufacturing of highly integrated micro/nano-electronic systems with ultrasmall devices and high-density interconnection. The other single crystal semiconductors such as III-V compounds have also played a key role in LSI development. More recently, this size and performance scaling has stalled economic reasons. However, state-of-the-art LSIs are still driven to the

**TABLE 9.1**

Comparison between Rigid Inorganic Electronics and Flexible Organic Electronics

| | Rigid inorganic electronics | Flexible organic electronics |
|---|---|---|
| Devices | Inorganic monocrystalline semiconductors (Si and III-V) | Organic semiconductors |
| Interconnects | PVD/plating-based metals | Specialty inks and pastes |
| Processing | Wafer-level | Sheet-level (roll-to-roll) |
| Manufacturing technology | Photolithography | Printing (screen- / inkjet- / gravure-) |
| Scalability | High | Low |
| Performance | High | Low |
| Flexibility | Low (rigid/flex PCBs) | High |

use of the single crystal semiconductors because their electronic properties as epitomized by a charge carrier mobility are superior to polycrystalline or amorphous semiconductors such as organic semiconductors, amorphous silicon (a-Si), low-temperature polycrystalline silicon (LTPS), and amorphous metal-oxide semiconductors (mainly indium-gallium-zinc-oxide, IGZO). The flexibility of thin-film transistors (TFT) formed with these non-single crystal semiconductors are much higher than the single crystalline Si and III-V compounds. Furthermore, the low-temperature deposition techniques allow or accelerated utilization of these materials due to their ability to stay within the thermal budget of polymeric flexible substrates. For example, polyimides (PI) show high heat resistance at 350°C, while PET is not thermally stable at above 120°C. Although the performance of the organic semiconductors has been recently improved [2, 3], the performance of inorganic single crystal semiconductors represented by Si and III-V compounds is unmatched by organic semiconductors. High-mobility device fabrication comparable to that of p-type single crystalline Si ($\mu_h$: ~500 cm$^2$/Vs) is still challenging even with amorphous metal-oxide semiconductors and LTPS [4, 5]. Sheet-level or roll-to-roll processes based on printing technology make it more difficult to enhance the size scaling of flexible electronics with the non-single crystalline semiconductors due to the limited thermal budgets. Although the applications that mix single crystal and polycrystalline semiconductors have been different so far, the capability of the use of single crystalline semiconductors in highly flexible device systems will be discussed.

## 9.2 Flexible Hybrid Electronics (FHE)

Flexible hybrid electronics (FHE) combine the flexibility of polymeric substrates with the performance of single crystal semiconductor devices to create a new category of flexible electronics [6, 7]. Traditional rigid/flex PCBs enable us to integrate thick Si dies on flexible substrates [8, 9]. However, these technologies are not based on wafer-level processing

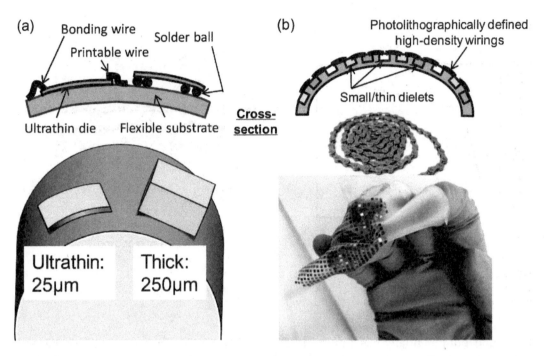

**FIGURE 9.1**
Comparison between conventional FHE with ultrathin dies [11] (a) and advanced FHE "FlexTrate™" with embedded rigid/thin/small dielets [13] (b).

and, in addition, the flexibility is limited by the rigid substrates. In order to enhance the flexibility of the rigid single crystalline semiconductors, ultrathin dies are mounted on flexible substrates [10, 11]. This is because such ultrathin dies can be more flexible and follow curved profiles with a bending radius of 5 mm when 25-μm-thick dies are employed [11], as shown in Figure 9.1a. The surface finishing of the mechanically thinned dies by grinding and CMP (chemical mechanical polishing) is an important factor to bend the ultrathin dies without damage. It is reported that a plasma treatment on ultrathin dies fabricated with dicing before grinding (DBG) process gives the dies higher repeated bendability down to 2 mm in curvature radius than saw dicing to singulate monocrystalline semiconductor wafers [12]. Another concern is the assembly/interconnection of the ultrathin dies onto flexible substrates. Low-stress processes are required for one-by-one die bonding and sequential wire printing with a wire pitch of >100 μm that is not scalable. As schematically shown in Figure 9.1a, thermocompression bonding with solder microbumps and wire bonding are gradually giving way to gentler assembly and interconnect the ultrathin dies onto flexible substrates. However, the ultrathin dies are very sensitive to applied mechanical stresses [10] by which both the performance degradation and property deviation would be induced by small bending radii. Lee et al. [13] have reported that the retention time of ultrathin dynamic random access memory (DRAM) chips having planar capacitors is degraded when the die thickness is less than 50 μm that is a critical point where the Young's modulus of single crystalline Si measured by nanoindentation is drastically lowered Figure 9.2a. In 2005 [14], Tohoku University has successfully stacked ultrathin dies with a thickness of 7 μm in layers and vertically interconnected with TSV (throughsilicon via) and solder microbump electrodes, as shown in Figure 9.2b. From the experience, ultrathin Si die fabrication is technologically possible by an advanced wafer-thinning

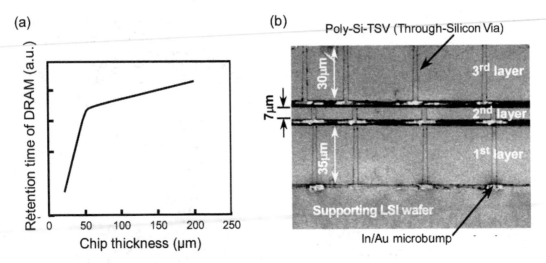

**FIGURE 9.2**
One example of property degradation with ultrathin DRAM dies [13] (a) and an SEM cross-section of 3D LSI with an ultrathin die stacked in Tohoku University [14] (b).

technique, but potentially has a property degradation issues given by thermomechanical stress. The seriousness of this issue depends on the applications. Single crystalline Si is profitably applied to stress sensor devices [15]. On the other hand, the property change by the stress is well controlled in strained Si application [16]. In an upcoming Internet-of-Everything (IoE) society with full-fledged artificial intelligent (AI), a large amount of memory is required in addition to high-performance logic and multifunctional sensors for the edge computing that is a distributed computing paradigm, which brings computation and data storage closer to the location where it is needed around us. The demands toward high-performance and multifunctional device systems with single crystalline semiconductors will be increasingly grown even in healthcare/biomedical flexible electronics for wearable and disposable applications with high reliability.

## 9.3 Advanced FHE: Concept and Fabrication

A new concept of advanced FHE is based on cutting-edge semiconductor packaging technology of FOWLP (fan-out wafer-level packaging) in which thin dies, not ultrathin dies, are embedded in an epoxy mold compound (EMC). The resulting epoxy resin wafers with embedded thin dies can realize wafer-level processing. The communication between the neighboring dies is given by wafer-level metallization called redistribution layers (RDL), as shown in Figure 9.3. It has been reported that not only the device package height can be reduced but the performance of the holistic system is also drastically improved by FOWLP even with old-generation chips are used although it is the technology governed by the price [17, 18]. This outstanding packaging technology can push the performance scaling in spite of recent Moore's law limitations. We, at UCLA and Tohoku University, are inspired by FOWLP and have tried to apply this technology to create high-performance FHE with embedded rigid/thin/small dielets [19, 20]. The dielets are not bent and the bending of flexible links

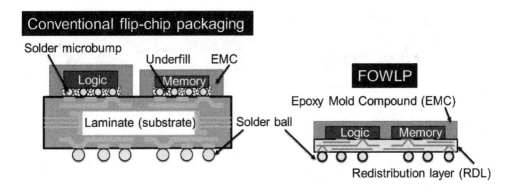

**FIGURE 9.3**
Structural comparison between conventional flip-chip packaging and FOWLP.

interconnecting the adjacent dielets gives high bendability to the advanced FHEs. The structural comparison between conventional FHE with an ultrathin/large die and advanced FHE with embedded rigid/thin/small dielets is shown in Figure 9.1b. Another advantage of our FHE is based on a modular "dielet" approach. In the dielet approach, instead of using one large application specific integrated circuits (ASIC) die, one can use multiple high-yielding smaller size dielets connected by lithographically defined metal interconnects at fine pitches (comparable to fat wire level) to allow for both higher mechanical flexibility and "on-chip"-like communication as shown schematically in Figure 9.1b.

A standard process flow is shown in Figure 9.4. First, Si or III-V (including passives, microelectromechanical system [MEMS], and small batteries – called battlets) dielets ranging in thickness from 50 to 400 μm and in size from hundreds micrometers to several millimeters in a side, typically 1 mm², were placed in a face-down configuration onto the 1st Si carrier wafer on which a thermally removable adhesive layer was laminated. A Teflon ring was used to make a retaining dam to keep the liquid raw material or precursor of a polymeric flexible substrate inside the ring. The height of this ring determines the total thickness of the advanced FHE. A biomedical grade PDMS (polydimethylsiloxane) is typically

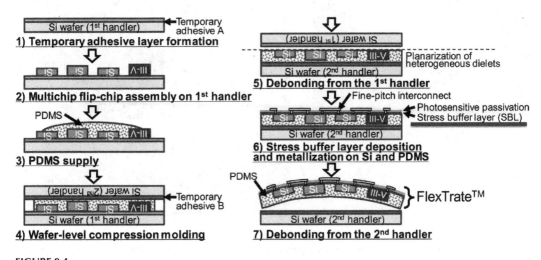

**FIGURE 9.4**
A fabrication process flow of advanced FHE called FlexTrate™.

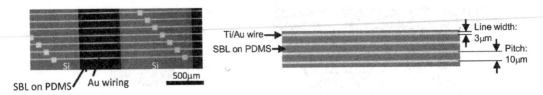

**FIGURE 9.5**
Fine-pitch interdielet wirings formed on Si dielets embedded in PDMS.

employed as a flexible substrate. In the subsequent step, the raw material of PDMS is poured on the dielets-on-wafer structure inward, and then, cured with the 2nd Si carrier wafers having another thermally removable adhesive layer that is stable at higher temperature than the 1st adhesive. After compression molding with the two Si carrier wafers, the thin/rigid/small dielets are embedded in the PDMS. The following step is debonding of the 1st Si carrier wafer at around 120°C to give a planarized surface of the dielets with various thicknesses on the 2nd Si carrier wafer without any mechanical thinning processes. Prior to the subsequent metallization processes, one or two dielectric layers as a stress buffer layer (SBL) is coated on the surface of the PDMS/dielets, followed by contact etching and metallization with using standard photolithography processes at the wafer level. Au or Cu wires are formed to interconnect the dielets embedded in the PDMS. Finally, the 2nd Si carrier wafer was debonded at around 150°C to give advanced FHE we call FlexTrate™. As seen from Figure 9.5, fine-pitch Au metallization can be performed by using a thin Ti as an adhesion layer. Figure 9.6 shows the three pictures of the advanced FHE with

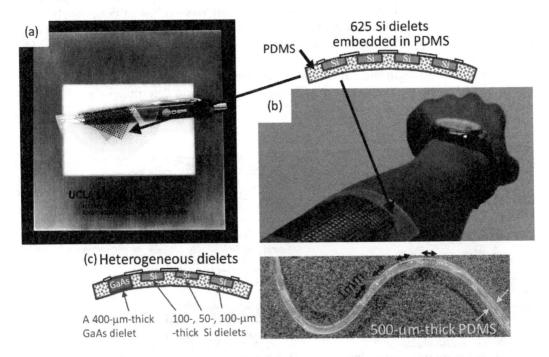

**FIGURE 9.6**
Pictures of FlexTrate™ demonstrators: rollable (a) and wearable (b) 100-μm-thick/1-mm² 625 Si dielets embedded in PDMS, and a cross-sectional image of PDMS embedding heterogeneous dielets composed of Si and GaAs with various thicknesses (c).

embedded dielets: FlexTrate™ with 625 (25 by 25) 1-mm² Si dielets are rollable to follow a pen with a curvature radius of 5 mm or less in Figure 9.6a and wearable on the human arm with a curvature radius of around 40 mm in Figure 9.6b, and heterogeneous dielets consisting of three Si dielets and a GaAs dielet are embedded in the PDMS in Figure 9.6c.

## 9.4  Advanced FHE: Characterization and Application

I-V behaviors are characterized by measuring the resistance of the Au wirings formed on the embedded dielets and the SBL in/on the PDMS. When a thin Au layer with a thickness of 200 nm is used, the resistance is pretty high, as shown in Figure 9.7. However, since the FlexTrate™ are fabricated in wafer-level processing, the resistance can be lowered by thickening the wire thickness using standard wafer-level electroplating. It is obvious that the resulting 5-μm-thick Au wire shows much lower resistance than the 200-nm-thick one. The low resistance can be kept after thermal debonding of the 2nd Si carrier wafer in the final step.

The bendability of the FlexTrate™ is characterized with an endurance testing system: tension-free U-shape folding tester (DLDMLH-FS/Yuasa). The resistances of FlexTrate™ having 600-nm-thick Cu wiring formed on the PDMS with 1-mm² embedded Si dielets with a thickness of 100 μm are evaluated with 4-point probe patterns. The interdielet distance is 0.8 mm and the Cu wirings are 15-mm long and 100, 40, 20, and 10 μm in width. The resistances are compared before and after bending with a bending radius of 10 mm and additional bending with the radius of 5 mm. As a result, both the Cu interconnects between the adjacent dielets embedded in the PDMS are still connected without severe wrinkling, micro-crack generation, and delamination. The resistance changes are within

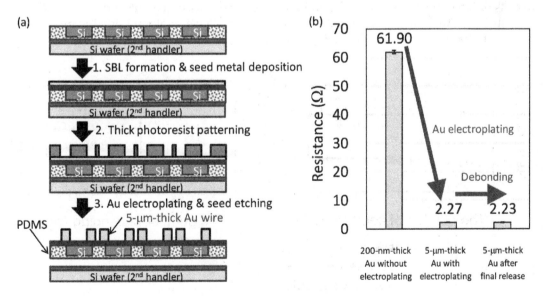

**FIGURE 9.7**
A wire thickening process in FlexTrate™ fabrication at the wafer level (a) and resistance measurement results for 200-nm-thick Au wires, 5-μm-thick Au wires before and after final debonding of the 2nd Si carrier wafer (b).

2% on average after the sequential bending cycle. Almost the same bending endurance results are obtained when 5-μm-thick Cu wirings are used. From these results, the wide ranges of Cu thicknesses are turned out to be applicable for the FlexTrate™. Here, a 1-μm-thick Parylene-C and 3-μm-thick SU-8 are employed as SBLs. Of course, surface modification such as a light plasma treatment with oxygen is required to enhance the adhesion between the PDMS and SBL.

The effect of SBLs are evaluated by using two types of SBLs: one has a low Young's modulus of 0.9 GPa and the other has a high Young's modulus of 2.3 GPa [21]. The Young's moduli of the PDMS substrate, Au metal wires, and single crystalline Si dielets are 0.0005, 79, and 190 GPa, respectively. SBL formed underneath the wirings can compensate the Young's modulus mismatch between the metal and PDMS. Figure 9.8 shows the comparison of bending property between the two SBLs. Since the PDMS is an elastic material, the Au wires are supposed to be deformed when high tensile or compressive stresses are applied to the PDMS. The small elongation at yield of Au leads to plastic deformation by which the wires are not restored to their original shape. The failure would be solved by geometrically and structurally optimizing SBL that can well control neutral plane and absorb the large stress resulted from PDMS deformation. As seen from Figure 9.8, 500-nm-thick Au wirings even with a width of 100 μm and a length of 10 mm are disconnected after 20 cycle bending with a bending radius of 20 mm when the soft SBL layer is used. In contrast, Au wirings formed on the hard SBL layer exhibit high bendability. After 1000 cycle bending, the 50-μm-width Au wirings are still connected. These bending results indicates that a hard SBL layer is essential for FlexTrate™ when we use extremely soft elastomer-based substrates such as PDMS.

Another approach is studied to obtain high bendability of FlexTrate™. As shown in Figure 9.9, corrugated wires vertically formed to a flexible substrate can further increase in bendability and give an opportunity to utilize FlexTrate™ in foldable and stretchable applications. Traditionally, serpentine U-shape wires are formed horizontally to the substrate, leading to high stretchability to the flexible electronics [22]. However, the high-density interconnections cannot be integrated by the XY-axis corrugated wires. UCLA has fabricated the Z-axis corrugated wires using wafer-level photolithographic processes with a photosensitive dielectric SU-8 [23]. Figure 9.10 shows the test results of 1000-cycle

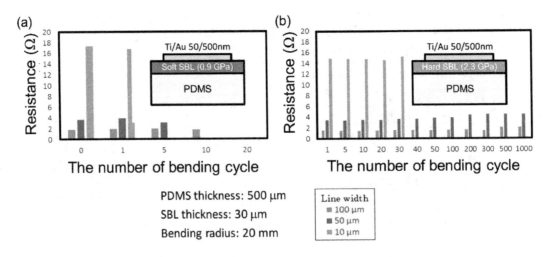

**FIGURE 9.8**
Bending property comparison between a soft SBL (a) and a hard SBL (b) [22].

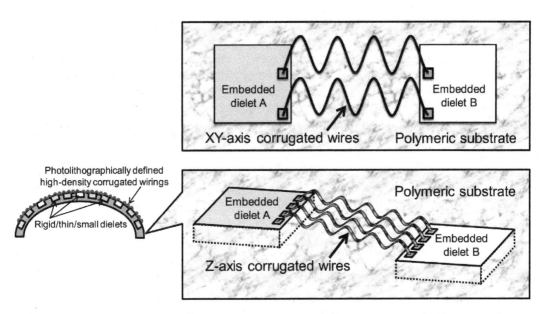

**FIGURE 9.9**
Schematic illustration of traditional XY-axis (a) and high-density Z-axis (b) corrugated wires [24].

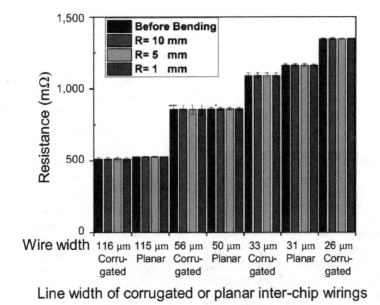

**FIGURE 9.10**
Interconnect reliability of FlexTrate™ after bending for 1000 cycles at bending radius of 10, 5, and 1 mm for planar and vertical corrugated wires [24].

bending with curvature radii of 10, 5, and 1 mm by the Yuasa system and lists the bending results with the Z-axis corrugated wires and planer straight wirings for comparison. The planar and corrugated wires were fabricated side-by-side. Surface profile demonstrating buckling of the corrugated wires with 3-μm peak-to-trough amplitude. As seen from this figure, high bendability is obtained from the corrugated wires after the additional bending cycle. The X-axis corrugated wires have shown less than 0.4% increase in average measured resistance per line width, which confirmed the reliability of the corrugated interconnects. These results are supported by FEM stress simulation by ANSYS®.

Several applications of the dielet-embedded FHE based on FOWLP are introduced in this section. The first one is flexible neural interfaces [24]. An opto-neural electrode system is fabricated in UCLA with an embedded micro-light-emitting diode (μLED) and a flexible internal coil to record the brain signals that are transferred to another flexible external coil through resonant magnetic coupling [25]. As shown in Figure 9.11, the green μLED is powered wirelessly with an efficiency >15% at 10-mm transmit distance. The implantable system is only 535 μm in thickness with a diameter of less than 20 mm. Such an ultra-flexible and thin wireless system can find immense application for subdermal implants where conforming to curvilinear surfaces and tight spots is a necessity. The standalone system with the single green μLED can be potentially applied to wireless optogenetics. Another opto-neural probe as a penetrating interface is also fabricated in Tohoku University for detecting local field potentials (LFP) in deep cerebral cortex, as shown in Figure 9.12 [26]. A 90-μm-thick blue μLED (125 μm by 225 μm) is embedded in a medical-grade flexible epoxy on which electrical wirings are formed for power supply to the μLED. The light is emitted to the other side on which recoding electrodes are formed in the same procedure. Wafer-level metallization was done on both the sides of

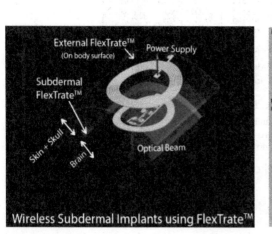
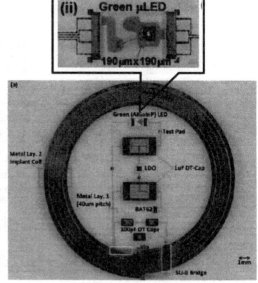

**FIGURE 9.11**
Schematic of wireless subdermal implant system using FlexTrate™ (a) and a top view of the fabricated wireless power transmission system with a bottom view of a green μLED embedded in PDMS (b) [25].

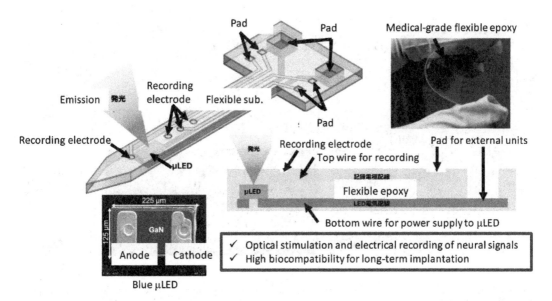

**FIGURE 9.12**
Schematic of opto-neural probe with a blue µLED embedded in flexible epoxy for optogenetic LFP detection from deep cerebral cortex [26].

the flexible epoxy with a total system thickness of 200 µm or less. When rigid Si probes are employed for deep brain stimulation, the probes push away the surrounding cells, and after that, glial encapsulation insulates the probes from the host tissues [27]. In addition, the cells/tissues are mechanically damaged and the immune reaction between the probes and cells/tissues are occurred. Therefore, biocompatible flexible neural probes alternative to rigid ones are a promising candidate to be neural interfaces for highly reliable long-term implantation.

The second example is a wearable display. As shown in Figure 9.13, an array of InGaN LED dielets are embedded in PDMS and successfully integrated with a microcontroller [28]. The basic operation of the foldable display is demonstrated by heterogeneous integration of 1-mm² Si and the µLED dielets on our platform FlexTrate™ in UCLA. This foldable display with the embedded dielets bent down to 1-mm bending radius for over 1000 bending cycles. Each segment of the 7-segment display has six µLEDs connected in parallel using Z-axis corrugated Cu interconnects at 40-µm pitch. The LEDs are powered using a power supply under current compliance of 200 mA. A display of "UCLA CHIPS" is shown in Figure 9.13. Furthermore, UCLA demonstrates complete folding of the display integrated with green LEDs, as shown in Figure 9.13. Two images of the foldable display during folding with a bending radius of 0 mm and in the range of 0–1 mm are shown in this figure. The green LEDs remain illuminated throughout the folding process.

The third application is a biomedical/healthcare sensor. Figure 9.14 shows the conceptual schematic of the trans-nail FHE system with photoplethysmographic (PPG) sensors [21]. 3D-printed nail chips are fabricated by OpenNail® technologies from Toshiba Digital Solutions Corporation. The tailor-made nail chips are designed to fit their own nail curves with various curvature radius with a 3D scanner. Underneath the Nail-Chip, an advanced FHE system composed of a red µLED and an LSI chip having photodiodes

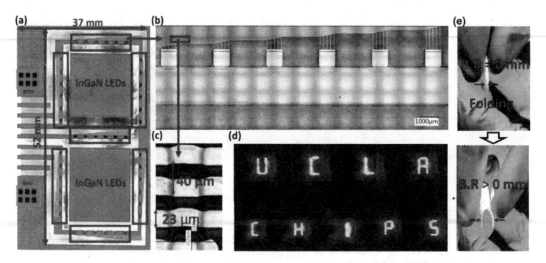

**FIGURE 9.13**
Foldable display with 42 µLEDs on FlexTrate™ (a), zoomed in image of a segment of the 7-segment display consisting of six µLEDs (b), zoomed in image of 40-µm-pitch-corrugated interconnects (c), programming of 7-segment display-to-display "CHIPS UCLA," as an example, on the foldable display (d), and images of foldable display during folding and after folding (e) [28].

and an LED driver is embedded in PDMS. Tohoku University has previously proposed the trans-nail PPG sensor system [29] and verified that the PPG recording is successfully monitored in both the reflection and transmission modes with an external LED chip. Our heart rate can be detected from the trans-nail PPG sensor. Nails are the only part not sweating on human body. Thus, the nails are the best part on which flexible devices are sitting.

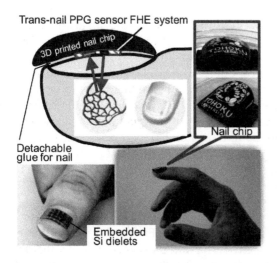

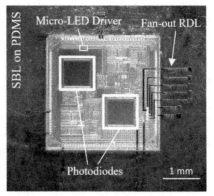

A photo of a FHE system with a PPG sensor dielet

**FIGURE 9.14**
Schematic of trans-nail FHE system with PPG sensors [29].

The forth example is retina prostheses. A well-known commercially available retina prosthesis is produced by Second Sight Medical Product, Inc., in the United States. Their epiretinal system, called Argus II, mainly consists of two parts: one is an external unit of camera module installed in an eyeglass and a video processing unit (VPU) with a signal amplifier [30]. The other is an internal unit of a wireless signal/power receiver and a 60 (6 by 10) Pt electrode array formed on a PI substrate. They only implant flexible stimulus electrodes behind the eyeball, but they have not yet implanted LSI chips. Tohoku University has proposed and demonstrated 3D LSI implantation behind the eyeball for restoring visual sensation to blind patients suffering from age-related macular degeneration (AMD) and retinal pigmentosa (RP) [31]. The 3D LSI has the layered structure of an image sensor system: the top and bottom layers have photodiodes and stimulus current generator, respectively. The biggest advantage of the 3D LSI is high aperture ratio. In conventional image sensors, photodiode and the other circuits such as logic devices occupy the chip area; while in 3D image sensor, only photodiodes can be designed on the top layer to give high resolution to the retina prosthesis. The 3D LSI can be embedded in a biocompatible flexible substrate, as shown in Figure 9.15 [32]. The advanced FHE with the embedded 3D LSI can further decrease the package size by skipping the encapsulation process and enhance the performance due to the elimination of microsoldering to the 3D LSI. In addition, the planar configuration of the advanced FHE with embedded 3D LSI will be expected to facilitate the surgery.

The final application is FHE embedding dielets in hydrogel [33]. Polyethylene terephthalate (PET), poly(ethylene 2,6-naphthalate (PEN), and polyimide are well known to be a typical flexible substrate; they cannot perfectly fit curved surfaces due to their rigidity. In contrast, PDMS has high adaptability to follow complicated 3D shapes and thermomechanical internal stress is less accumulated due to the low glass transition temperature ($Tg$) of –120°C. The above-mentioned three polymers have low permeability of substances such as water, tissue fluid, and gases (oxygen, etc.), and thus, perspiration will be trapped and lead to extreme discomfort when these flexible substrates are attached to the skin for a long time. On the other hand, hydrogels including a lot of water molecules are a network polymer and reveal higher biocompatibility (breathability) than PDMS and polyimides in addition to great adaptability. However, metallization on the hydrogels is very

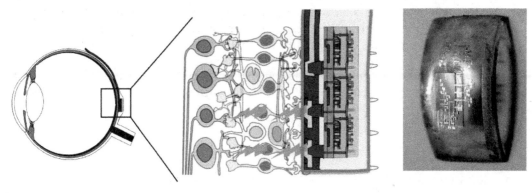

**FIGURE 9.15**
Schematic and photo of retina prosthesis system with 3D LSI embedded in PDMS [32].

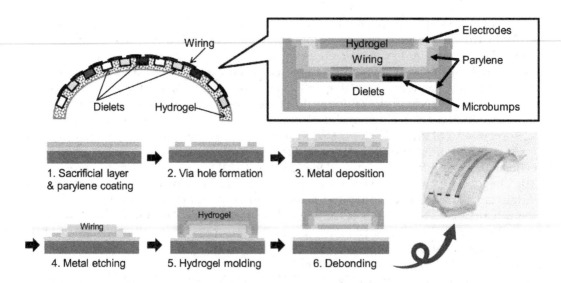

**FIGURE 9.16**
Schematic of hydrogel-based advanced FHE fabricated by RDL-first FOWLP technology [34].

challenging because heating and vacuum processes cannot be applied for the water-based hydrogels. Nishizawa et al. have reported conductive polymer wirings on a hydrogel [34] and Shimamoto et al. have described the formation of Au micro-patterns on a hydrogel [35]. Tohoku University has demonstrated high-density interconnect formation on hydrogel by using so-called RDL-first FOWLP technology with a transfer technique. As shown in Figure 9.16, the dielets can be embedded in the final step. Consequently, no heating processes are employed on the hydrogel substrate. This hydrogel based FHE is applicable to contact lens, biomedical patch, etc.

## 9.5 Summary

We have developed a novel technology for advanced FHE called FlexTrate™ with rigid/thin/small single crystalline dielets ranging in thickness from 50 μm to several hundred microns. Fine-pitch interconnects less than 20-μm pitch can be successfully formed on a large number of the Si chips embedded in PDMS by advanced WLP technologies. High bendability beyond 1 mm in curvature radius is enabled by vertical corrugated wires and SBLs that can mitigate bending stress. This new FHE can be used for various wearable and biomedical applications requiring high-performance and multifunctional heterogeneous integration with fine-pitch scalable interconnects.

## Acknowledgment

This work was supported in part by the Semiconductor Research Corporation (SRC), AFRL/NBMC, DARPA, UCLA CHIPS Consortium and the UC system. This work was partially supported by JSPS KAKENHI Grant-in-Aid for Scientific Research (A).

Grant Number: 18H04159, Challenging Research (Exploratory), Grant Number: 18K18841, and Promotion of Joint International Research (Fostering Joint International Research (B)), Grant Number: 19KK0101. This work was performed in UCLA CNSI and Nanolab cleanroom at UCLA and Micro/Nano-machining research and education Center (MNC) and Jun-ichi Nishizawa Research Center at Tohoku University.

## References

1. https://gigazine.net/news/20161221-airpods-teardown/https://www.ifixit.com/Teardown/AirPods+Teardown/75578
2. H. Iino, T. Usui, and J. Hanna, "Liquid Crystals for Organic Thin-Film Transistors", *Nat. Commun.*, Vol. 6, No. 6828, p. 8, April, 2015.
3. M. J. Kang, I. Doi, H. Mori, E. Miyazaki, K. Takimiya, M. Ikeda, and H. Kuwabara, "Alkylated Dinaphtho[2,3-b: 2′, 3′-f]Thieno[3,2-b]Thiophenes (Cn-DNTTs): Organic Semiconductors for High-Performance Thin-Film Transistors", *Adv. Mater.*, Vol. 23, pp. 1222–1225, August, 2011.
4. D. H. Lee, K. Nomura, T. Kamiya, and H. Hosono, "Diffusion-Limited a-IGZO/Pt Schottky Junction Fabricated at 200 degrees C on a Flexible Substrate", *IEEE Electron Device Lett.*, Vol. 32, pp. 1695–1697, 2011.
5. T. Serikawa, S. Shirai, A. Okamoto, and S. Suyama, "Low-Temperature Fabrication of High-Mobility Poly-SI TFTS for Large-Area LCDS", *IEEE Transactions on Electron Devices*, Vol. 36, pp. 1929–1933, 1983.
6. K. Jain, M. Klosner, M. Zemel, and S. Raghunandan, "Flexible Electronics and Displays: High-Resolution, Roll-to-Roll, Projection Lithography and Photoablation Processing Technologies for High-Throughput Production", *Proc. IEEE*, Vol. 93, pp. 1500–1510, August, 2005.
7. J. S. Chang, A. F. Facchetti, and R. Reuss, "A Circuits and Systems Perspective of Organic/Printed Electronics: Review, Challenges, and Contemporary and Emerging Design Approaches", *IEEE J. Emerging and Selected Topic in Circuit and Systems*, Vol. 7, pp. 7–26, March, 2017.
8. M. Fujiwara, Y. Shirato, H. Owar, K. Watanabe, M. Matsuyama, K. Takahama, T. Mori, K. Miyao, K. Choki, T. Fukushima, T. Tanaka, and M. Koyanagi, "Novel Optical/Electrical Printed Circuit Board with Polynorbornene Optical Waveguide", *Jap. J. Appl. Phys.*, Vol. 46, No. 4B, pp. 2395–2400, April, 2007.
9. F. Bossuyt, T. Vervust, and J. Vanfleteren, "Stretchable Electronics Technology for Large Area Applications: Fabrication and Mechanical Characterization", *IEEE Transactions on Components, Packaging and Manufacturing Technology*, Vol. 3, No. 2, pp. 229–235, February, 2013.
10. N. Wacker, H. Richte1, T. Hoang, P. Gazdzicki, M. Schulze, E. A. Angelopoulos, M. Hassan, and J. N. Burghartz, "Stress Analysis of Ultra-Thin Si Chip-On-Foil Electronic Assembly Under Bending", *Semicond. Sci. Technol.*, Vol. 29, No. 095007, p. 12, August, 2014.
11. R. L. Chaney, D. E. Leber, D. R. Hackler, B. N. Meek, S. D. Leija, K. J. DeGregorio, S. F. Wald, and D. G. Wilson, "Advances in Flexible Hybrid Electronics Reliability", *in Proc. 2017 IEEE Workshop on Microelectronics and Electron Devices (WMED)*, pp. 5–8, April, 2017.
12. DISCO Corporation 2017, Exhibition at the annual meeting of 2017 FLEX (June 19–22), Monterey.
13. K. Lee, S. Tanikawa, M. Murugesan, H. Naganuma, H. Shimamoto, T. Fukushima, T. Tanaka, and M. Koyanagi, "Degradation of Memory Retention Characteristics in DRAM Chip by Si Thinning for 3-D Integration", *IEEE Electron Device Lett.*, Vol. 34, pp. 1038–1040, August, 2013.
14. T. Fukushima, Y. Yamada, H. Kikuchi, and M. Koyanagi, "New Three-Dimensional Integration Technology Using Self-Assembly Technique", *IEEE International Electron Devices Meeting (IEDM) Technical Digest*, pp. 359–362, 2005.

15. K. Ikeda, H. Kuwayama, T. Kobayashi, T. Watanabe, T, Nishiwaka, T. Yoshida, and K. Harada, "Silicon Pressure Sensor Integrates Resonant Strain Gauge on Diaphragm", *Sensor Actuat.*, Vol. A21l–A23, pp. 146–150, 1990.
16. S. E. Thompson, G. Sun, Y. S. Choi, and T. Nishida, "Uniaxial-Process-Induced Strained-Si: Extending the CMOS Roadmap", *IEEE Transactions on Electron Devices*, Vol. 53, No. 5, pp. 1010–1020, May 2006.
17. http://www.appbank.net/2016/09/16/iphone-news/1253509.php
18. C.-F. Tseng, C.-S. Liu, C.-H. Wu, and D. Yu, "InFO (Wafer Level Integrated Fan-Out) Technology", in Proc. of the 66th Electronic Components and Technology Conference (ECTC), (2016), pp. 1–6.
19. T. Fukushima, A. Alam, S. Pal, Z. Wan, S. Jangam, G. Ezhilarasu, A. Bajwa, and S. Iyer, "'FlexTrate®'—Scaled Heterogeneous Integration on Flexible Biocompatible Substrates Using FOWLP," in Proc. of the 67th Electronic Components and Technology Conference (ECTC), (2017), pp. 1276–1284.
20. T. Fukushima, A. Alam, A. Hanna, S. C. Jangam, A. A. Bajwa, and S. S. Iyer, "Flexible Hybrid Electronics Technology Using Die-First FOWLP for High-Performance and Scalable Heterogeneous System Integration", *IEEE Transactions on Components, Packaging and Manufacturing Technology*, Vol. 8, pp. 1738–1746, 2018.
21. Y. Susumago, Q. Zhengyang, A. Jacquemond, N. Takahashi, H. Kino, T. Tanaka, and T. Fukushima, "Mechanical and Electrical Characterization of FOWLP-Based Flexible Hybrid Electronics (FHE) for Biomedical Sensor Application", in Proc. the 69th Electronic Components and Technology Conference (ECTC), (2019), pp. 264–269.
22. N. Lu, C. Lu, S. Yang, and J. Rogers, "Highly Sensitive Skin-Mountable Strain Gauges Based Entirely on Elastomers", *Adv. Funct. Mater.*, Vol. 22, pp. 4044–4050, 2012.
23. A. Hanna, A. Alam, T. Fukushima, S. Moran, W. Whithead, S. Jangam, S. Pal, G. Ezhilarasul, R. Irwin, A. Bajwa, and Subramanian S. Iyer, "Extremely Flexible (1mm bending radius) Biocompatible Heterogeneous Fan-Out Wafer-Level Platform with the Lowest Reported Die-Shift (<6 µm) and Reliable Flexible Cu-based Interconnects", in Proc. the 68th Electronic Components and Technology Conference (ECTC), (2018), pp. 1505–1511.
24. N. Lago, and A. Cester, "Flexible and Organic Neural Interfaces: A Review", *Appl. Sci.* Vol. 7, p. 1292 (27 pages), 2017.
25. G. Ezhilarasu, A. Hanna, R. Irwin, A. Alam, and S. S. Iyer, "A Flexible, Heterogeneously Integrated Wireless Powered System for Bio-Implantable Applications using Fan-Out Wafer-Level Packaging", *IEEE International Electron Devices Meeting (IEDM) Technical Digest*, pp. 683–686, 2018.
26. Shima, Y. 2019. Study of Flexible Neural Probe with Light Emitting Diode. Master diss., Tohoku Univ.
27. M. Welkenhuysen, A. Andrei, L. Ameye, W. Eberle, and B. Nuttin, "Effect of Insertion Speed on Tissue Response and Insertion Mechanics of a Chronically Implanted Silicon-Based Neural Probe", *IEEE Transaction on Biomedical Engineering*, Vol. 58, No. 11, pp. 3250–3259, November 2011.
28. A. Alam, A. Hanna, R. Irwin, G. Ezhilarasu, H. Boo, Y. Hu, C. W. Wong, T. S. Fisher, and S. S. Iyer, "Heterogeneous Integration of a Fan-Out Wafer-Level Packaging Based Foldable Display on Elastomeric Substrate", in Proc. the 69th Electronic Components and Technology Conference (ECTC), pp. 277–282, 2019.
29. Z. Qian, Y. Takezawa, K. Shimokawa, H. Kino, T. Fukushima, K. Kiyoyama, and T. Tanaka, "Development of Integrated Photoplethysmographic Recording Circuit for Trans-Nail Pulse-Wave Monitoring System", *Jap. J. Appl. Phys.*, Vol. 57, p. 04FM11.
30. https://www.secondsight.com/
31. T. Tanaka, K. Sato, T. Kobayashi, T. Watanabe, T. Fukushima, H. Tomita, H. Kurino, M. Tamai, and M. Koyanagi, "Fully Implantable Retinal Prosthesis Chip with Photodetector and Stimulus Current Generator", *IEEE International Electron Devices Meeting (IEDM) Technical Digest*, pp. 1015–1018, 2007.

32. S. Lee, Y. Susumago, Z. Qian, N. Takahashi, H. Kino, T. Tanaka, and T. Fukushima, "Development of 3D-IC Embedded Flexible Hybrid System", *IEEE International 3D System Integration Conference (3DIC)*, 2019.

33. N. Takahashi, Y. Susumago, Y. Miwa, H. Kino, T. Tanaka, and T. Fukushima, "RDL-First Flexible FOWLP Technology with Dielets Embedded in Hydrogel", in *Proc. of the 70th Electronic Components and Technology Conference (ECTC)*, pp. 811–816, 2020.

34. M. Sasaki, B. C. Karikkineth, K. Nagamine, H. Kaji, K. Torimitsu, M. Nishizawa, *Adv. Healthcare Mater.*, Vol. 3, pp. 1919–1927, 2014.

35. N. Shimamoto, Y. Tanaka, H. Mitomo, R. Kawamura, K. Ijiro, K. Sasaki, and Y. Osada, *Adv. Mater.*, Vol. 24, pp. 5243–5248, 2012.

# 10

## Metal-Laminated Fabric Substrates and Flexible Textile Interconnection

**Kyung-Wook Paik**

*Korea Advanced Institute of Science and Technology*


**Seung-Yoon Jung**

*Korea Advanced Institute of Science and Technology*

## CONTENTS

## 10.1 Introduction

Smart clothes using electronic textiles (e-textiles) technology are considered as one of the future wearable devices, because they combine electronic devices with the clothing which we are wearing on a daily basis. Although the goal of the e-textiles is to fabricate all fabric-based electronic devices (Cherenack and van Peterson 2012), current e-textiles integrate traditional electronic devices such as sensors, light emitting diode (LED) lighting, and thin film transistors (TFTs) into fabrics (Büscher et al. 2015, Cherenack et al. 2010, Bonderover and Wagner 2004, Buechley and Eisenberg 2009, and Krshiwoblozki, Linz, Neudeck, and Kallmayer 2012). Usually, electronic components were firstly assembled on conventional substrates such as printed circuit boards (PCBs) or flexible printed circuits (FPCs) using polyimide (PI) or polyethylene terephthalate PET films, and then inserted in the fabrics (Cherenack et al. 2010 and Bonderover and Wagner 2004) or surface mounted on the fabrics having electrical circuits already formed (Buechley and Eisenberg 2009 and Krshiwoblozki, Linz, Neudeck, and Kallmayer 2012).

Normally, electrical circuits in the fabrics were formed using conductive threads. They can be either pure metal wires (Cottet, Grzyb, Kirstein and Troster 2003), metallized polymeric fibers (Trindade, Martins, Miguel, and Silva 2014 and Mercier and Chandrakasan 2011), or conductive polymer-coated threads (Ding, Invernale, and Sotzing 2010). The threads were then sown or embroidered into the fabrics. Using conductive threads, these fabric substrates have advantages such as compatibility to the cloth production and flexibility. However, due to the limited width reduction, the conductive threads have disadvantages in achieving fine-pitch electrical circuits. In addition, only one or two contact points at the end of conductive threads can be interconnected when the threads were used. Because of these problems, a novel fabric substrate having fine metal patterns was introduced (Jung and Paik 2017). This method used B-stage adhesive films or non-conductive films (NCFs). Fine-pattern metal electrodes were fabricated on the metal foils laminated on the NCFs by conventional patterning processes, and then laminated onto the conventional fabrics. The NCFs are epoxy-based film type adhesives which are solid at room temperature. Therefore, fine-pitch metal circuits can be fabricated directly on the NCFs by a conventional lithography process. And during the lamination process with applying heat and pressure, the NCFs resin flowed into the fabrics during lamination resulting in the metal pattern on the fabrics. This method can provide fine-pitch electrical circuits on the fabric materials as conventional PCBs and FPCs.

In e-textiles, interconnection methods of the fabrics with the conventional electronic devices were commonly formed by connectors, stitching, or soldering (Trindade, Martins, Miguel, and Silva 2014, Mercier and Chandrakasan 2011, Cherenack et al. 2010, and Buechley and Eisenberg 2009). However, the button-type connector interconnection requires larger volume, which can be only applied for lower-density interconnections, and the soldering method may cause thermal damages on the polymer-based fabrics due to the high-temperature soldering process and also solder fatigue problems. In addition, both connector and soldering interconnection methods have flexibility problem at the joint area. On the other hand, the anisotropic conductive films (ACFs) interconnection can provide robust and flexible interconnection. The ACFs are film-type adhesive materials widely used in conventional electronic packaging such as flat-panel displays and flip-chip assemblies. ACFs consist of thermosetting resin and conductive particles randomly dispersed in the resin. The conductive particles can be either metal particle, metal-coated polymer balls, or solder particles. By applying heat and pressure, the ACFs resin flow occurred and filled the space between components being bonded, and the conductive particles were captured between electrodes forming electrical interconnection. After that, the ACFs resin started curing, and the cured resin provided the mechanical adhesion between two components. The ACFs interconnection has advantages over conventional interconnection methods such as lower temperature process, reduced assembly thickness, and fine-pitch capability. In addition to those advantages, recent studies have shown that the ACFs can be one of the interconnection solutions for flexible chip packaging such as chip on flex (COF) and chip in flex (CIF), because ACFs have compliance against bending and flexing environments (Kim et al. 2016, Kim, Kim and Paik 2019, and Kim et al. 2017).

In this chapter, the interconnection method of the Si chip and the fabric for e-textiles will be presented using metal-laminated fabric substrates and ACFs interconnection on the fabric substrates. First, the fine-pitch and flexible metal-laminated fabric substrates will be demonstrated, and the effects of the NCFs material properties on the lamination and bending properties of the fabric substrates will be discussed. And then, the interconnection of Si chip and fabric substrates, or chip on fabric (COFa) will be presented using ACFs.

## 10.2 Fine-Pitch Metal-Laminated Fabric Substrates Using B-Stage Non-Conductive Films (NCFs)

### 10.2.1 Materials and Fabrication Processes

To fabricate the metal-laminated fabric substrates, 12-μm thick Cu foil, 40-μm thick B-stage NCFs, and commercially available polyester/rayon-woven fabrics were used. The NCFs were prepared by coating NCFs resin on a releasing film. The NCFs resin consisted of epoxy resin, thermoplastic resin for film formability, curing agent, and elastomer to control the materials properties of the NCFs. Three types of NCFs having various elastomer contents were used and the materials properties of the NCFs were summarized in the Table 10.1. As the elastomer contents increased, the NCFs resin viscosity increased due to the high molecular weight of the elastomer; however, the modulus and peel adhesion strength decreased because the relative amount of the epoxy resin decreased.

Figure 10.1 shows the process steps to fabricate metal-laminated fabric substrates. Cu was firstly attached to the NCFs coated on a releasing film. And then, Cu was patterned using conventional photolithography and Cu wet etching processes. For better handling, Cu/NCF/releasing film was temporary attached to a 4-inch silicon wafer using a double-sided tape. Through the patterning processes, the NCFs remained as solid

**TABLE 10.1**

Materials Properties of the NCFs Used in This Study

| | Units | NCF A | NCF B | NCF C | Test methods |
|---|---|---|---|---|---|
| Elastomer contents | wt% | 40 | 30 | 0 | |
| Storage modulus[a] | MPa | 10 | 183 | 1000 | Dynamic mechanical analyzer (DMA) |
| Peel adhesion strength[b] | gf/cm | 583 | 841 | 1555.3 | 90-degree peel test |
| Minimum viscosity[c] | Pa•s | 14,470 | 4920 | 21 | Rheometer |

[a] At room temperature.
[b] Cu/Fabric laminates.
[c] Ramp up speed: 5°C/min.

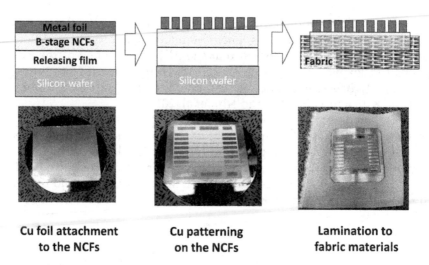

**FIGURE 10.1**
Fabrication of metal laminated fabric substrates using B-stage NCFs.

state, and the Cu patterns were successfully fabricated on the NCFs. Finally, the carrier wafer and releasing film were removed and the Cu electrodes on the NCFs were laminated onto the fabrics using a vacuum-lamination method (160°C, 60 min, and 0.14 MPa $N_2$ pressure). By applying heat and pressure, the NCFs resin flow due to the reduced NCFs viscosity caused NCFs permeation into the porous the fabrics which will be discussed later.

### 10.2.2 NCFs Curing Property Optimization

Normally, conventional patterning processes such as photoresist (PR) baking and metal wet etching are performed at elevated temperatures. As a result, the epoxy-based NCFs can be thermally pre-cured before laminating onto the fabrics. And if they are pre-cured too much, the lamination cannot be possible because of the lack of adhesion. To prevent the NCFs pre-curing, the NCFs curing reaction should take place above the maximum processing temperature, in this case, 110°C. On the other hand, if the curing reaction temperature was too high, the NCFs resin will not be sufficiently cured during the processing. Therefore, the NCFs curing onset temperature should be optimized to prevent NCFs precuring during metal patterning process and NCFs should be sufficiently cured during the lamination process at 160°C. Figure 10.2 shows the degree of cure of the NCFs after the patterning process measured by the Fourier transform infrared (FT-IR) spectroscopy and the peel adhesion strength of the Cu/fabric laminates using pre-cured NCFs by a 90-degree peel test. As the curing onset temperatures increased, the NCFs were severely pre-cured over 50%. However, the NCFs having higher curing onset temperatures did not show any resin pre-curing.

If severely pre-cured NCFs were laminated onto the fabrics, NCFs showed poor adhesion to the fabrics. As shown in the Figure 10.3, NCFs were clearly detached from the fabrics resulting in degraded peel adhesion strengths. In terms of resin pre-curing and lamination temperatures, the onset temperature of the NCFs was optimized at 150°C, which showed less than 10% resin pre-curing after the patterning process, and finally cured after the 160°C vacuum-lamination temperature.

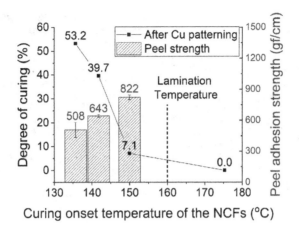

**FIGURE 10.2**
Degrees of pre-curing after the patterning process and the peel adhesion strength of Cu/fabric laminates using pre-cured NCFs.

### 10.2.3 Effects of NCFs Viscosities on the Fabric Substrates Morphology

The metal electrodes patterned on the NCFs were laminated onto the fabrics using a vacuum lamination method. During the lamination process, heat and pressure were applied and the NCFs viscosity gradually decreased with heating temperatures. As a result, NCFs resin flow occurred in two directions: 1) in-plane resin flow to fill the space between neighboring electrodes and 2) resin permeation into porous fabrics as shown in Figure 10.4.

Figure 10.5 shows the top-view optical microscope (OM) images of 100-μm-pitch Cu electrodes and cross-section SEM images of the 500-μm-pitch Cu electrodes on the fabric substrates using three types of NCFs. As the minimum viscosity of the NCFs decreased, the Cu electrodes were severely tilted, which could be explained by severe in-plane resin

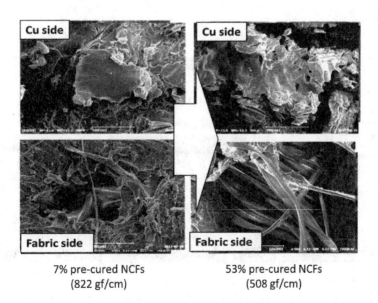

7% pre-cured NCFs
(822 gf/cm)

53% pre-cured NCFs
(508 gf/cm)

**FIGURE 10.3**
Top-view SEM images of fractured surfaces of Cu and fabrics after the peel test.

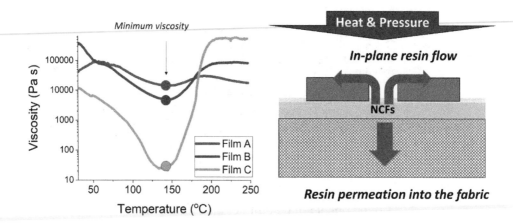

**FIGURE 10.4**
Viscosity behavior of the NCFs and NCFs resin flow behavior during the lamination process.

flow to push the electrodes away causing Cu electrodes' displacement. This result shows that the Cu electrodes cannot be laminated on the fabrics as desired.

On the other hand, when the NCFs viscosity increased, NCFs resin extruded between two neighboring Cu electrodes, which can be explained by the NCFs permeation behavior. The resin-permeation behavior through the porous fabric materials can be governed by the Darcy's law,

$$Q = -K \cdot \Delta P / \eta \tag{10.1}$$

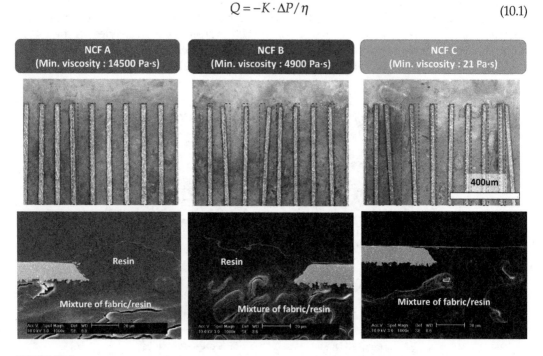

**FIGURE 10.5**
Top-view optical microscope (OM) images and cross-sectional SEM images of the fabric substrates using three types of NCFs (red dash line indicates the original metal patterns to be laminated, and red solid line indicates the NCFs resin area between the Cu electrodes).

**FIGURE 10.6**
Top-view OM images of the ENIG/Cu electrodes and voids on the NCFs.

where $Q$ is the volumetric permeation rate, $\Delta P$ the pressure difference, $\eta$ the viscosity, and $K$ the permeation constant depending on the porosity of the fabric substrates. Under Cu electrodes, NCFs resin was mixed with the fabric materials due to resin permeation. However, when higher-viscosity NCFs A and B were used, the extruded NCFs resin was observed between neighboring Cu electrodes as indicated as red lines in Figure 10.5.

### 10.2.4 Metal Surface Finish of Cu Electrodes on the B-Stage NCFs

In order to achieve a stable interconnection, Cu electrodes should be protected from oxidation. Therefore, electroless nickel immersion gold (ENIG) metal finish, which is one of the widely used for printed circuit boards (PCBs), was performed on the Cu electrodes after Cu patterning was completed on the NCFs. ENIG metal finish was also performed at elevated temperature (80°C); however, there was no NCFs resin pre-curing occurred, because the optimized onset temperature of the NCFs curing was higher than the ENIG-plating process temperature. However, another problem occurred during the electroless nickel plating process. Since the NCFs were exposed after Cu etching, the NCFs resin flow can be easily occurred at high temperature, as explained before. As a result, the low-viscosity film C showed severe voids in the NCFs, because severe resin flow caused resin agglomeration on the releasing film as shown in Figure 10.6. Therefore, to fabricate ENIG/Cu electrodes on the NCFs, NCFs viscosity should be as high as possible to prevent unstable NCFs morphology and Cu electrode pattern distortion.

## 10.3 Flexibility of the Fabric Substrates

Usually, metal electrodes may not be flexible compared to conductive threads made of polymeric fibers. Therefore, the flexibility of the metal patterns on fabric substrates needed to be investigated depending on the NCFs materials' properties. The effects of the NCFs modulus on the bending fatigue properties of metal electrodes were evaluated through simple theoretical bending stress analysis using the neutral axis theory and experimental dynamic bending tests. Since the ENIG process caused unstable NCFs morphology for lower-viscosity NCFs, the bending fatigue properties were evaluated only for bare Cu electrodes laminated on the fabric substrates.

### 10.3.1 Bending Stress Analysis of the Metal Pattern on Fabric Substrates

The neutral axis theory was used to predict the bending stress of the Cu electrodes on the fabric substrates. When the structure was bent, outer bent region was under a tensile stress while inner region under a compressive stress. The neutral axis is the plane or axis of the structure where no tensile or compressive stress is applied. The neutral axis position ($y_N$) of the multilayered composites can be calculated using the following equation,

$$y_N = \frac{\sum E_i \times y_i \times t_i}{\sum E_i \times t_i} \tag{10.2}$$

where $E$ is the modulus, $t$ the thickness, and $y$ the position of the center of each layer, which can be calculated using the thickness ($t$) variable. And the bending stress can be calculated as,

$$\sigma = E \times d/R \tag{10.3}$$

where $E$ is the modulus, $d$ the distance from the neutral axis, and $R$ the bending radius. In case of the fabric substrates, double-layer model consisting of Cu and NCF/fabric laminate was used because the NCFs resin permeated into the fabric materials. Therefore, mechanical properties of the Cu and NCF/fabric laminates need to be understood for the bending stress prediction using the neutral axis theory.

To measure the modulus of the NCF/fabric laminates, a 3-point bending test was used to measure the flexural modulus (ASTM D790). According to Lee et al. and Zweben et al., the flexural modulus of the polymer-reinforced fabric composites measured by a 3-point bending test depended on various factors such as 1) the polymer and fiber properties, 2) polymer distribution in the fabrics, 3) and test direction relative to the bending directions of the fabrics, which was characteristic of the fabrics (Zweben, Smith, and Wardie 1979 and Lee, Kim, Kim and Kim 2016). This study focused on the NCFs properties; therefore, other variables were assumed to be the same by fixing all the test directions (3-point bending test and dynamic bending test) relative to the fabric texture with the assumption that the surface resin amount was the same.

Table 10.2 summarized the modulus and thickness of Cu and various NCF-laminated fabrics. When the NCFs had higher modulus, the flexural modulus of the laminate increased,

**TABLE 10.2**

Modulus and Thickness Data for Bending Stress Calculation on the Fabric Substrates

| | Units | Cu foil[a] | NCF-laminated fabrics[b] | | |
|---|---|---|---|---|---|
| | | | NCF A | NCF B | NCF C |
| Modulus | GPa | 88.0 | $37.6 \times 10^{-3}$ | $80.6 \times 10^{-3}$ | $2512.9 \times 10^{-3}$ |
| Thickness | μm | 12 | 348 | 348 | 348 |

[a] ASTM standard E345 (Type B specimen).
[b] Flexural modulus (ASTM standard D790).

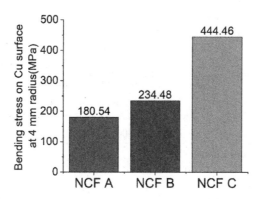

**FIGURE 10.7**
Calculated bending stress on the Cu surface at a 4-mm bending radius.

which can be explained by NCFs reinforcing effects. Based on the modulus and thickness data, the bending stresses on the Cu electrode surface was calculated using equation 10.2. As shown in Figure 10.7, the low bending stress at a 4-mm bending radius was achieved on the Cu electrodes when low-modulus NCFs were used.

## 10.3.2 Bending Fatigue Test Results

A bending fatigue test was conducted to evaluate the bending flexibility of the fabric substrates depending on the modulus of the NCFs. The test was performed as shown in the Figure 10.8. Fabric substrates were located between two parallel plates and the edge of the sample was gripped. While the upper plate was stationary, the lower plate was in cyclic linear motion using an actuator. The sliding length was 5 mm and the test speed was 200 cycles/min. The bending radius was fixed at 4 mm and the bending test was performed in convex bending direction, where tensile stress was applied on Cu electrodes. To find out the effects of bending stress as previously discussed, bending test was performed for three types of fabric substrates using three NCFs.

During the test, the linear motion of the bottom plates created two regions of the bending deformation. While the center of the fabric substrates was in static bent state, the region of the substrates nearby the plates (marked as red and blue dots in the Figure 10.8) were under cyclic bending deformation, which was fatigue region. As a result, Cu line resistance gradually increased and eventually failed by Cu line fracture at the fatigue region.

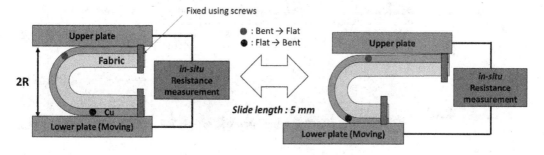

**FIGURE 10.8**
Test setup for the bending fatigue test of the Cu electrodes on fabric substrates.

**TABLE 10.3**

Failure Cycles after Dynamic Bending Test

| | Bending stress[a] (MPa) | Failure cycles of Cu lines[b] | | |
|---|---|---|---|---|
| | | 380-μm width | 180-μm width | 80-μm width |
| NCF A | 180.5 | 267,500 | 135,900 | 21,400 |
| NCF B | 234.5 | 109,700 | 52,200 | 4700 |
| NCF C | 444.5 | 1100 | 120 | 90 |

[a] At the 4-mm bending radius.
[b] Failure criteria: 20% of line resistance increase.

Table 10.3 summarized the failure cycles depending on the NCFs types and Cu line widths. Here, the failure cycle was defined as the cycle where the line resistance increased to 20% of the initial resistance. As a result, failure cycles increased when lower-modulus NCFs were used as expected. In addition, narrower Cu lines showed lower failure cycles.

Figure 10.9 shows the top and cross-section images of the 380-μm-width Cu lines after the bending test. The early failure of the narrow Cu lines might be explained by the Cu crack propagation behavior through the width direction. Under the Cu electrodes, debonding of the NCFs resin from the fibers was also observed for the film A and B (as shown in the red arrows in Figure 10.9). This delamination between fabric and resin under mechanical deformation was consistent with the other studies on epoxy reinforced fabric composites using glass fabric, braided carbon fiber, and flax fabrics (Yu et al. 2015, Lomov et al. 2008, and Mahboob et al. 2017). It was also found that Cu crack also initiated at the delaminated interface between fabric and Cu electrode (NCFs A in Figure 10.9). These findings suggested that the delamination might have an effect on the Cu crack initiation and the resulting Cu line failure. Normally, it was known that the interfacial delamination caused localized strain and even buckling (Kim et al. 2017 and Remmers and De Brost 2001). Therefore, to find out the effect of the delamination on the Cu crack behavior of the fabric substrates, bending fatigue behavior of the fabric substrates having high adhesion but also low bending stress was needed to be investigated.

### 10.3.3 Improvement of Bending Fatigue Life by Adhesion Enhancement

Surface modification of polymeric textile fiber is widely used to improve the adhesion between polymer and fabrics for laminates or coating applications. Among them, silane coupling agent was used to improve the adhesion strength of the fabric substrates without

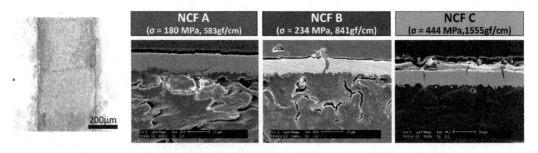

**FIGURE 10.9**
Top-view OM images of the failed Cu lines and cross-sectional SEM images of the fabric substrates after 500,000 cycles bending test.

**TABLE 10.4**

Materials Properties and Bending Test Results Using Coupling Agent-Treated Fabrics

| Types of fabrics[a] | Peel strength (gf/cm) | Bending stress[b] (MPa) | Failure cycles of Cu lines[c] | | |
|---|---|---|---|---|---|
| | | | 380-μm width | 180-μm width | 80-μm width |
| Untreated | 582.7 | 180.5 | 267,500 | 135,900 | 21,400 |
| Coupling agent-treated | 1152.0 | 171.2 | >500,000[d] | 170,400 | 75,900 |

[a] NCFs A.
[b] At 4-mm bending radius.
[c] Failure criteria: 20% of line resistance increase.
[d] Test cycle: 500,000 cycles.

affecting the mechanical properties of the fabric laminates. By the hydrolysis and condensation reactions, silane-derivate groups were chemically attached on the surface of the polymer fibers and then coupling reaction between organofunctional groups of the silane on the fibers and polymer resin matrix occurred to form chemical bonding between fiber and the polymer matrix (Luo and van Ooij 2002 and Xie et al. 2010).

Polyester/rayon fabrics were treated with the silane-coupling agent and the fabric substrates were fabricated using the low-modulus NCFs A, which showed the lowest bending stress and the highest failure cycles. When coupling agent-treated fabrics were used, peel adhesion strength drastically increased as shown in the Table 10.4. Although the adhesion strength was increased, the flexural modulus of laminates using coupling agent-treated fabrics was almost the same as that of the untreated fabric. This result might be because the coupling agent did not affect the mechanical properties of the fabrics.

The lower-modulus and higher-adhesion fabric substrates showed further improved bending fatigue life (Table 10.4). Especially, 380-μm-width line did not fail after 500,000 cycles for convex bending. After 300,000 cycles convex bending where Cu crack occurred for untreated fabrics, however, debonding of the coupling agent-treated fabric substrate was not severe, which also supported that the delamination affected the Cu crack formation. And the bending strain mapping using the digital image correlation (DIC) method (Figure 10.10) showed that the higher-adhesion fabric substrates showed lower bending strain at the 4-mm bending radius. And the localized strain up to 14% was observed at the interface between Cu and fabric laminates. This suggested that localized strain caused Cu crack on the interface, leading to early Cu failure. In summary, NCFs having lower modulus and higher adhesion properties showed the best bending performances, which showed no Cu electrode damages.

## 10.4 ACFs Interconnection of the Fabric Substrates: Chip On Fabric (COFa)

### 10.4.1 Materials and Test Vehicles

Si chip interconnection on the previously optimized fabric substrates was demonstrated by using ACFs. A 50-μm-thick Si chip was designed and there were Cu/Ni/Au bumps on the peripheral array of I/O pads of the chip. Fabric substrates were designed to have a daisy chain test pattern to measure continuity of the ACFs joint and a 4-point Kelvin structure to measure single ACFs joint contact resistances after Si chip bonding.

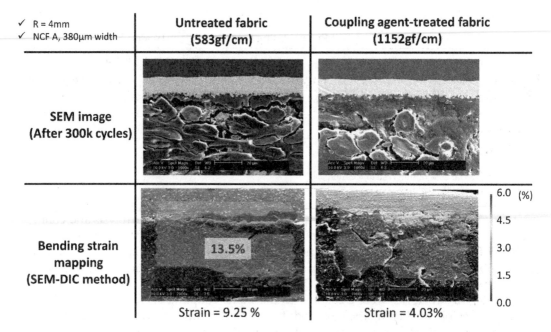

**FIGURE 10.10**
Cross-sectional SEM images after 300,000 cycles bending test and the SEM-DIC results of the Cu electrode on fabric substrates.

To fabricate the fabric substrates, Cu circuits were firstly fabricated on the NCFs followed by ENIG metal finish. And the ENIG/Cu electrodes were laminated onto the fabrics using a vacuum-lamination method. Figure 10.11 shows the test chip and the fabric substrates.

ACFs consisted of epoxy-based thermosetting polymer resin and Au/Ni-coated polymer balls as conductive particles. Polymer balls had 20-μm diameter and the content was 20 wt%. Also, as cover layer materials, 135-μm-thick PI film and epoxy-based cover adhesive films were used. The thickness of the cover adhesive films varied from 15 to 60 μm.

### 10.4.2 Fabrication and Evaluation of COFa Using ACFs

To fabricate the flip-chip COFa assemblies, ACFs were laminated on the Si chip and bonded onto the fabric substrates using a thermo-compression (T/C) bonding method. The bonding condition was 1 MPa, 210°C, for 10 sec. After T/C bonding, fabric substrates on the bonding area were highly compressed because high temperature and pressure were applied on the fabric substrates, as shown in Figure 10.12.

Figure 10.13 shows the ACFs joint properties of the COFa in comparison to conventional chip on flex (COF). In the ACFs joint, conductive polymer balls were well captured between bumps on the chip and electrodes on fabric substrates. As a result, stable daisy chain resistances and contact resistances were obtained. Compared to conventional COF, the single joint resistance using the same 4-point Kelvin structure was lower when the fabric substrates were used.

Higher peel strength was obtained when the fabric substrates were used than using flex substrates. While COF failure occurred between ACFs resin and flex substrates (Kim et al. 2017), COFa showed cohesive failure of the fabric substrates and ENIG/Cu electrodes were detached from the fabric substrates as shown in Figure 10.14.

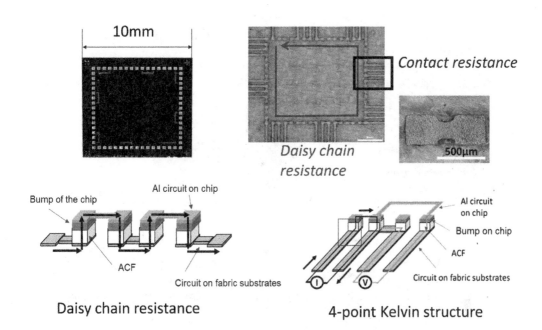

Contact resistance

Daisy chain resistance

Daisy chain resistance

4-point Kelvin structure

**FIGURE 10.11**
Test chip and fabric substrates with a daisy chain resistance and the 4-point Kelvin structures for COFa assembly.

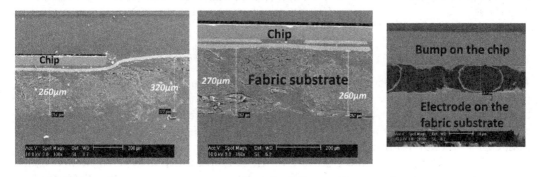

**FIGURE 10.12**
Cross-sectional SEM images of the COFa assemblies after T/C bonding.

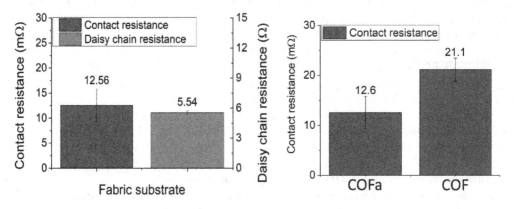

**FIGURE 10.13**
Electrical daisy chain resistances and contact resistances of the COFa and COF.

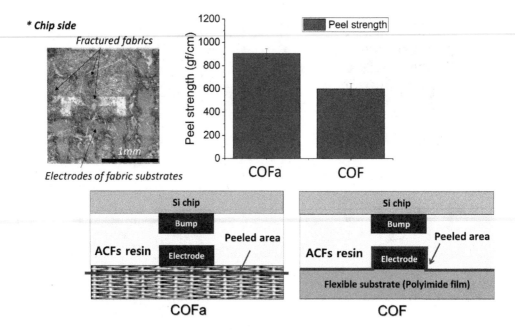

**FIGURE 10.14**
90-degree peel adhesion test results of COFa and COF and schematic diagrams of the locus of failure during the peel test.

### 10.4.3 Flexibility of COFa with the Cover Layer Structure

Although the flexible substrates and thinned Si chip were used, it was well known that the COF had low flexibility, because Si chip can be easily fractured under a bending condition. To solve this problem, the chip in flex (CIF) package structure was introduced, whose cover layer structure was applied on top of the COF. CIF showed much more bendable than COF without chip fracture and this was theoretically explained by the neutral axis theory. The bending stress on the surface of the Si chip ($\sigma$) can also be calculated using equation 10.3; whereas in this case, $E$ is elastic modulus of the Si chip, $d$ the distance from the neutral axis position, and $R$ the bending radius. By applying cover layer structure on the chip, the neutral axis position shifted toward the chip center and the theoretical bending stress on the chip was reduced (Kim et al. 2016 and Kim, Kim and Paik 2019).

This cover layer structure concept was also applied to the COFa to improve the static bending flexibility. After COFa was fabricated, the cover layer structure consisting of 135-µm-thick PI film and epoxy-based cover adhesive films was laminated onto the flip-chip surface of the COFa using a vacuum-lamination method. The lamination condition was 110°C, 5 min under nitrogen pressure of 0.14 MPa (25 psi).

Figure 10.15 shows the calculated "$d$" value of the COFa structure depending on the total cover layer thickness. Cover PI film thickness was fixed at 135 µm and cover adhesive film thickness was varied from 15 to 60 µm. As the thickness of cover layer on the chip increased, the "$d$" value gradually decreased, which suggested that thicker cover layer structure was necessary for better flexibility without chip crack.

The 4-point bending test was performed to evaluate the static bending flexibility of COFa. Test speed was 10 µm/sec and the minimum bending radius was measured where Si chip was fractured. Since the bending can be either convex (tensile stress on

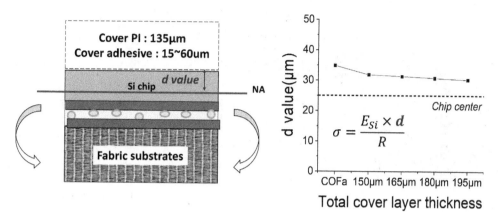

**FIGURE 10.15**
Theoretical neutral axis position of COFa with various cover layer structures.

the chip surface) or concave (compressive stress on the chip surface), the 4-point bending test was conducted for both bending directions. Figure 10.16 shows the minimum bending radius of the COFa depending on the total cover layer structure thickness. When the tensile stress was applied on the chip (convex bending), Si chip without cover layer structure was easily fractured at 25-mm bending radius. However, as the thickness of cover layer structure on chip increased, minimum bending radius drastically reduced to 7.4 mm. On the other hand, for concave bending, slight minimum bending radius changes were observed because in this case, compressive stress was applied on the chip. Based on the results, COFa structure was optimized as 135-μm cover PI film and 60-μm cover adhesive films.

## 10.4.5 Reliability of the Optimized COFa

Reliability of the optimized COFa was evaluated under dynamic bending, 85°C/85% relative humidity (RH) (85/85) and washing environment.

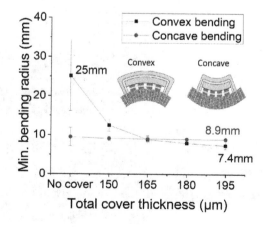

**FIGURE 10.16**
4-point bending test results of COFa with various cover layer structures under convex and concave bending directions.

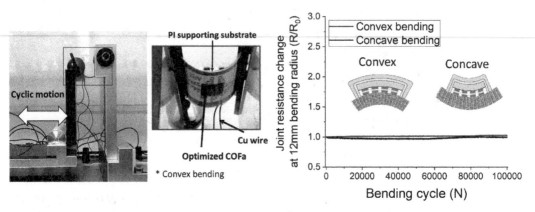

**FIGURE 10.17**
Dynamic bending test setup of COFa and bending results of the optimized COFa.


First, dynamic bending test was conducted using 2-axis bending test machine. During the bending test, daisy chain resistances were measured in situ by connecting pads to the electrical resistance measuring system using Cu wires as shown in Figure 10.17. After 100,000 cycles bending at 12-mm radius, there was no changes in the daisy chain resistances. This stable joint resistance can be explained by that the ACFs had compliance against bending deformation (Kim et al. 2016). In addition, there was no Si chip fracture under cyclic bending deformation.

The 85°C/85% RH (85/85) test is one of the reliability test methods for traditional ACFs flip-chip assemblies. Figure 10.18 shows the 85/85 test results of the optimized COFa. Contact resistances gradually increased during the test; however, no open failure was observed after 1000 h. Cross-section SEM images showed that ACFs resin and fabric substrates were swollen after 85/85 test. Especially, for fabric substrates, NCFs resin swelling caused interfacial delamination between ENIG/Cu electrode and fabric substrates. Under high temperature and humid environment, both NCFs resin in the fabric

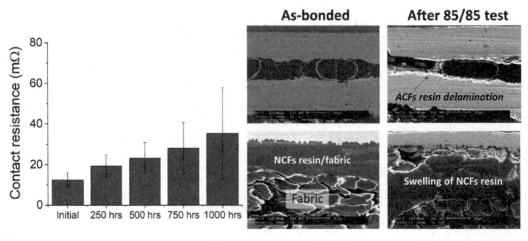

**FIGURE 10.18**
85°C/85% RH test results of the optimized COFa.

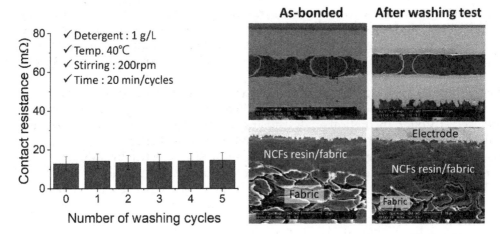

**FIGURE 10.19**
Washing test results of the optimized COFa.

substrates and the ACFs can absorb moisture permeated through the porous fabric substrates. This moisture uptake of two resins might cause hygroscopic swelling (Ray 2006, Scida, Assarar, Poilane, and Ayad 2013, and Kim and Seo 2006), resulting in increased joint resistances.

Figure 10.19 shows the washing test results of the optimized COFa. To duplicate actual laundry washing, COFa samples were immersed into the neutral detergent-added tap water and stirred at 200 rpm for 20 min per cycle. The test temperature was 40°C, which was consistent with the washing standards. After five cycles or total 100 min washing, stable ACFs joint resistances were obtained, which proved that the optimized COFa using ACFs was totally washable. Similar to moisture, liquid water can also penetrate and act as a plasticizer to cause interfacial damages of polymer-reinforced fabric laminates (Wan et al. 2006 and Chilali et al. 2017). However, cross-section SEM images showed that the ACFs joint and fabric substrates remained stable without any delamination or swelling compared with 85/85 test. Since the washing test was performed at low temperature (40°C), the water-induced damage did not occur compared with that of 85/85 test.

## 10.5 Conclusion

Fabric electronics interconnection using ACFs and metal pattern laminated fabric substrates was successfully demonstrated. Metal pattern laminated fabric substrates were successfully fabricated using B-stage NCFs with patterned metal circuits and laminating onto the fabrics. This method has several advantages such as design flexibility, fine-pitch capability, and stable electrical properties of the metal electrodes compared with conventional fiber-based electrodes.

The effects of the NCFs properties on the lamination and bending properties of the Cu-laminated fabric substrates were fully investigated. First, Cu-patterning process on the NCFs resin caused NCFs pre-curing, resulting in poor adhesion with the fabrics after the laminating onto the fabrics. Therefore, the NCFs curing onset temperature was optimized as 150°C. In terms of the lamination morphology, fine-pitch fabric substrates can be fabricated without any pattern distortion as the NCFs minimum viscosity was high. However, this could lead to extruded NCFs resin due to the slow NCFs resin permeation into the porous fabrics.

Regarding the bending fatigue properties, the best performance was obtained when lower-modulus NCFs and coupling agent-treated fabrics with higher adhesion strength were used. Through theoretical analysis using the neutral axis theory and experimental results, it was found that the lower-modulus NCFs showed better bending fatigue properties of the Cu electrodes due to the lower bending stress on Cu. In addition, adhesion-improved fabric substrates using coupling-agent treatment on the fabrics showed better bending properties, because the delamination between NCFs resin and fabrics was restricted, reducing possibility of strain localization at the interface between NCFs and Cu electrodes.

Using the optimized NCFs, fabric substrates and COFa assemblies were fabricated. The fabric substrates were fabricated with additional ENIG metal finish on Cu electrodes using NCFs; and after T/C bonding, stable ACFs joint properties such as electrical resistances and peel adhesion strength were obtained. In addition, static bending flexibility was optimized by applying the cover layer structure on top of the flip-chip COFa and the optimized COFa showed minimum bending radius down to 7.4 mm without chip fracture, excellent dynamic bending reliability up to 100,000 cycles at 12-mm bending radius, washing reliability up to 100 min washing at 40°C, and no open failure after 1000 h 85/85 test.

As a result, COFa can replace the conventional chip-integrated e-textiles using inserted plastic substrates, because direct chip assemblies can be possible. Fabric electronics using COFa assemblies can open the way to integrate highly functional semiconductors directly on the textiles.

## References

Bonderover, Eitan, and Sigurd Wagner. 2004. "A Woven Inverter Circuit for E-Textile Applications." IEEE Electron Device Letters 25 (5): 295–97.

Buechley, Leah, and Michael Eisenberg. 2009. "Fabric PCBs, Electronic Sequins, and Socket Buttons: Techniques for e-Textile Craft." Personal and Ubiquitous Computing 13 (2): 133–50. https://doi.org/10.1007/s00779-007-0181-0.

Büscher, Gereon H., Risto Kõiva, Carsten Schürmann, Robert Haschke, and Helge J. Ritter. 2015. "Flexible and Stretchable Fabric-Based Tactile Sensor." Robotics and Autonomous Systems 63 (January): 244–52. https://doi.org/10.1016/j.robot.2014.09.007.

Carl H. Zweben, W. Smith, and M.W. Wardle, "Test Methods for Fiber Tensile Strength, Composite Flexural Modulus, and Properties of Fabric-Reinforced Laminates," in Composite Materials: Testing and Design (Fifth Conference), ed. S. Tsai (West Conshohocken, PA: ASTM International, 1979), 228–262. https://doi.org/10.1520/STP36912S.

Cherenack, Kunigunde, and Liesbeth Van Pieterson. 2012. "Smart Textiles: Challenges and Opportunities." Journal of Applied Physics 112 (9). https://doi.org/10.1063/1.4742728.

Cherenack, Kunigunde, Christoph Zysset, Thomas Kinkeldei, Niko Münzenrieder, and Gerhard Tröster. 2010. "Woven Electronic Fibers with Sensing and Display Functions for Smart Textiles." Advanced Materials 22 (45): 5178–82. https://doi.org/10.1002/adma.201002159.

Chilali, Abderrazak, Wajdi Zouari, Mustapha Assarar, Hocine Kebir, and Rezak Ayad. 2017. "Effect of Water Ageing on the Load-Unload Cyclic Behaviour of Flax Fibre-Reinforced Thermoplastic and Thermosetting Composites." Composite Structures 183 (1): 309–19. https://doi.org/10.1016/j.compstruct.2017.03.077.

Cottet, Didier, Janusz Grzyb, Tünde Kirstein, and Gerhard Tröster. 2003. "Electrical Characterization of Textile Transmission Lines." IEEE Transactions on Advanced Packaging 26 (2): 182–90. https://doi.org/10.1109/TADVP.2003.817329.

Ding, Yujie, Michael A. Invernale, and Gregory A. Sotzing. 2010. "Conductivity Trends of PEDOT-PSS Impregnated Fabric and the Effect of Conductivity on Electrochromic Textile." ACS Applied Materials and Interfaces 2 (6): 1588–93. https://doi.org/10.1021/am100036n.

Jung, Seung Yoon, and Kyung Wook Paik. 2017. "A Study on the Fabrication of Electrical Circuits on Fabrics Using Cu Pattern Laminated B-Stage Adhesive Films for Electronic Textile Applications." In Proceedings of the Electronic Components and Technology Conference, 2145–50. https://doi.org/10.1109/ECTC.2017.100.

Kim, Hyo Jin, and Seo Do Won. 2006. "Effect of Water Absorption Fatigue on Mechanical Properties of Sisal Textile-Reinforced Composites." International Journal of Fatigue 28 (10 SPEC. ISS.): 1307–14. https://doi.org/10.1016/j.ijfatigue.2006.02.018.

Kim, Ji Hye, Ik Lee Tae, Won Shin Ji, Soo Kim Taek, and Kyung Wook Paik. 2016. "Bending Properties of Anisotropic Conductive Films Assembled Chip-in-Flex Packages for Wearable Electronics Applications." IEEE Transactions on Components, Packaging and Manufacturing Technology 6 (2): 208–15. https://doi.org/10.1109/TCPMT.2015.2513062.

Kim, Ji Hye, Ik Lee Tae, Soo Kim Taek, and Kyung Wook Paik. 2017. "The Effect of Anisotropic Conductive Films Adhesion on the Bending Reliability of Chip-in-Flex Packages for Wearable Electronics Applications." IEEE Transactions on Components, Packaging and Manufacturing Technology 7 (10): 1583–91. https://doi.org/10.1109/TCPMT.2017.2718186.

Kim, Ji Hyun, Ji Hye Kim, and Kyung Wook Paik. 2019. "Effects of the Types of Anisotropic Conductive Films on the Bending Reliability of Chip-in-Plastic Packages." IEEE Transactions on Components, Packaging and Manufacturing Technology 9 (3): 405–11. https://doi.org/10.1109/TCPMT.2019.2893979.

Krshiwoblozki, Malte von, Torsten Linz, Andreas Neudeck, and Christine Kallmayer. 2012. "Electronics in Textiles – Adhesive Bonding Technology for Reliably Embedding Electronic Modules into Textile Circuits." Advances in Science and Technology 85 (September): 1–10. https://doi.org/10.4028/www.scientific.net/AST.85.1.

Lee, Tae Ik, Cheolgyu Kim, Min Sung Kim, and Taek Soo Kim. 2016. "Flexural and Tensile Moduli of Flexible FR4 Substrates." Polymer Testing 53: 70–76. https://doi.org/10.1016/j.polymertesting.2016.05.012.

Lomov, S. V., D. S. Ivanov, T. C. Truong, I. Verpoest, F. Baudry, K. Vanden Bosche, and H. Xie. 2008. "Experimental Methodology of Study of Damage Initiation and Development in Textile Composites in Uniaxial Tensile Test." Composites Science and Technology 68 (12): 2340–49. https://doi.org/10.1016/j.compscitech.2007.07.005.

Luo, S, and W J van Ooij. 2002. "Surface Modification of Textile Fibers for Improvement of Adhesion to Polymeric Matrices: A Review." Journal of Adhesion Science and Technology 16 (13): 1715–35. https://doi.org/Doi 10.1163/156856102320396102.

Mahboob, Zia, Ihab El Sawi, Radovan Zdero, Zouheir Fawaz, and Habiba Bougherara. 2017. "Tensile and Compressive Damaged Response in Flax Fibre Reinforced Epoxy Composites." Composites Part A: Applied Science and Manufacturing 92: 118–33. https://doi.org/10.1016/j.compositesa.2016.11.007.

Mercier, Patrick P., and Anantha P. Chandrakasan. 2011. "A Supply-Rail-Coupled Etextiles Transceiver for Body-Area Networks." IEEE Journal of Solid-State Circuits 46 (6): 1284–95. https://doi.org/10.1109/JSSC.2011.2120690.

Ray, B. C. 2006. "Temperature Effect during Humid Ageing on Interfaces of Glass and Carbon Fibers Reinforced Epoxy Composites." Journal of Colloid and Interface Science 298 (1): 111–17. https://doi.org/10.1016/j.jcis.2005.12.023.

Remmers, J. J.C., and R. De Borst. 2001. "Delamination Buckling of Fibre-Metal Laminates." Composites Science and Technology 61 (15): 2207–13. https://doi.org/10.1016/S0266-3538(01)00114-2.

Scida, Daniel, Mustapha Assarar, Christophe Poilâne, and Rezak Ayad. 2013. "Influence of Hygrothermal Ageing on the Damage Mechanisms of Flax-Fibre Reinforced Epoxy Composite." Composites Part B: Engineering 48: 51–58. https://doi.org/10.1016/j.compositesb.2012.12.010.

Trindade, Isabel G, Frederico Martins, Rui Miguel, and Manuel S Silva. 2014. "Design and Integration of Wearable Devices in Textiles." Sensors & Transducers 183 (12): 42–47.

Wan, Y. Z., Y. L. Wang, Y. Huang, H. L. Luo, F. He, and G. C. Chen. 2006. "Moisture Absorption in a Three-Dimensional Braided Carbon/Kevlar/Epoxy Hybrid Composite for Orthopaedic Usage and Its Influence on Mechanical Performance." Composites Part A: Applied Science and Manufacturing 37 (9): 1480–84. https://doi.org/10.1016/j.compositesa.2005.09.009.

Xie, Yanjun, Callum A.S. Hill, Zefang Xiao, Holger Militz, and Carsten Mai. 2010. "Silane Coupling Agents Used for Natural Fiber/Polymer Composites: A Review." Composites Part A: Applied Science and Manufacturing 41 (7): 806–19. https://doi.org/10.1016/j.compositesa.2010.03.005.

Yu, B., R. S. Bradley, C. Soutis, P. J. Hogg, and P. J. Withers. 2015. "2D and 3D Imaging of Fatigue Failure Mechanisms of 3D Woven Composites." Composites Part A: Applied Science and Manufacturing 77: 37–49. https://doi.org/10.1016/j.compositesa.2015.06.013.

# 11

## Flexible and Stretchable Systems for Healthcare and Mobility

**Kai Zoschke, Thomas Löher, Christine Kallmayer, and Erik Jung**
*Fraunhofer IZM*

## CONTENTS

## 11.1 Introduction

The growing need to shrink not only the footprint of an electronic system but also to limit its thickness has led to increased interest in ultrathin circuits. Namely, flexible substrates and consequently ultrathin ICs have become a pivotal technology for demanding systems. With the capability to conform towards nonplanar bodies, even to support organically shaped surfaces and offer some dynamics, flexibility has eventually expanded to flex-stretch and –finally– conformal systems.

Such properties not only fuel the drive of miniaturization in general [Beica 2018] but also open up new application areas especially in in the sector of healthcare, mobility, robotics, arts, and fashion.

Among these emerging application areas, we find patches to monitor vital parameters such as ECG, temperature and activity, prosthetic enhancements restoring haptic sensing, or robotic systems capable of manipulating sensitive goods. Also hybrid devices such as connected pieces of art in a smart home or jewelry (e.g., rings), allowing to interact with the environment as building blocks of a "smart-X" connected world are enabled by flex-stretch systems, which merge imperceptibly into the main functionality of the goods [Dudhe 2017, Chen 2016]. The latter is key to customer acceptance. Lastly, implanted devices strongly benefit from flexible and soft systems, rendering rigid circuit boards enclosed in a titanium housing obsolete.

## 11.2 Core Technologies for Flexible/Stretchable System Integration

Recent developments to realize high-density flexible circuits have demonstrated the capability to improve production volume, cost, and reliability significantly. Since the conception of ultrathin substrates, the challenge to obtain such (<25 µm) substrates was the release of such flexible circuits from the production panel. Especially with the material of main interest, polyimide, substantial improvements in the production of high-density circuits have been achieved. In contrast to the cast polyimide sheets or rolls with RA (rolled-annealed) or ED (electro-deposited) copper subtractively structured, more delicate circuits can be created by multilayered processes, including the integration of thin chips into the dielectric material itself. The concept of thin chip integration, albeit not a new idea by itself [Christiaens 2010], has since found its implementation into the additive realization of complex circuitry.

Although polyimide is regarded as a highly versatile, robust, and capable material, flex circuit realization is not limited to this. Niche materials like parylene or LCP (liquid crystal polymers) have been sided with materials targeting volume markets, e.g., PU (polyurethanes) or PET (polyethylenterephthalat). Thermoplastic polyurethanes (TPUs) have come to the attention of the industry, as processing of these as substrate material not only results in flexible products but also in highly pliable products and – with proper track material and geometry – are even stretchable. In combination with thermosetting or thermoformable backings, these stretchable properties even allow to build electronic systems conforming to complex, biomorphic surfaces.

### 11.2.1 Ultrathin Flexible Circuits and Systems by Laser-Assisted Debonding

For ultrathin, albeit high-density flexible substrates, spin-on polyimide is selected as material of choice. Mandatory for the creation of reliable flex circuits is the proper material selection for the polymer, which is deposited in a process well understood in the wafer-level redistribution industry, offering manufacturing sizes up to 300-mm diameter [Kuo 2010]. Also, a capable release technology of the polymer/metal multilayer stack from the surface of the process wafer is key to a successful implementation. Among release options such as mechanical substrate removal or usage of anti-sticking layers, a laser-assisted release of the polymer/metal multilayer stack through a transparent carrier wafer has been demonstrated to be a very promising approach [Zoschke 2016b, Woehrmann 2016]. Figure 11.1 shows the schematic flow for a typical flow for high-density flex fabrication process, based on glass carrier wafers with a thickness <1 mm conforming to semiconductor wafer standards [SEMI M01].

In the first step, a release layer of polymer is deposited on the glass surface. The release layer is only temporary present and required for the final detach of the multilayer stack from the glass carrier. It will not be part of the final flex circuit. In the second step, the backside pads of the later flex circuits are generated by semi-additive metal structuring. The sequence includes the sputtering of an adhesion and seed layer followed by photoresist lithography and electroplating to the desired metal. Finally, the resist defining the ED structures is removed and the respective seed layer and adhesion layer are etched. With the need for subsequent surface mounting, in this first layer, the plating sequence of, e.g., pad finishes like Ni/Au, has to be reversed. As with additive high-density structuring, a sputtered adhesion- and an additional seed-layer are also required to the temporary

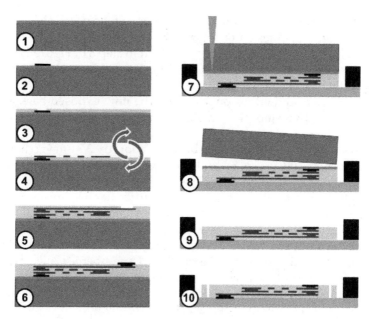

**FIGURE 11.1**
High-density flex technology based on standard wafer-level multilayer redistribution and laser-assisted release.

release layer, proper processes have to be put in place to remove this thin layer after flex detachment from the carrier wafer.

In order to realize multilayered flex substrates, vias are defined by lithography and subsequently, the respective wiring layers are built up semi-additively.

The multilayer wiring stack is now created by the sequential repetition of the steps 3 and 4 as of Figure 11.1. The maximum number of routing layers, which can be processed is limited by the warpage of the carrier wafer. Due to the coefficient of thermal expansion (CTE), mismatch between the carrier wafer and the deposited materials, each layer will increase the warpage after processing. Reduction of metal load and thicknesses of both, metal and polymer layers, as well as introduction of interruptions in the polymer layers (like scribe lines) can reduce the warpage issue. In order to optimize process yield, a multilayer stack with four routing layers and a total thickness of 60 μm could be reliably processed by the proposed sequence.

After finishing the multilayer stack with deposition and structuring of the final polymer layer in step 5, the topside I/O pads are created by using again a semi-additive metal structuring technology, allowing at this point to even include interconnect elements such as gold or solder finish.

The process wafer is now flipped onto a dicing tape on film frame carrier (Figure 11.1, step 7), exposing the backside of the glass wafer and thus the interface glass-polyimide to the debonding laser. The exposure for the laser-assisted debonding is done using a 248-nm excimer laser stepping system, focusing at the polymer-glass interface. With shot repetition rates of 50 Hz and a spot size of >6 × 6 mm², a 300-mm wafer can be debonded within approx. 60 sec [Zoschke 2016a].

During the laser exposure, the release layer absorbs the laser radiation completely within a layer thickness of several 100 nm behind the glass surface. Above a certain threshold, the energy intake into the polymer caused by the laser exposure leads to a physical damage of the material at molecular level, meaning the destruction of the materials' molecular bonds. As a result, the remaining adhesion of the material to the glass wafer is close to zero, so that the glass carrier can be detached easily as shown in step 8 of Figure 11.1, assisted also by the tensile stress built into the multilayer circuit. The carrier wafer can be cleaned and reused subsequently.

The remaining release and adhesion layers on the now-exposed backside of the circuit are removed by a two-step plasma etch process using reactive ion etching (RIE) plasma.

Typically, such an ultrathin circuit can be built with approx. 35-µm via diameter (limited by stray light reflection and warpage) and approx. 12-µm lines/space. When additional process efforts to minimize the influence of stray light to the via formation are considered, the design rule guidelines for the ultrathin high-density flex are improved to allow for a 10-µm via diameter, and 7-µm L/S, allowing for an I/O pitch of approx. 27 µm.

With the availability of thin bare dice (less than 25 µm in thickness), driven by the mobile devices industry, the proposed process flow lends itself also to the creation of not only passive substrates but also active systems [Hassan 2013, Landesberger 2016]. The main difference to the previously described sequence is the placement of the chips on the polyimide dielectric and their embedding into an additional ~25-µm-thick polyimide layer, which subsequently serves as slightly thicker dielectric to receive the next metallization layer. Due to the topography introduced, minimum via diameters acceptable for the next interconnect layer are limited to ~40 µm [Wolf 2014].

As such, highly integrated systems will seldom find their use as standalone systems, interconnect structures to the higher-level system partition can be provided for in the same sequence. A partial debonding, leaving parts of the circuit still attached to the carrier wafer, can be useful for subsequent handling in the next level of integration. Figures 11.2–11.4 show some technical details of the intermediate process steps, with Figure 11.5 showing the integration of a circuit built for condition monitoring, i.e., with stress and vibration sensors integrated in this ultrathin flex circuit technology, partly debonded and affixed to the carrier element for the next-level system integration.

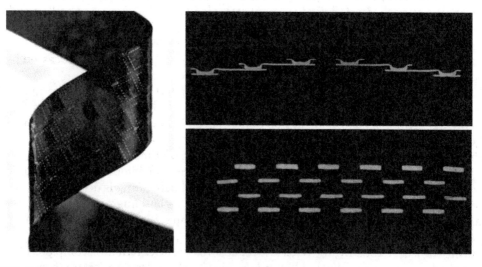

**FIGURE 11.2**
Flex substrate after release from carrier (left), cross-sectional view of staggered vias (right/top), and multilayer wiring (right/bottom).

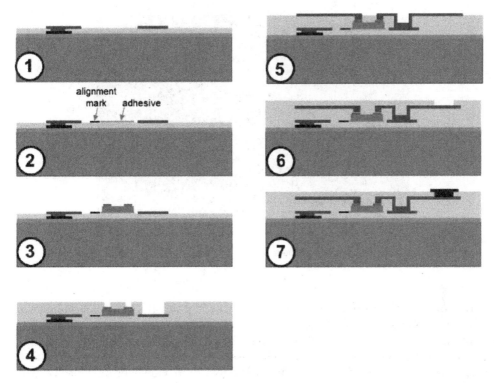

**FIGURE 11.3**
Schematic process flow for the embedding of thin ICs into the flex build-up layers.

## 11.2.2 Stretchable Circuits and Systems

While the polyimide-based circuits described in section 11.2 have shown a substantial step forward with respect to thickness and integration density, their mechanical properties and, thus, dimensional stability are less favorable when the target application requires a "soft" touch or even a certain amount of dynamic flexing without offering space for a freely moving flexure. Here, elastic materials could offer an adequate solution. However,

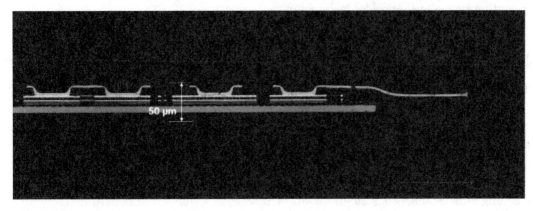

**FIGURE 11.4**
Cross-sectional view of I/O pad connections to 20-μm thin 24GHz transceiver IC embedded in polymer layers.

**FIGURE 11.5**
Stress/vibration sensor module with thin embedded ICs after singulation from wafer and partial substrate release.

the intrinsic elasticity conflicts with the circuit tracks, e.g., mostly made of metal. Advances in conductive polymers as well as geometrically compliant structures have since overcome this conflict to a certain extent. With a combination of elastic materials and corresponding compliance in the conductors, e.g., silicone-based substrates have been demonstrated to stretch for 3× their original dimensions in $x/y$ [Rogers 2019]. However, silicone is a material not favored by the industry, as processing and curing as well as its raw material cost are often incompatible with industry's capabilities and needs. Alternatively, TPU has emerged as a promising material for stretchable circuits [Loeher 2014].

It offers large area processing of circuitry in a process flow compatible with the PCB industry's manufacturing processes in combination with a high dynamic stretchability. Additionally, its dielectric properties suitable for a range of applications render TPU material of choice for stretchable circuits. Single layer and multilayer substrates can be realized with this material and – with proper adaption of the process details – an integration of active circuits with state-of-the-art surface mount technology (SMT) is possible [Ostmann 2016]. Notably, the thermoplasticity of the TPU limits the maximum process temperatures to work with to ~190°C, which will have a significant influence on SMT-mounting process using solder technology.

Figure 11.6 depicts the process flow as implemented on a 457 × 610 mm sized rigid carrier (e.g. 3-mm-thick epoxy-glass material), supporting the flexibly and pliable TPU substrate throughout the processing steps.

After laminating a copper foil to the TPU, subsequently a photoresist is applied and structured, e.g., by laser direct imaging. Exposing the copper film through this patterned mask to an etchant, the circuit design is etched into the foil and either an additional TPU layer or a solder-resist layer is coated on the surface, again opening in a lithography-based process the interconnect pads for subsequent processing. For a solderable finish (also suitable for adhesive interconnect), electroless silver is chosen, as it offers both a perfect finish as well as compatibility to the two aforementioned interconnect methods.

All process steps are fully compatible with respect to structuring, etching, rinsing, and cleaning with established, state-of-the-art PCB processes; thus, circuit board manufacturers have an easy way to adapt their process flows to produce stretchable circuits also in volume.

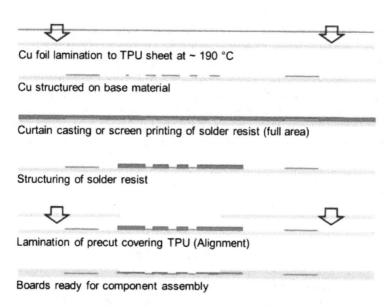

Cu foil lamination to TPU sheet at ~ 190 °C

Cu structured on base material

Curtain casting or screen printing of solder resist (full area)

Structuring of solder resist

Lamination of precut covering TPU (Alignment)

Boards ready for component assembly

**FIGURE 11.6**
Process flow of TPU substrate manufacturing (single layer; handling panel not shown).

While the TPU by itself is elastic and thus stretchable, the interconnecting metal tracks are not. Some polymer material innovations have since shown that additively deposited tracks using elastomeric conductive inks or pastes can be used to achieve good stretchability. However, such additive concepts cannot be easily adopted by the PCB manufacturer. In order to ensure the compatibility to established manufacturing lines, not the material or process, but the structure of the conductors need to enable the stretch capabilities. This is done by meandering geometries of the tracks implemented already in the design phase. Figure 11.7 shows this schematically; the resulting horseshoe structure allows for

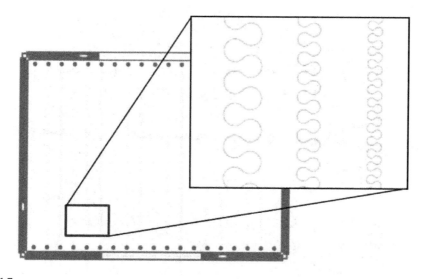

**FIGURE 11.7**
Horseshoe structure design in a repetitive pattern to allow for high stretchability of the circuit, using copper as track material.

substantial stretchability without exerting too much strain to the material. This is due to two separate, highly efficient effects – a), the geometry distributes in-plane stress due to the effective length of the meander and b) three-dimensional buckling effect of the meanders adds to lowering the stress introduced into the track material (Figure 11.8, [Loeher 2014]). With the PCB-compatible design rules, the radii of the meanders are in the range of 1.2 mm with line width of the tracks down to 75 µm. With smaller radii, the force to deform the circuit as a whole is increased due to the larger metal load of the substrate, rendering designs thus less useful for the intended purpose for high compliance to external complex geometries [Joshi 2018].

A notable drawback of this approach, clearly, is the limitation of achievable track densities as well as the high-frequency capabilities, which are anyway limited by the dielectric properties of the TPU ($\varepsilon \sim 4.4$, $\tan \delta \sim 0.04$).

Iteratively repeating the process flow of the principal steps as depicted in Figure 11.6, including through-and blind-vias by laser drilling and subsequent additive metallization, complex multilayer designs can be realized (Figure 11.8).

Lastly, to build a functional system using the stretchable platform, active and passive components can be assembled and interconnected in surface mount technology. Here, due to the inherent properties of TPU, a processing temperature of 190°C cannot be surpassed – typically, low melting point solders are thus used for the interconnect with typical components such as LED (light-emitting diodes), µC (microcontrollers) in, e.g., QFN (quad flat nonlead) package or passive R-C-L chips. Alternatively, silver-filled adhesive can be used to form the interconnect, albeit limiting the suitable surface finishes of the selected components. Notably, the components as well as the interconnect areas offer no elasticity/stretchability anymore. Thus, any mechanical load applied will result in a stress maximum at exactly these critical interfaces, leading to early failure in operation. To compensate for this, either local stiffening structures (Figure 11.9a) or appropriate encapsulation is used to distribute the stress over a larger area, shifting the stress maximum to less critical regions of the system. As a rigid encapsulant (e.g., epoxy) would

**FIGURE 11.8**
Active test structure built with a two-layer TPU-based stretchable circuit board.

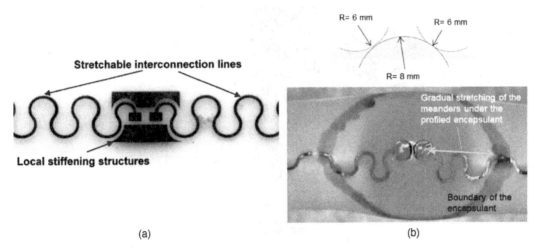

**FIGURE 11.9**
Concepts for stress relocation from the critical component interconnect area.

only shift the region of maximum stress impact, preferably soft (by choice of material) and graded (by choice of encapsulant geometry, see Figure 11.9b) encapsulation is used, which not only shifts the position of maximum stress impact but also spreads the strain applied to a larger affected area, effectively minimizing the reliability impact of the load under operation.

Implementing these design and process considerations, stretchable systems with elongation capabilities up to 300% can be obtained (Figure 11.10a). At lower deformation levels, several 1000 (@10% strain) dynamic stretch cycles have been well demonstrated (Figure 11.10b).

Such stretchable systems can either be used in a standalone setting, where their stretchability offers the expected robustness against mechanical dynamic load (Figure 11.11a shows an example of a miniature rapid eye moving detector mounted on the eye lid of the patient, allowing the monitoring of the moving eye via a miniature acceleration sensor), or laminated to supporting structures like textiles. In the latter case, the properties of the textile (i.e., drapeability, breathability, crinkling) are not compromised by adding the electronic functionality (Figure 11.11b, showing a sensing system, which can also be integrated into a T-shirt (Figure 11.14), to enable the shirt to serve as a HMI device).

### 11.2.3 Conformally Integrated Electronics

With the lamination capabilities of the stretchable circuits mentioned in Section 11.2.2, such systems can also be integrated in thermoformed shapes. Thermoforming is a well-established technique to bring thermoplastic materials with rigid properties at their typical operating condition into a specific shape. This has both been employed for low-cost disposables, like food containers, or high-value products, like avionics interiors or medical prosthetics. Until now, adding electronic functionalities has been difficult and a labor-intensive process. With the advent of stretchable systems, offering the electronic functionality in conjunction with their intrinsic property to follow a given complex geometric shape, such integration processes have now become much more efficient.

(a)

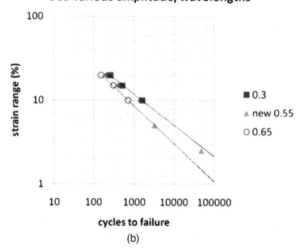

(b)

**FIGURE 11.10**
Stretchability of TPU-based substrates.

The process flow for thermoforming is modified in that way, that, e.g., the foils used for state-of-the-art in-mold decoration is replaced or supplemented by the stretchable system. The layer stack (Figure 11.12) is then placed over the master mold of the thermoforming unit, heated and – by applying vacuum and pressure – conformed to the surface geometry of the mold.

After cooling below the $T_g$, the mold clamp is opened and the product in its final shape is taken out. Figure 11.13 shows a typical mold insert, a test structure that has been designed to assess the local distortion and a demonstrator of a custom-shaped LED lamp interior.

In order to limit the local deformation to protect the active circuits on the stretchable system, the temperature distribution in the heater can be tailored to specifically modify the flow properties of the thermoform substrate at a given site. With this approach, both protruding as well as intruding shapes can be electronically enhanced, shaping complex geometries with sidewall angles up to 87°.

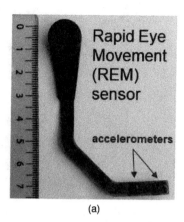

(a)

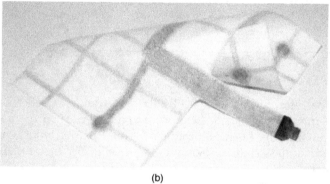

(b)

**FIGURE 11.11**
(a) Sensor for in-sleep monitoring of rapid eye movement (REM) episodes, built as standalone stretchable system and (b) haptic sensor sheet with the stretchable system laminated on a highly breathable textile (nonwoven).

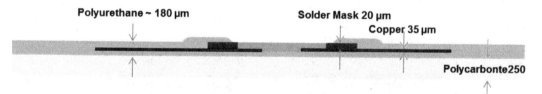

**FIGURE 11.12**
Layer stack of thermoform polymer (e.g., polycarbonate) and TPU-based stretchable system.

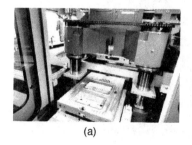

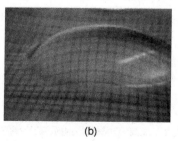

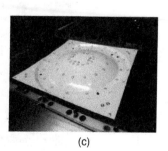

(a)  (b)  (c)

**FIGURE 11.13**
(a) Thermoforming equipment with master mold, (b) test shape, and (c) product demonstrator with protruding shape.

**FIGURE 11.14**
Stretchable PCB colaminated on textile carrier, integration into child's romper suit.

## 11.3 Application Examples

From the vast number of applications conceivable with the described technologies, the domains of highest interest are clearly healthcare and mobility. In the past years, several hundred publications have been reported successful integration of sensors, readout-electronics, and even wireless transmission into stretchable circuits placed, e.g., on skin, notably for wound care.

Here, stretchability, breathability, and biocompatibility are the keys to realize wound care patches with high acceptance in the field. Similar requirements are in demand for pediatric care, either in a clinical setting or in a home monitoring situation to prevent sudden child death by colaminating stretchable circuits with the main textile body [Seckel 2013] (see Figure 11.14).

**FIGURE 11.15**
Interior car ceiling illumination panel conformally shaped to a dome geometry.

In the field of advanced mobility scenarios, conformally adaptable circuits have demonstrated their capability to be integrated into car dashboards [Ostmann 2016], vehicle ceilings [Figure 11.15], and aircraft sidewall covers [Hahn 2019].

As obviously these two application areas are only a small excerpt of the range where conformal and stretchable electronics can be used, it is expected that with the maturity of technology now being ready for use in products (i.e., up to Technology Readiness Level - TRL- 8) numerous innovations will hit the market in the forthcoming years.

## 11.4 Summary and Outlook

With the advent of technologies to realize advanced high-density multilayer flexible substrates, event with stretchable properties, and the availability of thin bare dice, the level of integration into flexible systems has reached a maturity level, which opens up the perspective for a multitude of products outside the traditional perspective. In the era of digitalization, of Internet of Things (IoT) and distributed, connected smart devices, this technology options now can be expected to serve as hardware platform for our future "smart x" societies. Engineers and artists creativity can now be extended to nontraditional representations with improved interfaces to the user and use cases.

## References

R. Beica, *"Enabling Information Age through Advanced Packaging Technologies and Electronic Materials"*, 2018 Pan Pacific Microelectronics Symposium (Pan Pacific), 5–8 Feb. 2018, Waimea, HI, USA.

W. Chen, *"Heterogeneous Integration for IoT Cloud and Smart Things: A Roadmap for the Future"*, 2016 International Symposium on 3D Power Electronics Integration and Manufacturing (3D-PEIM), 13–15 June 2016, Raleigh, NC, USA.

W. Christiaens, E. Bosman, J. Vanfleteren, *"UTCP: A Novel Polyimide-Based Ultra-Thin Chip Packaging Technology"*, IEEE Transactions on Components and Packaging Technologies 33(4), Dec. 2010, pp. 754–760.

P. V. Dudhe, N. V. Kadam, R. M. Hushangabade, M. S. Deshmukh, *"Internet of Things (IOT): An Overview and its Application"*, 2017 International Conference on Energy, Communication, Data Analytics and Soft Computing (ICECDS), 1–2 Aug. 2017, Chennai, India, pp. 2650–2653.

R. Hahn et al., *"Abheben mit Gedruckter Elektronik"*; Press Release on innovations during LOPEC 2019, Munich.

M. U. Hassan, C. Schomburg, C. Harendt, E. Penteker, J. N. Burghartz, "Assembly and Embedding of Ultra-Thin Chips in Polymers", 9–12 Sept. 2013, Grenoble, France.

S. Joshi, *"Free Standing Interconnects for Stretchable Circuits"*, Thesis, TU Delft, 2018, https://doi.org/10.4233/uuid:20ba0c91-198a-4334-bb6d-a7d99d76d32b

T. Y. Kuo, Z. C. Hsiao, Y. P. Hung, W. Li, K. C. Chen, C. K. Hsu, C. T. Ko, Y. H. Chen, *"Process and Characterization of Ultra-thin Film Packages"*, 2010 5th International Microsystems Packaging Assembly and Circuits Technology Conference, 20–22 Oct. 2010, Taipei, Taiwan.

C. Landesberger, N. Palavesam, W. Hell, A. Drost, R. Faul, H. Gieser, D. Bonfert, K. Bock, C. Kutter, *"Novel Processing Scheme for Embedding and Interconnection of Ultra-thin IC Devices in Flexible Chip Foil Packages and Recurrent Bending Reliability Analysis"*, 2016 International Conference on Electronics Packaging (ICEP), 20–22 Apr. 2016, Sapporo, Japan, pp. 473–478.

T. Loeher et al., "Stretchable and Deformable Electronic Systems in Thermoplastic Matrix Materials", IEEE CPMT Symposium Japan, 2014, Kyoto, DOI: 10.1109/ICSJ.2014.7009639.

A. Ostmann et al., *"Large Area Processes for 3D Shaped Electronics"*, ESTC 2016, Grenoble, DOI:10.1109/ESTC.2016.7764463.

J. A. Rogers, *"Materials for Biointegrated Electronic and Microfluidic Systems"*, MRS Bulletin 44(3), 2019, pp. 195–202.

M. Seckel et al., *"Romper Suit to Protect Against Sudden Infant Death"*, FhG Research News, Mar. 2013, Permalink: https://www.fraunhofer.de/en/press/research-news/2013/january/romper-suit-to-protect-against-sudden-infant-death.html.

SEMI M01, *Specification for Polished Single Crystal Silicon Wafers*, Sep. 2018, https://store-us.semi.org/products/m00100-semi-m1-specification-for-polished-single-crystal-silicon-wafers.

M. Woehrmann, O. Wuensch, K. D. Lang, R. Gernhard, K. Hauck, K. Kroehnert, K. Zoschke, N. Juergensen, M. Toepper, T. Braun, H. Hichri, M. Arendt, *"New Excimer Laser-Based Dual Damacene Process for High I/O Applications with Ultra-Fine Line Routing"*, Süss Report, 2016, pp. 4–10.

J. Wolf, J. Kostelnik, K. Berschauer, A. Kugler, E. Lorenz, T. Gneiting, C. Harendt, Z. Yu, *"Ultra-thin Silicon Chips in Flexible Microsystems"*, ECWC 13, 13th Electronic Circuits World Convention, Nürnberg, DE, 7–9 May 2014, pp. 1–5.

K. Zoschke, K. D. Lang, *"Evolution of Structured Adhesive Wafer to Wafer Bonding Enabled by Laser Direct Patterning of Polymer Resins"*, 18th Electronics Packaging Technology Conference, Singapore, 2016a, pp. 223–228.

K. Zoschke, M. Wegner, T. Fischer, K. D. Lang, *"Temporary Handling Technology by Polyimide Based Adhesive Bonding and Laser Assisted De-bonding"*, 6th Electronics System-Integration Technology Conference, 13–16 Sept. 2016b, Grenoble, France.

# 12

## Fabrication of Transparent Antennas on Flexible Glass

**Jack P. Lombardi III, Darshana L. Weerawarne, Robert E. Malay, and Mark D. Poliks**
*Binghamton University*

**James H. Schaffner and Hyok Jae Song**
*HRL Laboratories, LLC*

**Ming-Huang Huang and Scott C. Pollard**
*Corning Research and Development Corporation*

**Timothy Talty**
*General Motors*

## CONTENTS

## 12.1 Introduction

Flexible electronics is an emerging technology that holds promise to revolutionize functionality of electronics and how they are manufactured. With a projected growth from $29 billion in 2017 to $73 billion in 2027, the flexible electronics industry shows potential for

**TABLE 12.1**

Properties of Commonly Used R2R Substrates

| Property | Substrate | | | | |
|---|---|---|---|---|---|
| | **PET** | **PEN** | **PI** | **Flexible Glass** | **Stainless Steel** |
| Thickness (μm) | 12–300 | 25–300 | 25–300 | 50–200 | 100 |
| Weight (g/m²) | 120 | 120 | 120 | 250 | 800 |
| Max processing temperature (°C) | 125 | 175 | 350 | 700–750 | 1000 |
| Linear CTE (ppm/°C) | 20–25 | 18–20 | 30–60 | 3–5 | 10 |
| Visible spectrum transmission (%) | ~85 | ~90 | <50 | >95 | 0 |
| Water absorption (%) | 0.14 | 0.14 | 3 | 0 | 0 |
| Young's modulus (GPa) | 5 | 6 | 2.5 | 70–80 | 200 |
| Tensile strength (MPa) | 225 | 275 | 230 | 70 | 485–515 |
| Thermal conductivity (W/m°C) | 0.15–0.4 | 0.15 | 0.5 | 1 | 161 |

huge growth with efficient, low-cost, and environmentally friendly applications in automobile [1, 2], aerospace [3], healthcare [4], and many more. The flexibility and conformability of the substrates used in flexible electronics and roll-to-roll (R2R) manufacturing provide unique capabilities for manufacturers to fit and conform devices into confined spaces [5, 6]. Flexible substrates are often based on polymers such as polyethylene naphthalate (PEN), polyethylene terephthalate (PET), and polyimide (PI), but thin flexible glass substrates play a critical role. While maintaining the high transparency and flexibility that polymer substrates provide, flexible glass presents an ultra-smooth, high-temperature, and high-vacuum processable surface, which makes it attractive in conventional micro-/nano-fabrication. Corning® Willow® Glass is one such flexible glass substrate that provides lower coefficient of thermal expansion (CTE) and better surface quality [7–9] (see Table 12.1) and is extensively used in this report for fabrication of transparent antennas. The flexibility of these glass substrates also makes them compatible with high-throughput R2R manufacturing, which further lowers the cost and allows for large area fabrication, an advantage for photovoltaics [10–12].

As shown in Table 12.1, flexible glass substrates, such as Corning® Willow® Glass, possess a unique set of properties attractive to flexible electronics manufacturers [13]. The inherent flexibility of electronic devices also allows for R2R processing as it helps electronics to be rolled up during processing just like newspaper printing. As a promising substrate for flexible electronics and R2R processing, here we develop fabrication processes on flexible glass wafer and sheet level that are R2R compatible. By doing this, we aim to reduce complications and waste that could be generated in developing a process completely R2R. Furthermore, vacuum coating techniques such as sputtering and various lithographic and etching techniques have also been modified and expanded to R2R processing [13–16]. The selection of the fabrication method depends on the material to be deposited, the pattern resolution needed, and flexible substrate used. In the R2R work conducted here, the focus is on replicating the same photolithography, subtractive, and semi-additive processes usually available in wafer scale for large area using tools similar to those shown in Figure 12.1.

For applications in radio frequency (RF) antennas, the dielectric properties of the flexible substrate contribute to the overall design and resulting performance. Table 12.2 shows the measured dielectric properties of Willow Glass and polymer film substrates typically

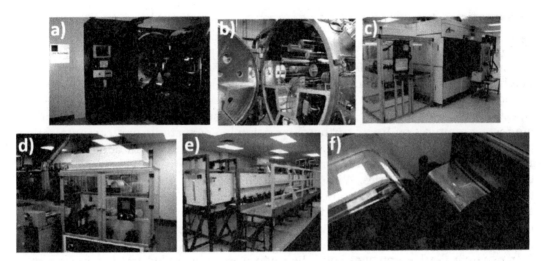

**FIGURE 12.1**
Photos of R2R processing at CAMM (Binghamton University), with (a) CHA sputtering system, (b) GVE sputtering system, (c) Azores stepper photolithography, (d) wet processing and developing, (e) Cu electroplating, and (f) an example of Corning® Willow® Glass in R2R processing. (Adapted from binghamton.edu/camm/.)

used in flexible electronic applications [13]. Willow Glass shows a higher $D_k$ and lower $D_f$ in the frequency range from 1 KHz to 1 MHz, making it as preferable flexible substrate for RF antenna applications.

In this chapter, we investigate the development of transparent antennas on flexible glass. Corning® Willow® Glass is extensively studied for process development. Fabricated antennas showed enhanced connectivity while being unobtrusively mounted on glass surfaces, such as building or automotive windows. Corning® Willow® Glass also has added advantages such as high transparency, excellent surface quality, high thermal and chemical stability, and a refractive index matched to other glasses [13, 17–19]. The fabrication and process development are done on wafer or sheet scale with the goal of transitioning to fully R2R processing.

**TABLE 12.2**

Dielectric Properties of Corning® Willow® Glass and Commonly Used R2R Polymer Substrates

| Material | $D_k$ | | | | $D_f$ | | | |
|---|---|---|---|---|---|---|---|---|
| | 1 kHz | 10 kHz | 100 kHz | 1 MHz | 1 kHz | 10 kHz | 100 kHz | 1 MHz |
| Corning® Willow® Glass | 5.21 | 5.20 | 5.19 | 5.16 | 0.0015 | 0.0012 | 0.0013 | 0.0008 |
| PC | 3.15 | 3.13 | 3.11 | 3.05 | 0.0031 | 0.0040 | 0.0075 | 0.0123 |
| PEN | 3.06 | 3.04 | 3.00 | 2.94 | 0.0046 | 0.0069 | 0.0099 | 0.0092 |
| PET | 3.31 | 3.27 | 3.21 | 3.11 | 0.0052 | 0.0104 | 0.0163 | 0.0175 |
| PI | 3.50 | 3.49 | 3.46 | 3.40 | 0.0024 | 0.0043 | 0.0077 | 0.0091 |
| PMMA | 3.52 | 3.33 | 3.15 | 2.99 | 0.0421 | 0.0366 | 0.0354 | 0.0304 |

## 12.2 Background

Advancements in automobile technology and ever-growing need for communication have made radio communication systems a necessary part of our day-to-day life. Capacity increase in connectivity and sensing demands for vehicles including GPS, entertainment, satellite radio, and systems for collision avoidance [20, 21] are pushing the communication limits further. Expansion in connectivity and communication promised by 5G technology, which will greatly increase vehicle communications, both vehicle to overhead or terrestrial transceivers and vehicle-to-vehicle [22–24] will further widen these boundaries. With growing demand for radio frequency communication, antenna design and fabrication have gained new momentum. While monopole and shark fin antennas have been dominant [20], the increasing need, along with aesthetics and cost, is driving other alternatives. One such alternative is transparent flexible antenna.

Transparent flexible antennas could be unobtrusively mounted to the windows of a vehicle, matching the curve of the glass and located where terrestrial and satellite signals can be received. The substrates used for antenna fabrication therefore need to be high transparent and have an index of refraction that matches with other glass to provide an uninterrupted view through vehicle windows. From a fabrication point of view, these substrates should have good surface quality and high thermal and chemical stability. Corning® Willow® Glass studied in this chapter provides aforementioned qualities, thus providing an excellent substrate for transparent antenna fabrication [13, 17–19]. Furthermore, low cost and high volume that are essential in automotive industry drive the selection of R2R manufacturing.

R2R has been utilized in industries such as document printing and can be applied for the fabrication of antennas and structures for RFID to reduce cost and increase throughput [10, 11, 25]. R2R manufacturing allows provisions to use glass rolls instead of using glass sheets, which are difficult to handle [16, 19]. At Center for Advanced Microelectronics Manufacturing (CAMM) at Binghamton University, we have worked extensively on R2R processing of Willow Glass [26, 27]. In this chapter we do not fully develop R2R manufacturing. Alternatively, we use 100-μm thick glass in 100-mm diameter wafers in initial tests and 150 mm × 150 mm sheets for the antennas and develop processes in small scale that are fully compatible with R2R manufacturing. Here, we will discuss the use of low-cost etched or semi-additive copper and weak acid etchants compatible with R2R fabrication to make conductive meshes and transparent antennas. Transparent antennas, both rigid and flexible, have been demonstrated using meshed conductors. Refs. [28] and [29] demonstrated transparent antennas using freestanding wire mesh screens patterned and attached to a dielectric (Lexan polycarbonate or polydimethylsiloxane) to form patch antennas. Guan et al. [30] have demonstrated the use of mesh conductors for the fabrication of single-layer antennas by utilizing printed meshes of silver paste with lines as small as 2.5 μm. Furthermore, Hong et al. [31] and Lee and Jung [32] used copper meshes to create microstrip patch antennas and receivers for wireless power transfer, with lines approximately 20-μm wide. This work demonstrates that the initial prototyping on flexible glass can be done with conventional lithographic tools and plating rigs, allowing processes to be developed in wafer and sheet scale that are compatible with R2R systems.

## 12.3 Indium Tin Oxide Transparent Antenna on Flexible Glass

### 12.3.1 Fabrication

The materials used in this study primarily consisted of Corning® Willow® Glass wafers of 100-µm thickness as the substrate, sputtered indium tin oxide (ITO) as a conductor layer, and an Al-SiO$_2$ adhesion layer between ITO and glass as a barrier layer. In most cases, the flexible glass wafers were attached to a polyimide wafer-on-web (web) and these thin-film coatings were applied using our R2R GVE sputtering system.

### 12.3.2 Design

The goal was to design a single-sided and single-layer antenna that could be directly connected to the co-planer waveguide (CPW) of a GPPO connector. Additionally, the antenna performance had to be optimized for 100-µm-thick glass. Losses in the ITO could be minimized if a small antenna size was used. With these requirements in the building and the work done in [33, 34], a single-sided CPW-fed antenna design serves as the basis for further design and optimization. Optimizations include the additional design frequency of 5.8 GHz and changing the initial design from a split-ring type antenna to a patch antenna to minimize resistive losses and improve performance. Antenna design and simulations were carried out using ANSYS HFSS and CST Microwave Studio software. This allowed for tuning and optimization of the antenna, including compensating for differences in the sheet resistance of the ITO as shown in Figure 12.2. Optimized antenna designs for 2.4 and 5.8 GHz are shown in Figure 12.3.

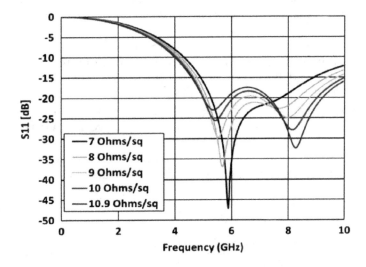

**FIGURE 12.2**
Measured S parameters for different ITO sheet resistances. Note that both the magnitude and location of the radiation peak are affected by this value.

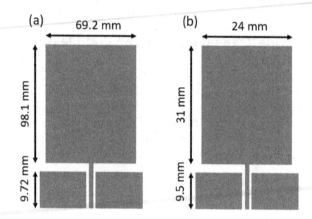

**FIGURE 12.3**
Schematic antenna designs and dimensions for (a) 2.4 GHz antenna and (b) 5.8 GHz antenna. Note that the 5.8 GHz design is smaller in size as indicated on the drawing.

### 12.3.3 Fabrication Method

A Corning® Willow® Glass wafer of 100-μm thickness with 675 nm of ITO layer was deposited using a KDF (panel) or the GVE (R2R) sputter systems. A 100-nm Al-SiO$_2$ barrier layer was deposited between the glass and the ITO layer. The wafer was then annealed at 500°C for 5 min with a ramp rate of 5°C/s in Argon using a rapid thermal processing system. The measured sheet resistance of the wafer with 675 nm of ITO was 3–4 Ω/sq and the transparency was 75–85%. The wafer showed some amount of bowing which was manageable during further fabrication steps.

Since the antennas are based on ITO, prior knowledge on ITO standard recipes could be revised and reused. Standard lithography recipes were assessed to assure the compatibility with flexible glass. Furthermore, an ITO etch recipe has to be compatible with R2R fabrication. Our standard process involved exposing the patch antenna design using a Suss MJB-4 mask aligner, etching with heated 50 g/L concentration oxalic acid mixed from anhydrous crystals, and thermal annealing. In the revised antenna-fabrication process, a Heidelberg Instruments PG 101 direct write laser lithography tool was used to expose the design directly from a GDSII design file. The design file was generated using KLayout [35], where the DXF designs produced from electromagnetic simulations were directly imported and manipulated to form a virtual reticle for the whole wafer.

During the lithography process, first the ITO-coated wafer was primed with MicroChem MCC Primer by spin coating. Spin coating parameters are tabulate in Table 12.3. Next, the

**TABLE 12.3**

Primer Spin Coating Parameters

| Parameter | Step 1 | Step 2 |
|---|---|---|
| Velocity (RPM) | 0 | 3500 |
| Ramp (RPM/s) | 0 | 30,000 |
| Time (s) | 60 | 60 |

**TABLE 12.4**

S1813 Spin Coating Parameters

| Parameter | Value |
|---|---|
| Velocity (RPM) | 4400 |
| Ramp (RPM/s) | 30,000 |
| Time (s) | 60 |

primed wafer was coated with Microposit S1813 photoresist to a 1-μm nominal thickness using the parameters shown in Table 12.4. The wafers were baked for 90 sec at 115°C to further solidify the resist before exposure.

A Heidelburg Instruments PG 101 direct write laser lithography tool in Write Mode III was used to directly write the design. A standard S1813 exposure energy of 40 mW @ 27% was used in writing. The puddle development process tabulated in Table 12.5 was used to develop the resist in an automated spinner. The ITO was then etched with gentle agitation using an oxalic acid solution created from anhydrous crystals dissolved in deionized (DI) water at 50°C. Solution concentrations from 50 to 80 g/L were tested for etching. However, there was little to no effect of the concentration on etch time which was about 40 min for 675-nm ITO layer.

It was also very common to see flakes of undercut ITO, especially along the edges of the patterned structures, that were not directly removed by etching. Physical removal, i.e., a gentle wipe with a cleanroom wiper moistened with DI water, worked well and improved the etched wafer quality. Annealing was performed in a rapid thermal processing system with a nitrogen environment. The annealing parameters are tabulated in Table 12.6. After fabrication, these antennas were singulated using a diamond scribe and mounted onto a 3D printed ABS plastic frames for testing. We used a cyanoacrylate-based adhesive to secure the antenna on the plastic frame. Silver paste was used to provide electrical contact between the connector and the antenna.

### 12.3.4 Testing and Evaluation

Diced and packaged antennas were tested for S parameters. Measured S parameters are shown in Figures 12.4 and 12.5, for 2.4 and 5.8 GHz, respectively. These antennas showed good performance, especially for 5.8-GHz antenna, where smaller size antenna was less affected by the relatively high (compared to metals) resistivity of ITO. Radiation efficiencies of 92 and 49% were observed for the 2.4 and 5.8 GHz antennas, respectively.

**TABLE 12.5**

Puddle Development Process Parameters

| Parameter | Step 1 | Step 2 | Step 3 | Step 4 | Step 5 | Step 6 | Step 7 | Step 8 |
|---|---|---|---|---|---|---|---|---|
| Velocity (RPM) | 50 | 50 | 0 | 350 | 50 | 0 | 600 | 3000 |
| Ramp (RPM/s) | 5000 | 5000 | 5000 | 5000 | 5000 | 5000 | 5000 | 5000 |
| Time (s) | 3 | 4 | 9 | 2 | 4 | 90 | 15 | 40 |
| Dispense | None | MF-26A | None | None | MF-26A | None | DI Water | None |

**TABLE 12.6**

ITO Annealing Parameters

| Parameter | Value |
|---|---|
| Temperature (°C) | 500 |
| Ramp rate (°C/s) | 5 |
| Time (min) | 5 |
| Gas Flow (SLM) | 30 |

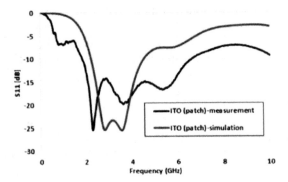

**FIGURE 12.4**
Plot of simulated and measured S parameters for the 2.4 GHz design. Simulations in red color and measurements in black color.

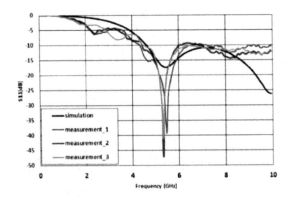

**FIGURE 12.5**
Plot of simulated and measured S parameters for the 5.8 GHz design. Simulations in black color and measurements in color.

## 12.4 Copper Mesh Antennas on Flexible Glass

So far, in Sections 12.1–12.3 we have used transparent conductive materials to fabricate antennas. In Section 12.4, we make use of the concept of achieving transparency by opening holes in a continuous and opaque conductor layer. The advantage is that an antenna designer is not now limited to select transparent (and often less) conductive materials. A meshed conductor with a certain ratio between the width of the conductor lines (w) and the period of the lines (p), as shown in Figure 12.6, shows higher transparency than a continuous conductor plane. The transparency and the sheet resistance of the meshed conductor depends on the radio of w and p. We used FEKO electromagnetic analysis software to study the effect of w and p on the sheet resistance and the transparency. As shown in Figure 12.6, for a 2–2.5-μm-thick layer of Cu, a period to width ratio (p/w) between 10 and 15 is needed to achieve 80% transmission and a sheet resistance of less than 0.3 Ω/sq.

Following the calculations, designs consisting of 25.4 mm × 25.4 mm blocks of grids were created with width-period variations (in micrometers) of 5–50, 5–75, 5–100, 10–100, 10–150, 10–200, and 30–450. The transmission and the sheet resistance were measured for the design variations after fabrication. Initial small-scale tests were conducted on wafers using a design that consisted of 5 × 5 mm blocks of each grid pattern. These samples provided an initial verification that a 1-μm copper layer could be wet etched with the needed precision, and the dry (plasma) etching of the Ti adhesion layer would work. Based on the confirmation, the designs were scaled up for a 100-mm wafer size.

The antenna design that will be discussed here is based on an automobile window glass-mounted multiband antenna for cellular AMPS/PCS reception [36]. The antenna was scaled in size for operation in 4G LTE (fourth generation broadband cellular network based on the long-term evolution communication standard) bands down to 690 MHz and up to 2200 MHz. Figure 12.7 illustrates a drawing of this design. The antennas were fabricated using two types of processes: 1) etched-back process where deposited conductor layer was etched to form the mesh pattern and 2) a semi-additive process where the Cu was plated on a seed layer with the aid of a photomask. Figure 12.8 shows a flowchart and a schematic representation of two types of processes.

Even though we focus on transparent antenna in this chapter, solid designs are also fabricated to serve as baselines for performance comparisons. During the grid creation,

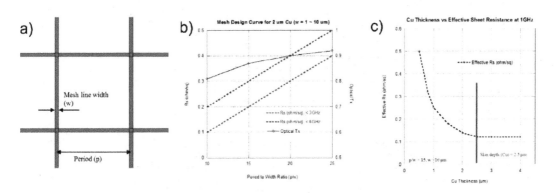

**FIGURE 12.6**

(a) A diagram of mesh layout showing line width and period, (b) calculated sheet resistance and transparency for different period/width ratio, and (c) Cu thickness vs effective sheet resistance at 1 GHz.

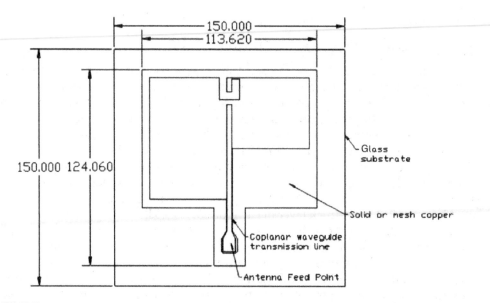

**FIGURE 12.7**
AutoCAD screen capture of the antenna design on 150 mm × 150 mm flexible glass substrate. All dimensions are in millimeters.

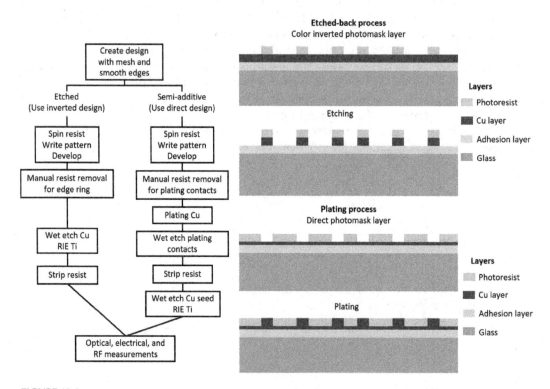

**FIGURE 12.8**
Flowchart and schematic diagram of the antenna fabrication using both etched and semi-additive processes.

AutoCAD and KLayout [35] software packages were used to create a meshed version of the original design, while keeping smooth edges and producing an output that can be used for lithography. The meshing process started with the original antenna in AutoCAD in .DXF format. Next, an offset of the edges of the antenna was created. This was to make a smooth edge around the antenna and create a boundary for the meshed area. The design was then imported to KLayout and meshing was done via a Boolean operation between an array of squares representing the holes in the mesh and the solid of the antenna resulting in the desired mesh. A subsequent Boolean operation was done to merge the edge created in AutoCAD with the meshed pattern and a final Boolean may be done to reverse the field of the design for an etched-back process. Figure 12.9 illustrates this software manipulation process. After the design was finalized, it was exported to .GDSII file format. This format is used by the laser lithography system and provided an unambiguous check of the design to be written.

Before fabricating a full antenna design, it was important to ensure the quality to which a copper mesh can be created since the eventual antenna performance is based on the mesh. Therefore, first 100-mm wafers consisting of 25.4-mm squares of Cu mesh were fabricated. Once the meshes were fabricated and characterized, the rest of the antenna fabrication proceeded. Corning® Willow® Glass was compatible with all the conventional fabrication processes such as spin coating of resist, development, electroplating, etc. The challenges were in the wafer and sheet handling compared to a standard silicon wafer. Sharp edges

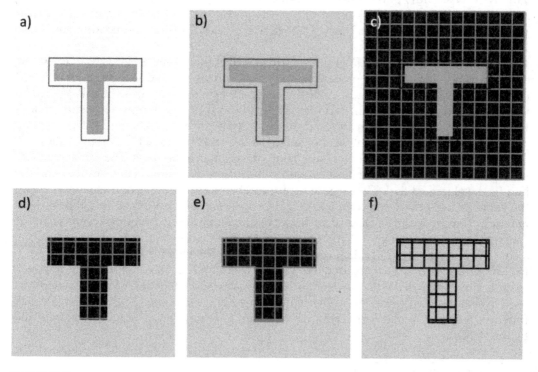

**FIGURE 12.9**
Software manipulation process for mesh creation. (a) Design with an offset edge shown in red, (b) solid design on the write area, and (c) design with blocks (black) superimposed for mesh creation. (d) Mesh created from Boolean operation with blocks and (e) offset edge merged with the created mesh. (f) Boolean operation to create mesh design for writing positive resist.

and point loads, such as those produced by metal wafer tweezers, are not recommended as they introduce defects that quickly propagate. Therefore, handling had to be done with tools that have smooth and wide contacts to avoid unnecessary point loading. To help handling in certain steps such as the wet etching, polyimide tape tabs (in the shape of handles) were put on the wafers and sheets to provide ways to use tweezers without breakage.

Due to the limitations of the tools used, manual removal of photoresist was used to complete the antennas. This consisted of removing the excess resist outside of the laser lithography write area and creating windows for contact with the seed layer for plating in the semi-additive process. While multiple exposures could have been used to expose these areas, it was much faster and easier to manually remove the resist with solvent. It should be noted that the use of R2R could eliminate the need for these additional manual steps. Furthermore, all the process steps demonstrated here can be ported to R2R facilities available in our laboratory.

### 12.4.1 Fabrication – Etch Process

In etch process, first a 100-nm Ti adhesion layer and a 1-µm copper layer were deposited on as-received Corning® Willow® Glass using a KDF in-line sputter system. A 1-µm copper layer was easy to work with due to relatively low bowing of the glass. Furthermore, the etch time for a copper layer thicker than 1 µm would result in some form of denaturing of the photoresist. The substrate was primed with MicroChem MCC primer and standard Microposit S1813 photoresist was spin coated to 1-µm nominal thickness. Following the spin coating, a 90-sec baking at 115°C was performed to further solidify the resist. The same set of parameters tabulated above were used in prime and coating steps (see Tables 12.3 and 12.4).

A Heidelberg Instruments µPG 101 laser direct write lithography tool was used to expose the resist. Since this is an etch process, as shown in Figures 12.8 and 12.9, the color inverted image of the design was used to expose the resist. An exposure energy of 40 mW @ 25–27% was used with write mode III. Write mode III allowed for the largest exposure area of 125 mm × 125 mm with 5-µm resolution. Three writes were used: one for the antenna design, and two rectangles that were offset to the top and bottom of the design to allow for easier removal of the excess photoresist "ring" at the edge of the sheet. The same exposure settings tabulated above were used to expose the wafers in an automated spinner and the puddle development tool. The large size of the substrate sometimes necessitates running the same process twice for complete removal of the resist. Any excess photoresist was manually removed using a cleanroom wiper with acetone and isopropyl alcohol. Next, the Cu layer was etched using ammonium persulfate (APS) mixed from anhydrous crystals at a concentration of 20 g/200 mL at room temperature. The etch took approximately 2 min to clear all Cu. After Cu was cleared, the sample was rinsed in DI water using an in-bench cascade rinse and dried with nitrogen. A Nano Master NRP-4000 PECVD/RIE system was used to etch the Ti adhesion layer. The etch parameters are tabulated in Table 12.7. The etch time was 5 min. A final solvent cleaning was used to remove any residual resist followed by the etching.

### 12.4.2 Fabrication – Semi Additive

Difficulties and challenges in a subtractive process for creating fine features motivated us to develop a semi-additive process for antenna fabrication. This process started with depositing a 100-nm Ti adhesion layer on Corning® Willow® Glass using a KDF in-line

**TABLE 12.7**

Ti RIE Etch Parameters

| Parameter | Value |
|---|---|
| Time (min) | 5 |
| RF power (W) | 200 |
| Pressure (torr) | $3 \times 10^{-1}$ |
| $CF_4$ flow (sccm) | 80 |
| $O_2$ flow (sccm) | 10 |

sputter tool followed immediately by a 100-nm Cu seed layer. This seed layer will be used in electroplating copper in further steps. Next, the substrate was primed with MCC primer. MicroChem SPR 220 photoresist was spin coated to a 4-μm nominal thickness. The same parameters used in Section 12.4.1 were used for priming and the parameters used for resist coating are tabulated in Table 12.8. A hotplate baking step at 115°C for 90 was performed to ensure further solidifying of the resist.

As shown in Figures 12.8 and 12.9, for semi-additive process, the direct meshed antenna design was used and exposed using the Heidelberg Instruments μPG 101 laser direct write lithography system using an exposure of 30 mW @ 90%. Next, the resist was developed using the same set of parameters outlined previously in an automated puddle development tool. Similar to the etching process, the development step had to be run twice to ensure full exposure and resist removal. Copper plating was done in a different facility and special care was taken to eliminate possibilities of glass breakage during loading, unloading, and plating. For example, clips used to hold the glass to plating fixtures were polished to remove any sharp edges that could exert point loads on the glass. During the plating process, the substrates were immersed in a sulfuric acid solution for 30 sec for cleaning prior to plating. They were then plated in a copper sulfate bath using 5-A current and 5-V potential. A 100-mm wafer took 3 min for plating. A solid antenna and a meshed antenna took 5.30 and 4.30 min, respectively, to plate. The plated samples were rinsed in DI water to remove any residual plating solution left on the substrates and dried with canned air to prevent corrosion of the plated Cu and the sticking of the interleaf foam to the glass. After stripping the resist, the seed copper layer was etched using APS etchant. The etch time for the seed layer was ~4.5 min. The Ti adhesion layer was removed by using RIE process described in Section 12.4.1. At the end of the fabrication process, antennas with 2- to 2.5-μm-thick copper conducting layers were obtained.

## 12.4.3 Testing and Evaluation

First, the meshes fabricated to test the concept of meshing to create transparent antennas were characterized for their electrical and optical properties. A Jandel Model RM3000

**TABLE 12.8**

SPR 220 Spin Coating Parameters

| Parameter | Value |
|---|---|
| Velocity (RPM) | 2200 |
| Ramp (RPM/s) | 30,000 |
| Time (s) | 60 |

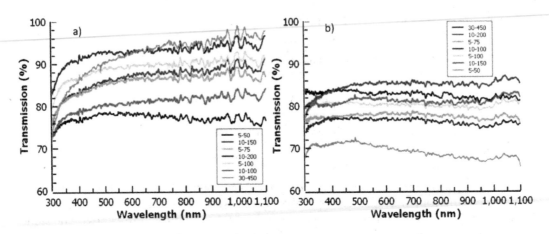

**FIGURE 12.10**
The measured transmission spectra for 25.4-mm blocks of mesh made by (a) etching and (b) semi-additive processing.

four-point probe tool was used to determine the sheet resistance and a Filmetrics F20-UV reflectometer configured for transmission was used to measure the optical transparency. Figures 12.10a and 12.10b show the transmission as a function of wavelength for meshes fabricated using different linewidth-period combinations for etched meshes and plated meshes, respectively. Furthermore, the changes in the sheet resistance and the optical transmission as a function of the mesh type and fabrication method are shown in Figure 12.11. Based on these results, it was decided to use a 10–100 linewidth-period combination for further processing as this provided a good result for both fabrication processes with transmission near 80% and sheet resistance less than 0.2 Ω/sq. Antennas fabricated with 10–100 linewidth-period combination using etch process and plating process are shown together with a solid antenna in Figure 12.12.

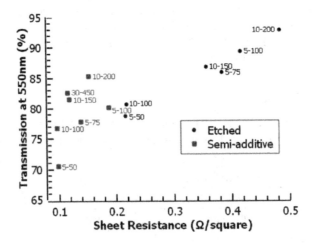

**FIGURE 12.11**
Scatter plot of sheet resistance and transmission for mesh samples fabricated by etching (in black) and semi-additive (in red) processes.

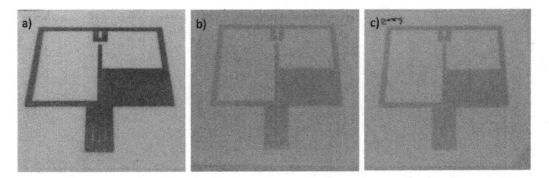

**FIGURE 12.12**
Photographs of antennas that were RF tested. (a) A solid copper antenna, (b) an etched copper mesh antenna, and (c) a plated copper mesh antenna.

### 12.4.3.1 On-Vehicle Testing

A selected antenna from each type was used for on-vehicle testing at the antenna range located at Oakland University. This range had the roll-on roll-off capability where a vehicle was spun on a turntable while a goniometer-mounted antenna measured the emitted field. Figure 12.13 shows a photograph of a fabricated antenna on a vehicle and other details related to antenna measurement setup. Antennas were slot-die coated with an adhesive to allow for easy attachment and detachment from the automobile window. A coaxial pigtail with SMA connectors were attached with conductive epoxy. Our prior work suggests that the conductive adhesive provided better connectivity to perform the testing. Measurements were carried out for LTE frequencies in the range of 1700–2200 and 700–900 MHz, for both vertical (V) and horizontal (H) polarizations, respectively.

The linear average gain of the antennas taken over a 360° azimuth sweep with at 5° elevation increments, with 0° perpendicular to the ground is shown in Figure 12.14 for 1700–2200 and 700–900 MHz frequency ranges in LTE band. The linear average gain gives an indication of the antenna efficiency at each elevation angle. Here, the average gain is the ratio of the power measured at an elevation angle to the total power radiated isotropically.

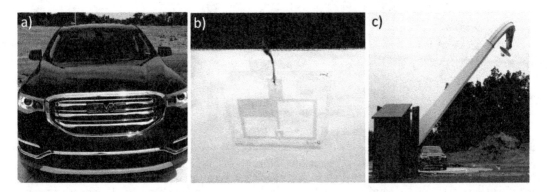

**FIGURE 12.13**
(a) A photograph of a fabricated antenna on a vehicle, (b) a zoomed in view of the windscreen, and (c) a photograph of the vehicle in the antenna range at Oakland University.

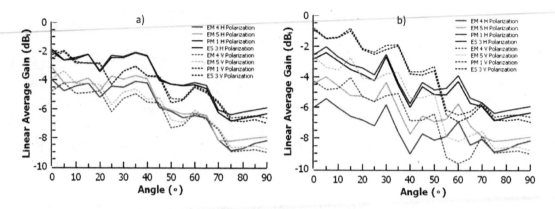

**FIGURE 12.14**

Plot of linear average gains (a) in the high band (1700–2200 MHz) and (b) in the low band (700–900 MHz) for on-vehicle antenna tests. ES – etched solid antenna, EM – etched mesh antenna, and PM – plated mesh antenna.

The etched mesh antennas show similar performance in both polarizations. Furthermore, the high-band and low-band measurements closely overlap with each other with less than 2 dB$_i$ difference. A semi-additive mesh antenna and a solid antenna show very similar performance. These observations are also highlighted in 3D visualizations of the data shown in Figure 12.15. Both solid and plated mesh antennas have gains about 2–3 dB$_i$ higher than an etched mesh antenna. The disparity can be partially due to the

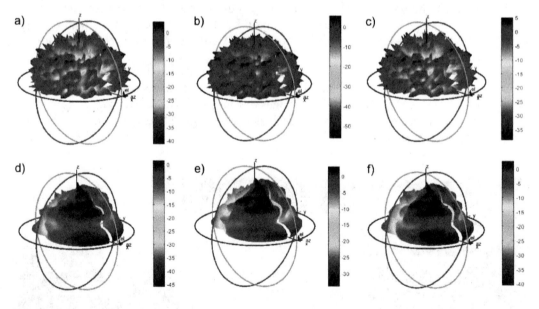

**FIGURE 12.15**

3D visualizations of the observed antenna gain (dB$_i$) for the H polarization at a high-band frequency of 2110 MHz and low-band frequency of 717 MHz, with (a) etched mesh antenna high band, (b) plated mesh antenna high band, (c) etched solid antenna high band, (d) etched mesh antenna low band, (e) plated mesh antenna low band, and (f) etched solid antenna low band. Note that the visualizations are oriented such that the front of the vehicle is pointing to the lower right. One fabricated antenna from each type was selected for visualization.

differences in copper thickness between a semi-additive/solid and an etched antenna. A semi-additive mesh antenna had ~2-μm-thick copper whereas an etched mesh antenna only had ~1-μm-thick copper. A thicker conductor usually mitigates RF losses due to skin depth. Overall, all the antennas fabricated had gains ~2 dB$_i$ less than a conventional roof mounted LTE antenna. Therefore, the direct usability of the antenna for wireless communication was limited. However, these antennas can be implemented as a part of a multiple input multiple output (MIMO) system where antennas such as these have demonstrated LTE data throughput as good or even better than the LTE throughput of roof mounted antennas [37].

## 12.5 Discussion and Conclusion

We demonstrated fabrication and packaging of transparent antenna using ITO on Corning® Willow® Glass. Antennas were fabricated for 2.4 and 5.8 GHz microwave frequencies. Design and simulation of these antennas were completed using ANSYS HFSS and CST Microwave Studio, which allowed for the tuning and optimization of the antenna. These antennas were improved upon work previously done in our group. Improvements in packaging allowed for more consistent and accurate measurements of antenna performance and increased durability of the glass. Furthermore, the design thickness provided higher conductivity. As shown in Section 12.3, the antennas performed well, especially at 5.8 GHz. Further improvements in the performance of the antenna can be done by reducing the ITO resistance.

Furthermore, design, fabrication, characterization, and on-vehicle testing of transparent mesh antennas on Corning® Willow® Glass were discussed. These copper-based antennas were fabricated using a subtractive and a semi-additive process. Both these processes were demonstrated on sheet level and are compatible with R2R processing. The meshing technique introduced here to achieve transparency proved to be effective with providing greater than 70% transparency with ~0.2 Ω/sq sheet resistance. On-vehicle testing demonstrated that the antennas performed well and that a solid antenna and a semi-additively fabricated mesh antenna had similar performance. Further improvements can be made to increase the usability of these antennas for vehicle wireless communication.

One of the main limiting factors that hindered further experimentation was the time taken for direct writing of the antenna using the laser direct write lithography system. The use of a large area-projection lithography system would significantly reduce the exposure time. This would greatly enhance the throughput of the process. Once the design is optimized and finalized, a photomask can be used to reduce the exposure overhead as well. Many packaging and connection-related aspects of these antennas are still open questions. For example, when antennas are laminated onto surfaces, signal feed lines have to be redesigned to meet lamination requirements. Developments need to be made in leaving some flexibility in the way that RF signals are fed to the antenna. For example, flexible transmission lines made on polymer substrates could be used. The connection between a conventional connector on one end and the flexible transmission lines on the other end could be joined with an anisotropic conductive film (ACF) adhesive providing an improved, compliant, and low-loss connection.

# References

[1] "Printed and Flexible Electronics in Automotive Applications 2016–2026," Tech. rep. (2016).

[2] "In-Mold Electronic Technology | DuPont USA," http://www.dupont.com/products-and-services/electronic-electrical-materials/printed-electronics/products/in-mold-electronic-technology.html.

[3] "Potential for printed electronics in aerospace components," https://www.insidecomposites.com/potential-for-printed-electronics-in-aerospace-components/

[4] T. Q. Trung and N. Lee, "Flexible and stretchable physical sensor integrated platforms for wearable human-activity monitoring and personal healthcare," Advanced Materials 28, 4338–4372 (2016).

[5] A. Nathan, A. Ahnood, M. T. Cole, S. Lee, Y. Suzuki, P. Hiralal, F. Bonaccorso, T. Hasan, L. Garcia-Gancedo, A. Dyadyusha, S. Haque, P. Andrew, S. Hofmann, J. Moultrie, Daping Chu, A. J. Flewitt, A. C. Ferrari, M. J. Kelly, J. Robertson, G. A. J. Amaratunga, and W. I. Milne, "Flexible electronics: The next ubiquitous platform," Proceedings of the IEEE 100, 1486–1517 (2012).

[6] N. R. Council, The Flexible Electronics Opportunity (The National Academies Press, Washington, DC, 2014). DOI: 10.17226/18812.

[7] A. Shorey, P. Cochet, A. Huffman, J. Keech, M. Lueck, S. Pollard, and K. Ruhmer, "Advancements in fabrication of glass interposers," in "2014 IEEE 64th Electronic Components and Technology Conference (ECTC)," (2014), pp. 20–25.

[8] A. Shorey, J. Keech, G. Piech, B. K.Wang, and L. Tsai, "Glass substrates for carrier and interposer applications and associated metrology solutions," in "ASMC 2013 SEMI Advanced Semiconductor Manufacturing Conference," (2013), pp. 142–147.

[9] A. Shorey, S. Pollard, A. Streltsov, G. Piech, and R. Wagner, "Development of substrates for through glass vias (TGV) for 3ds-IC integration," in "2012 IEEE 62nd Electronic Components and Technology Conference," (2012), pp. 289–291.

[10] N. Savage, "Roll to Roll Electronics Manufacturing Rolls On," IEEE Spectrum: Technology, Engineering, and Science News (2016).

[11] Accrington, "On a roll," The Economist (2016).

[12] M. Sullivan, "Printed Electronics: Global Markets to 2022," Tech. Rep. IFT066C, BCC Research (2017).

[13] S. M. Garner, ed. Flexible Glass: Enabling Thin, Lightweight, and Flexible Electronics (John Wiley & Sons, 2017).

[14] S. M. Garner, S. C. Lewis, and D. Q. Chowdhury, "Flexible glass and its application for electronic devices," in "2017 24th International Workshop on Active-Matrix Flatpanel Displays and Devices (AM-FPD)," (2017), pp. 28–33.

[15] N. Kooy, K. Mohamed, L. T. Pin, and O. S. Guan, "A review of roll-to-roll nanoimprint lithography," Nanoscale Research Letters 9, 320 (2014).

[16] H. Tamagaki, Y. Ikari, and N. Ohba, "Roll-to-roll sputter deposition on flexible glass substrates," Surface and Coatings Technology 241, 138–141 (2014).

[17] M. Junghähnel and S. Garner, "Glass meets flexibility," Vakuum in Forschung und Praxis 26, 35–39 (2014).

[18] S. M. Garner, K. W. Wu, Y. C. Liao, J. W. Shiu, Y. S. Tsai, K. T. Chen, Y. C. Lai, C. C. Lai, Y. Z. Lee, J. C. Lin, X. Li, and P. Cimo, "Cholesteric liquid crystal display with flexible glass substrates," Journal of Display Technology 9, 644–650 (2013).

[19] M. Junghähnel, S. Weller, and T. Gebel, "P-65: Advanced processing of ITO and IZO thin films on flexible glass," SID Symposium Digest of Technical Papers 46, 1378–1381 (2015).

[20] B. D. Pell, E. Sulic, W. S. Rowe, K. Ghorbani, and S. John, "Advancements in automotive antennas," in New Trends and Developments in Automotive System Engineering (InTech, 2011).

[21] K. Solbach and R. Schneider, "Review of antenna technology for millimeter wave automotive sensors," in "Microwave Conference, 1999. 29th European," (IEEE, 1999), vol. 1, pp. 139–142.

[22] "Experience the Future of Mobility - 5G Automotive Association," http://5gaa.org/5g-technology/experience-the-future/.

[23] "Paving the way towards 5G - 5G Automotive Association," http://5gaa.org/5g-technology/paving-the-way/.

[24] "A look ahead at 5G's impact on the automotive industry," https://www.qualcomm.com/news/onq/2017/05/22/look-ahead-5gs-impact-automotive-industry (2017).

[25] R. R. Søndergaard, M. Hösel, and F. C. Krebs, "Roll-to-Roll fabrication of large area functional organic materials," Journal of Polymer Science Part B: Polymer Physics 51, 16–34 (2013).

[26] R. Malay, A. Nandur, J. Hewlett, R. Vaddi, B. E. White, M. D. Poliks, S. M. Garner, M.-H. Huang, and S. C. Pollard, "Active and passive integration on flexible glass substrates: Subtractive single micron metal interposers and high performance IGZO thin film transistors," in "Electronic Components and Technology Conference (ECTC), 2015 IEEE 65th," (IEEE, 2015), pp. 691–699.

[27] M. D. Poliks, Y. L. Sung, J. Lombardi, R. Malay, J. Dederick, C. R. Westgate, M. H. Huang, S. Garner, S. Pollard, and C. Daly, "Transparent antennas for wireless systems based on patterned indium tin oxide and flexible glass," in "2017 IEEE 67th Electronic Components and Technology Conference (ECTC)," (2017), pp. 1443–1448.

[28] E. R. Escobar, N. J. Kirsch, G. Kontopidis, and B. Turner, "5.5 GHz optically transparent mesh wire microstrip patch antenna," Electronics Letters 51, 1220–1222 (2015).

[29] H. A. Elmobarak Elobaid, S. K. Abdul Rahim, M. Himdi, X. Castel, and M. A. Kasgari, "A transparent and flexible polymer-fabric tissue UWB antenna for future wireless networks," IEEE Antennas and Wireless Propagation Letters 16, 1333–1336 (2017).

[30] N. Guan, H. Tayama, S. Kaushal, and Y. Yamaguchi, "A wire-grid type transparent film UWB antenna," in "2017 IEEE-APS Topical Conference on Antennas and Propagation in Wireless Communications (APWC)," (2017), pp. 166–169.

[31] S. Hong, Y. Kim, and C. Won Jung, "Transparent microstrip patch antennas with multilayer and metal-mesh films," IEEE Antennas and Wireless Propagation Letters 16, 772–775 (2017).

[32] H. H. Lee and C. W. Jung, "Magnetic resonant-wireless power transfer for transparent laptop applications using metal mesh film," Microwave and Optical Technology Letters 59, 2781–2785 (2017).

[33] H. Khaleel, Innovation in Wearable and Flexible Antennas (WIT Press, 2014). Google-Books-ID: _x0RBQAAQBAJ.

[34] H. Khaleel, H. Al-Rizzo, and A. Abbosh, "Design, fabrication, and testing of flexible antennas," in "Advancement in Microstrip Antennas with Recent Applications," Ahmed Kishk, ed. (InTech, 2013). DOI: 10.5772/50841.

[35] "KLayout Layout Viewer And Editor," http://www.klayout.de/index.php.

[36] H. J. Song, C. R. White, J. H. Schaffner, A. Berkaryan, and E. Yasan, "Multifunction antenna," (2010). US Patent 8,704,719 B2.

[37] J. H. Schaffner, H. J. Song, A. Bekaryan, T. Talty, D. Carper, and E. Yasan, "Scanner based drive test LTE capacity measurements with MIMO antennas placed inside the vehicle," in "2015 IEEE International Conference on Microwaves, Communications, Antennas and Electronic Systems (COMCAS)," (2015), pp. 1–4.

# 13

# Testing and Reliability Characterization Methods for Flexible Hybrid Electronics

Pradeep Lall, Jinesh Narangaparambil, and Kartik Goyal

*Auburn University*

## CONTENTS

Flexible electronics in wearable application may be subjected to twisting, folding, or flexing depending on the form factor and the location of use. The trend toward weight reduction and miniaturization of electronics is driving the emergence of flexible electronics. A number of additive printed electronics methods are available including aerosol-jet printing (AJP), inkjet printing, screen printing, and gravure printing. Reduction in size and weight are especially important for applications focusing on sports, leisure, healthcare, military and security apparel, fashion, and wearable consumer electronics [Tao 2017]. Mechanical stresses are induced on flexible electronics when subjected to bending, which can lead to delamination, cracking, or shearing of interconnects and additively printed layers [Dai 2015].

An understanding of the shear strength of the deposited layer is important in addition to the effect of process conditions on the reliability and survivability of the printed structures.

The AJP is a popular method in flexible electronics because of its high-resolution in comparison with inkjet, screen printing. It has also gained much popularity as a low-volume, custom print electronics manufacturing technique. The vulnerability of the conductors for mechanical stresses needs to be investigated thoroughly as they are an essential block in flexible electronics [Happonen 2015]. From a bendability point of view, high mechanical strength and stability are key parameters required from the substrate material. A common choice for the substrate is plastic film, such as polyethylene terephthalate (PET), polyethylene naphthalate (PEN), or polyimide (PI) [Yakimets 2009; Lacerda 2013]. The relationship between microstructural evolution, such as particle growth, porosity, and electrical conductivity during post heat treatment has been widely studied. Reducing the porosity and forming a dense microstructure can improve the electrical conductivity, which can be attained by controlling annealing temperature [Lee 2005; Park 2006; Jeong 2006], annealing time [Greer 2007], heat treatment method [Kim 2011], and ambient atmosphere [Yia 2010]. According to the end application, the process parameters can be varied in order to get a lower resistance but, at the same time, enough shear strength to suffice the need. Flexing is one of the common body joint motions. Wearable application in flexible electronics being the major area, it is important to analyze and improve reliability.

## 13.1 Flexure Reliability of Flexible Hybrid Electronics (FHE)

Figure 13.1 shows the flexure setup developed at Auburn University's CAVE3 Research Center. The setup includes a data acquisition unit connected to the test vehicle to measure the resistance across each individual trace, which records the data into the computer. A stepper motor, power supply source, stepper motor driver, microcontroller, lead screw actuator, and hinge setup are used to produce flexure.

The acquisition unit measures real-time data and records it into the software, which helps in analyzing the fatigue failure point more conveniently. Figure 13.2 shows the

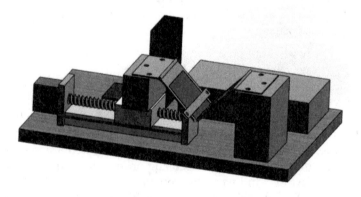

**FIGURE 13.1**
Experimental setup (flexure).

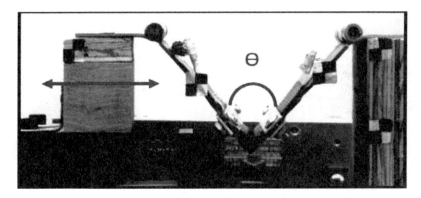

**FIGURE 13.2**
Motion of the test board (flexure).

motion of setup. The test setup replicates the v-bend encountered in body motions like folding hands at elbow or folding of legs. These bends allow the testing of local features in the line allowing for applying stress at a region or over a region of line and not on the entire test board [Lall 2018]. The wires are connected to the traces on the substrate using silver conductive epoxy. Figure 13.2 shows how the actuator moves in the horizontal direction imparting flexure to the test-specimen. The samples are fixed with the help of two metal sheets of width 4 mm at 10 mm from the center of the hinge. Figure 13.3 shows the experimental setup with the flexible electronics test vehicle.

Figure 13.4 shows the experimental stepwise flow. Each end of the traces is connected to the data-acquisition system wire using silver conductive epoxy. The wire connection is routed to the multiplexer (34901A) which is connected to the Agilent Data Acquisition Unit (34970A). The acquisition unit is connected to a computer through general purpose interface bus-to-universal serial bus (GPIB-USB) cable. Cycle time for the setup is 2 sec (0.5 Hz). The acquisition unit takes readings at an interval of 1 sec such that it takes reading at the start of the cycle and at the bent position similarly in twist state. The data recorded by the software is saved with '.csv', an extension, which can be imported to the excel for analysis.

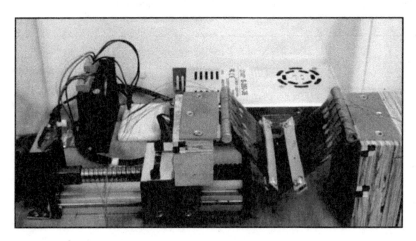

**FIGURE 13.3**
Experimental setup with a test vehicle (flexure).

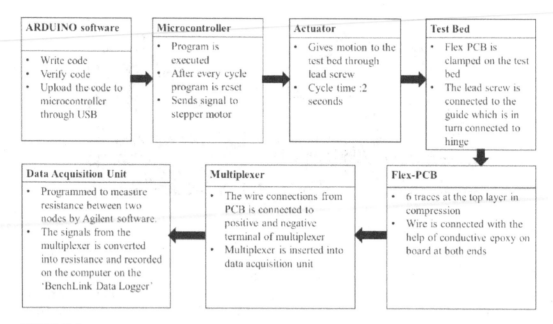

**FIGURE 13.4**
Block diagram of the experimental setup.

Figure 13.5 shows the actual graph plot using the high-speed camera for its position versus time graph to measure frequency and velocity.

### 13.1.1 Specimen Dimensions and Test Conditions

A typical substrate dimension of 132 mm × 76 mm was used. The printing is done to achieve the line width of 0.15 mm. The pneumatic atomizer is used for the deposition of ink on the substrate. The samples are printed using a single pass.

The trace patterns for twist and flexure are shown in Figure 13.6. The printing parameters used for fabrication of the specimen are shown in Table 13.1. Each geometrically shaped trace has two identical traces in order to measure the consistency within the same sample.

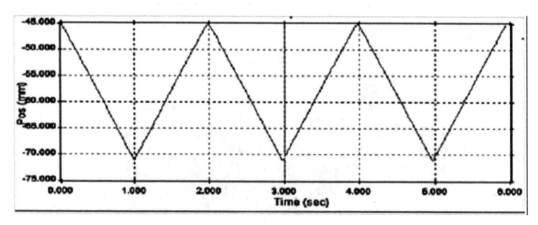

**FIGURE 13.5**
Graph plot for frequency.

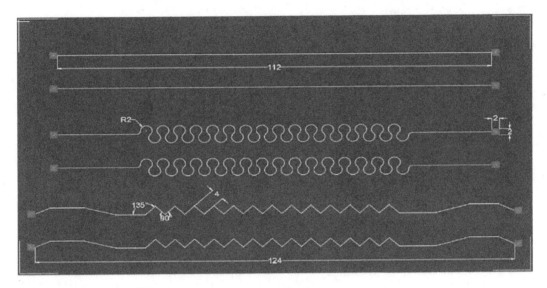

**FIGURE 13.6**
Trace pattern for flexure.

To study the effect of sintering temperature on the electrical and mechanical properties, after the printing is done the samples are sintered at four different temperatures. The samples are sintered for 1 h at sintering temperatures 150°C, 200°C, 250°C, and 300°C. The pressure of the nitrogen ($N_2$) gas, mass flow rate of ink is been established using a detailed study of the process parameters. The optimal set of process parameters has then been used for fabrication of the samples used for twisting and flexing experiments. Figure 13.7 shows nomenclature for each of the additively printed trace morphologies.

The flexure was between an initial angle of 100° and a final angle of 80° at a frequency of 0.5 Hz. The reliability in flexure can be used to identify the desirable manufacturing parameters of the printed lines. Table 13.2 shows the sintering conditions used for the data-collection in the test matrix.

## 13.1.2 Effect of Sintering Temperature on Flexure Reliability

The flex-PCBs (printed circuit boards) sintered at different temperatures are tested on the flexure machine. The traces on the substrate undergo compression stress state during

**TABLE 13.1**

Printing Parameters

| Process Parameters | Value |
|---|---|
| PA MFC | 800 ccm |
| Sheath flow rate | 40 ccm |
| Exhaust MFC | 750 ccm |
| Speed | 7 mm/s |
| Platen temperature | 25°C |
| Ink temperature | 25°C |
| Print height | 3 mm |
| Nozzle diameter | 300 μm |

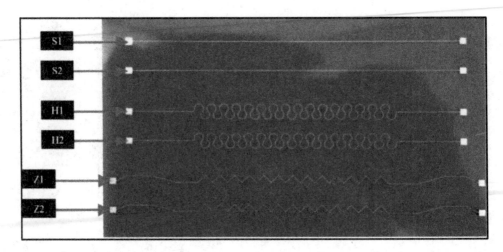

**FIGURE 13.7**
Nomenclature of the test vehicle for/flexure.

flexure. The samples are initially fixed on the test bed at 100°. The flexing cycle is from 100° to 80°; $\Delta\theta = 20°$. The failure criteria of the test board considered are 20% increase in resistance from its initial value.

Figure 13.8 shows the effect of varying sintering temperature on resistance keeping the sintering time identical. It can be observed from Figure 13.8 that as the temperature of sintering is increased, the resistance of the traces decreases. During sintering, the nanoparticles of the ink are compacted due to evaporation of carrier solvents and the surfactants in the ink. The reduction in the resistance reduces with the increase in sintering temperatures with the temperatures higher than 250°C yielding marginal benefits. The trend holds true for straight, zigzag, and horseshoe trace geometries.

### 13.1.3 Effect of Sintering Temperature on Shear Load to Failure

Figure 13.9 shows the relation of sintering temperature and shear load to failure. With an increase in sintering temperature, the shear load to failure tends to increase till a sintering temperature of 250°C. Increase in the sintering temperature above 250°C causes a reduction in the shear strength of the printed traces. While the precise temperature at which the shear strength peaks may vary from ink-to-ink, the reversal of the trend in the shear strength with the increase of sintering temperature is significant. The reported values of shear have been measured on a Dage shear tester machine.

**TABLE 13.2**

Test Matrix for Flexural Test

| Flexure Test | | | | |
|---|---|---|---|---|
| $\Delta\theta = 20°$ | Sintering Temperature @ 1 hour | | | |
| | 150C | 200C | 250C | 300C |

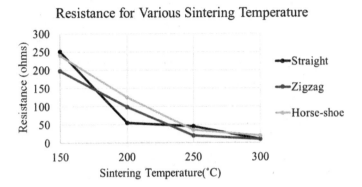

**FIGURE 13.8**
Effect of various sintering temperature on resistance.

### 13.1.4 Cyclic Flexure Accelerated Tests

Each sample was tested for 100,000 cyclic flexing. The change in resistance was recorded with the help of the data acquisition unit. For the test conditions, it is observed that the resistance of the traces gradually increases over the number of cycles. The resistance of the traces increases due to crack formation in the region of the bend. The resistance fluctuations in the graphs (Figures 13.10–13.33) are due to the change in the cross-sectional area of the printed trace during flexure. Continuous flexing leads to wrinkles on the flexible substrate thereby accelerating the crack-formation process. The resistance of the traces has an increasing trend with the number of cycles to failure. Figures 13.10–13.21 show the test results for the testing condition of initial angle of 100° and final angle of 80°.

#### 13.1.4.1 Samples Sintered at 150°C for 1 H

In this section, the trend of resistance change with the increase in the flexure cycles is reported for samples sintered at 150°C for 1 h. The failure criterion of 20% resistance change

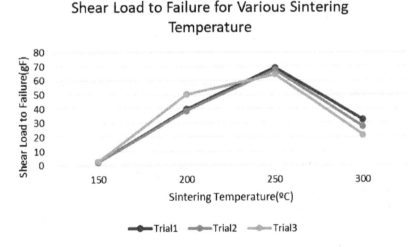

**FIGURE 13.9**
Effect of various sintering temperatures on shear load to failure.

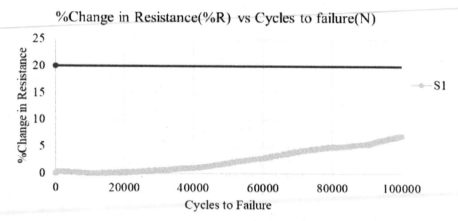

**FIGURE 13.10**
Data on trace S1 for cyclic flexing.

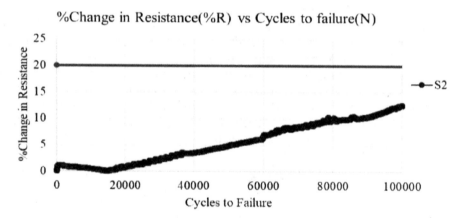

**FIGURE 13.11**
Data on trace S2 for cyclic flexing.

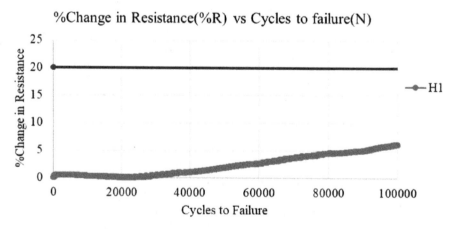

**FIGURE 13.12**
Data on trace H1 for cyclic flexing.

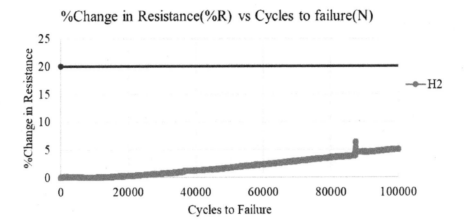

**FIGURE 13.13**
Data on trace H2 for cyclic flexing.

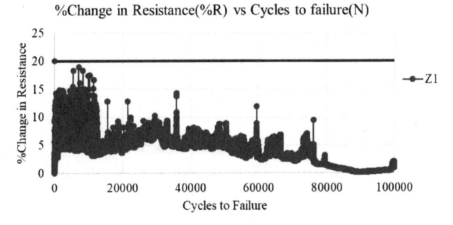

**FIGURE 13.14**
Data on trace Z1 for cyclic flexing.

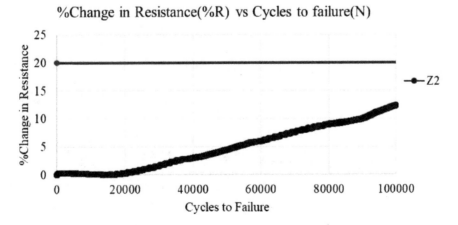

**FIGURE 13.15**
Data on trace Z2 for cyclic flexing.

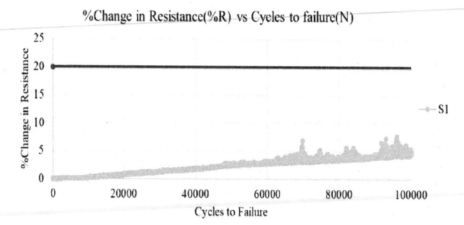

**FIGURE 13.16**
Data on trace S1 for cyclic flexing.

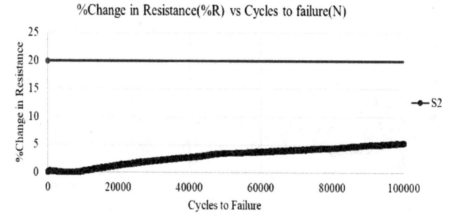

**FIGURE 13.17**
Data on trace S2 for cyclic flexing.

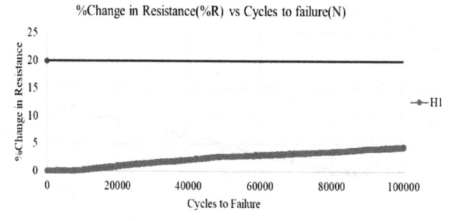

**FIGURE 13.18**
Data on trace H1 for cyclic flexing.

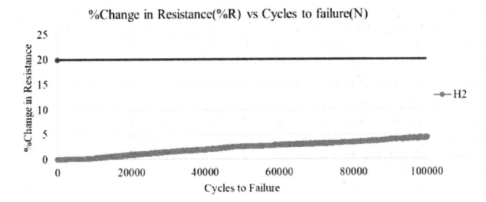

**FIGURE 13.19**
Data on trace H2 for cyclic flexing.

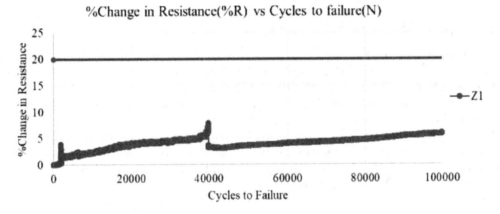

**FIGURE 13.20**
Data on trace Z1 for cyclic flexing.

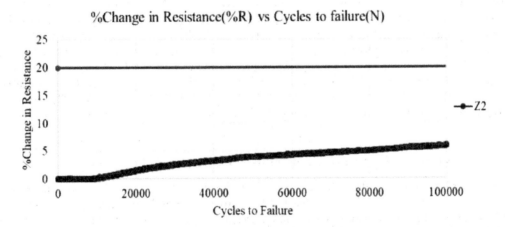

**FIGURE 13.21**
Data on trace Z2 for cyclic flexing.

is indicated with a horizontal line on each of the graphs. None of the sample morphologies exhibits failure after 100,000 cycles, even though the zigzag traces show the highest resistance increase with the increase in the flexure cycles.

### 13.1.4.2 Samples Sintered at 200°C for 1 H

In this section, the trend of resistance change with the increase in the flexure cycles is reported for samples sintered at 200°C for 1 h. The failure criterion of 20% resistance change is indicated with a horizontal line on each of the graphs. None of the line morphologies exhibits failure after 100,000 flexure cycles.

By comparing the results of samples sintered at 150°C and 200°C, it can be noticed that the slope of the resistance graph reduces with increasing sintering temperature in each geometry. The improved flexure fatigue life can be attributed partly to the reduction in the line porosity with the increase in sintering temperature and partly to the increase in the adhesion with the underlying substrate. Failure is generally accompanied with deadhesion of the printed traces from the underlying substrate.

### 13.1.4.3 Samples Sintered at 250°C for 1 H

In this section, the trend of resistance change with the increase in the flexure cycles is reported for samples sintered at 250°C for 1 h. The failure criterion of 20% resistance change is indicated with a horizontal line on each of the graphs. Figures 13.22 and 13.23 show the data for straight traces. Figures 13.24 and 13.25 show the data for horseshoe traces. Figures 13.26 and 13.27 show the data for zigzag traces. A comparison between the 250 and 200°C sintering conditions in terms of the resistance change with the flexure cycles shows that the resistance change is muted for the higher sintering condition in the case of straight and horseshoe traces. However, the zigzag traces in the 250°C sintering condition exhibit failure in contrast with the zigzag traces in the 200°C sintering condition, which do not exceed 20% resistance change after 100,000 cycles (Figures 13.24 and 13.25).

### 13.1.4.4 Samples Sintered at 300°C for 1 H

In this section, the trend of resistance change with the increase in the flexure cycles is reported for samples sintered at 300°C for 1 h. The failure criterion of 20% resistance change is indicated with a horizontal line on each of the graphs (Figure 13.28–13.33). Figures 13.28 and 13.29 show the data for straight traces. Figures 13.30 and 13.31 show the data for horseshoe traces. Figures 13.32 and 13.33 show the data for zigzag traces. The increase in the sintering temperature to 300°C from the initial sintering temperature of 200°C does not show a significant effect on the resistance change versus cycles. Similar to the sintering condition of 250°C, a noticeable increase in the resistance is observed after 100,000 cycles for the zigzag traces. The effect is expected owing to the normal forces along the length of the zigzag interconnect.

### 13.1.5 Comparison of the Flexure Reliability versus Sintering Conditions

Figure 13.34 shows the data of the samples tested at initial angle of 100° and final angle of 60°. In this condition, the samples tend to fail before 8000 flexure cycles; whereas in the current test, conditions it was observed that there are no samples failed in the 100,000 testing cycles. Thus, the reduction of angle variation by 20° improves the fatigue-life significantly. One more major observation, which also relates to the prior

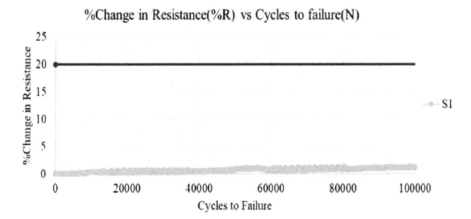

**FIGURE 13.22**
Data on trace S1 for cyclic flexing.

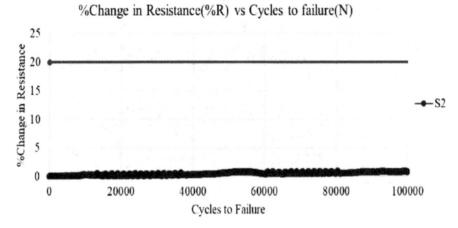

**FIGURE 13.23**
Data on trace S2 for cyclic flexing.

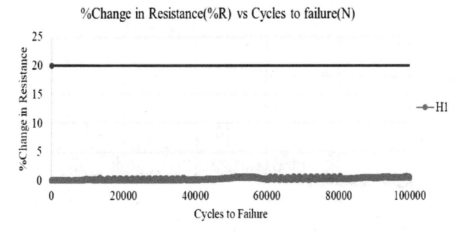

**FIGURE 13.24**
Data on trace H1 for cyclic flexing.

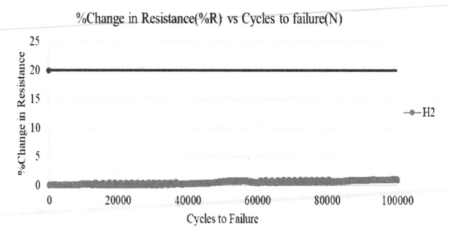

**FIGURE 13.25**
Data on trace H2 for cyclic flexing.

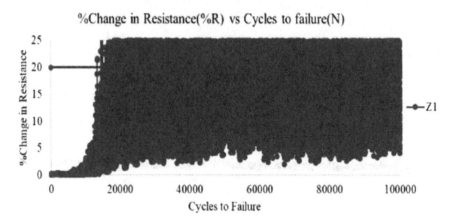

**FIGURE 13.26**
Data on trace Z1 for cyclic flexing.

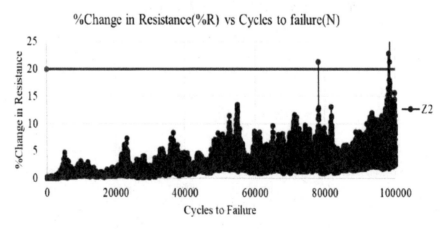

**FIGURE 13.27**
Data on trace Z2 for cyclic flexing.

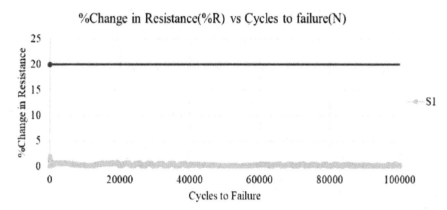

**FIGURE 13.28**
Data on trace S1 for cyclic flexing.

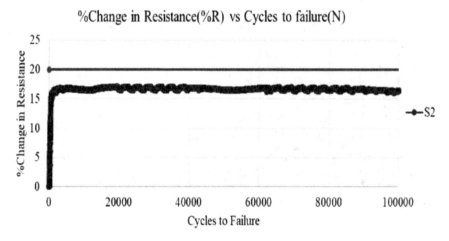

**FIGURE 13.29**
Data on trace S2 for cyclic flexing.

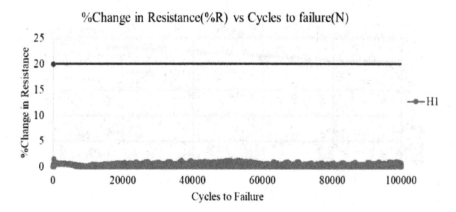

**FIGURE 13.30**
Data on trace H1 for cyclic flexing.

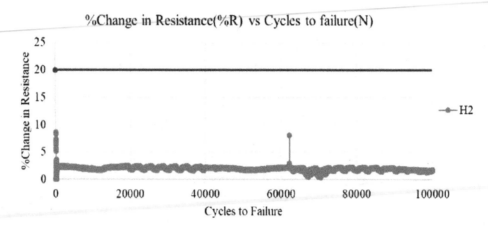

**FIGURE 13.31**
Data on trace H2 for cyclic flexing.

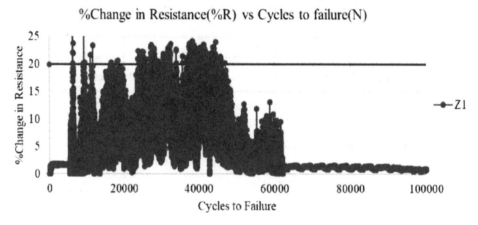

**FIGURE 13.32**
Data on trace Z1 for cyclic flexing.

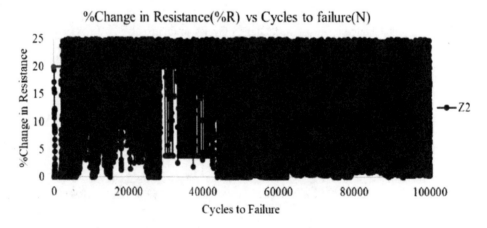

**FIGURE 13.33**
Data on trace Z2 for cyclic flexing.

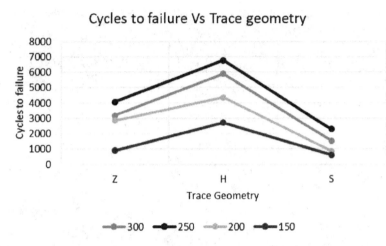

**FIGURE 13.34**
Comparison of trace geometries.

study that the horseshoe bend geometry is less susceptible to failure as compared to rest of the geometries. Further test cycles need to be run to quantify the cycles to failure for flexing condition.

### 13.1.6 Optical Images for Various Sintering Conditions

Figures 13.35–13.37 show that the trace begins to stabilize with increasing sintering temperature accompanied with the evaporation of the solvent from the printed line. Similar results are observed in zigzag and horseshoe geometrical traces.

Figures 13.38–13.40 show the optical micrographs of printed traces for straight, horseshoe, and zigzag traces, respectively. Locations closer to the curvature and bend-angles have been included in the micrographs.

**FIGURE 13.35**
150°C for 1 h.

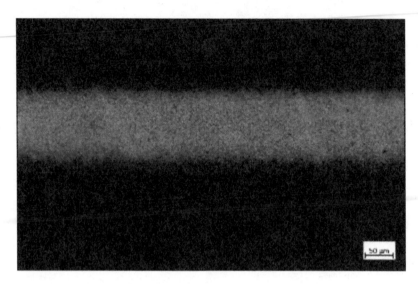

**FIGURE 13.36**
200°C for 1 h.

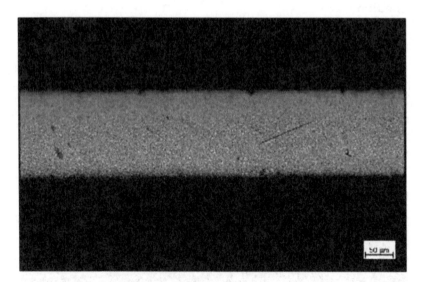

**FIGURE 13.37**
250°C for 1 h.

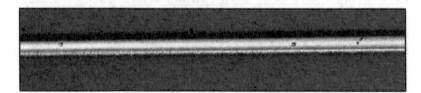

**FIGURE 13.38**
Pristine image of straight geometry trace.

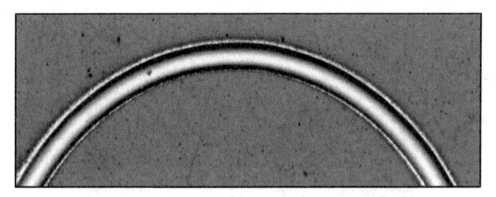

**FIGURE 13.39**
Pristine optical image of horseshoe trace.

The samples after cyclic flexing undergo deformation and micro-cracks that lead to an increase in the resistance of the trace have been imaged as well. Figures 13.41–13.43 show the location of micro-cracks at a number of locations in the straight, horseshoe, and zigzag traces.

### 13.1.7 Scanning Electron Microscope (SEM) images

Figures 13.44–13.49 show the effect of sintering temperature on printed silver nanoparticle. The SEM images are all taken at the same specification for comparison.

Comparison of the results as seen in Figures 13.44–13.49 demonstrates that there is particle transformation till 250°C. The evolution of the line morphology is muted for increase in temperature above 250°C. The variation till 250°C is due to removal of the solvent content in the printed line. Once the solvent is removed, additional exposure to higher temperatures results in the incidence of micro-cracks or increase in the constriction resistance.

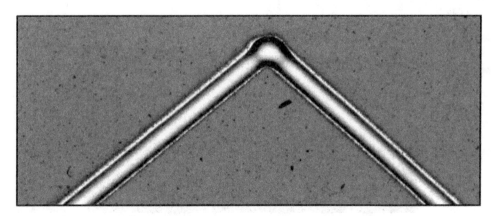

**FIGURE 13.40**
Pristine optical image of zigzag geometry.

**FIGURE 13.41**
Crack locations in straight trace after failure.

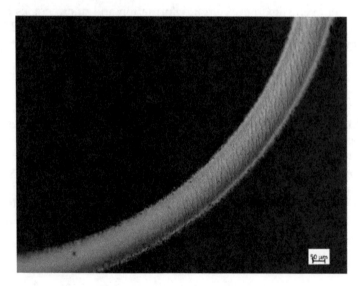

**FIGURE 13.42**
Crack locations in horseshoe trace after failure.

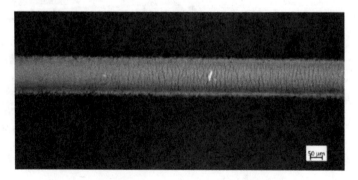

**FIGURE 13.43**
Crack locations in zigzag trace after failure.

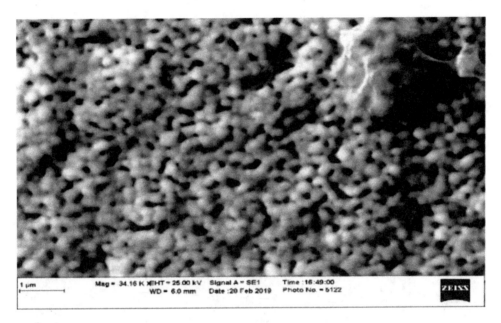

**FIGURE 13.44**
SEM image of one of the samples sintered at 120°C.

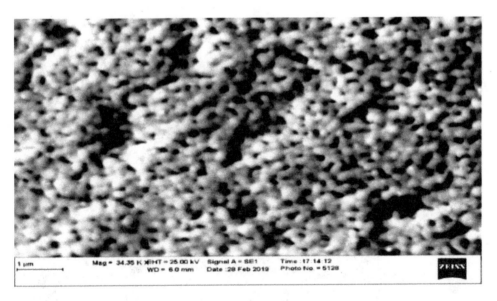

**FIGURE 13.45**
SEM image of one of the samples sintered at 150°C.

**FIGURE 13.46**
SEM image of one of the samples sintered at 200°C.

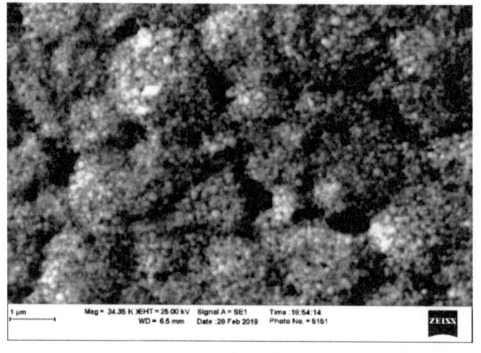

**FIGURE 13.47**
SEM image of one of the samples sintered at 250°C.

**FIGURE 13.48**
SEM image of one of the samples sintered at 270°C.

**FIGURE 13.49**
SEM image of one of the samples sintered at 300°C.

## 13.2 Stretching Reliability of FHE

The experimental setup is designed to mechanically stretch the sample for different levels of stretching. The setup consists of a power supply and a microstep driver to send the current to the stepper motor attached to the linear actuator. The microcontroller coding is done in Arduino to control the parameters such as amount of travel and actuator speed.

The setup is shown in Figure 13.50. Data-acquisition unit (Agilent Benchlink) is used to record the in situ resistance measurements during the stretch cycle. Wires are attached to the two pads on each trace, which are routed through to the multiplexer housed inside the data acquisition unit. The unit is connected to the computer through GPIB-USB interface, which feeds the real-time resistance values across each trace to the computer and displays it on the software as a reading as well as a graph. The data acquisition unit is set to take readings at two different positions, i.e., one at unstretched and another at stretched. The data recorded can be exported as a .csv file, which can be used for analysis purposes.

### 13.2.1 Test Vehicle

The test vehicle is designed to investigate the effect of stretching on the electrical resistance of a stretchable substrate. The substrate is supplied from DuPont, composed of a TE-11C polyurethane with PE874 as a conductive ink with PE671 carbon encapsulant on top. The vehicle is shown in Figure 13.1.

The test vehicle consists of two wavy traces with pad-to-pad distance of 4 inches and two circular pads of 5-mm diameter for each trace (Figure 13.51). The initial resistance for each trace was around 5 Ω. The silver-conductor paste, PE874, and the carbon-bearing encapsulant, PE671 by DuPont are both stretchable conductive paste, compatible with wide

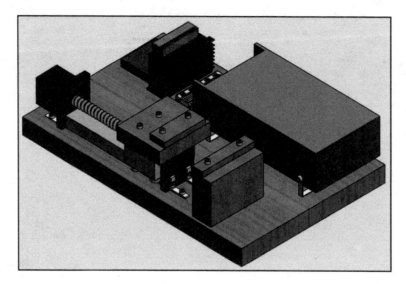

**FIGURE 13.50**
3D drawing of experimental test setup.

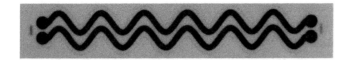

**FIGURE 13.51**
Stretch test sample.

variety of fabric and film substrates. As indicated in Table 13.1, when printed on Melinex ST505 Polyester film, at 150°C, PE874 consists of 60–65% solids while PE671 30–34% solids. Some of the other typical properties are listed in Tables 13.3 and 13.4.

## 13.2.2 Data on Stretch Testing of FHE

Stretch testing results for screen-printed samples with silver conductive ink in addition to encapsulant on top are presented for different stretch levels ranging from 5% to 20%. The stretch levels were decided based on the general stretch level that may accrue if incorporated into a wearable technology. The stretch-level was monitored with the actuator data-acquisition interface and the line resistance measured during the test continuously.

### 13.2.2.1 Stress Relaxation Test

Stress relaxation test was performed to understand the substrate behavior when subjected to stretching. Stress relaxation is a material property to study the decrease in stress over a period of time. In this study, instead of stress, we will be looking at the decrease in resistance over a finite period of time. The sample was stretched to 8% and left for 14 h. Figure 13.52 (top) shows the change in resistance for the whole period of time. We can see an initial spike in first few minutes, and over a period time, it stabilizes to 100% change, which can be regarded as a plastic deformation. To analyze the initial spike, a zoomed-in graph is plotted in Figure 13.52 (bottom). When the sample is stretched to 8% level, the highest change in resistance is around 250%, which plateaus out eventually. This decrease may be attributed to the change in properties of the stretchable substrate when it is in a stretch state. Polyurethane undergoes multiple changes while it is stretched such as microphase separation, change in the polymer bonds, which may be the reason for the decrease in the resistance after a period of time. It can also be argued that this decrease is the stretch recoverable property of the PE874 conductive silver paste used on the substrate.

**TABLE 13.3**

PE874 Typical Properties

| Test | Value |
| --- | --- |
| Solids (%) @150°C | 60–65 |
| Viscosity (PaS) Brookfield RVT, #14 spindle 10 rpm, 25°C | 50–80 |
| Density (g/cc) | 2 |
| Coverage (cm$^2$/g @5μm) | 350 |
| Coverage (cm$^2$/g @10μm) | 175 |
| Dried Print Thickness (microns) | 8–12 |

**TABLE 13.4**

PE671 Typical Properties

| Test | Value |
|---|---|
| Solids (%) @150°C | 30–34 |
| Viscosity (PaS) Brookfield RVT, #14 spindle 10 rpm, 25°C | 40–75 |
| Density (g/cc) | 1.6 |
| Coverage (cm²/g @5μm) | 400 |
| Coverage (cm²/g @10μm) | 200 |
| Dried Print Thickness (microns) | 8–12 |

### 13.2.2.2 Stretch Results

In the case of stretchable electronics, the traditional failure criteria of 20% increase in the resistance cannot be applied because 20% increase in resistance occurs in a very short period of time, and sometimes right after the stretch cycle is started. Thus, failure criterion still needs to be developed for stretchable and conformal electronics. Stretch results are presented for different levels of stretch ranging from 5% to 20%. All of the stretchable samples were subjected to 10,000 cycles. Zoomed-in versions of the graphs are also plotted for each of the condition. A closer look for about 90-cycles is looked at in the zoomed-in versions. Figures 13.53–13.56 show the change in resistance for the stretchable samples when subjected to 5, 10, and 15% strain. For the 5% strain, we see the change in resistance plateaus out at about 100%. For 10%, the resistance change is in the range of 350%. For 15% strain, the resistance change is in the neighborhood of 600%. We can also see the general trend of saw-tooth type waveform of the resistance as the number of cycles go up, which is due to way reading are taken, i.e., one at stretched state and one at unstretched.

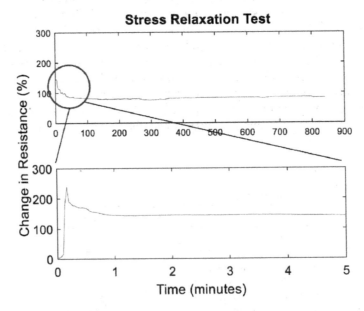

**FIGURE 13.52**
Stress relaxation test (top) for 14 h (bottom) zoomed-in to 5 min.

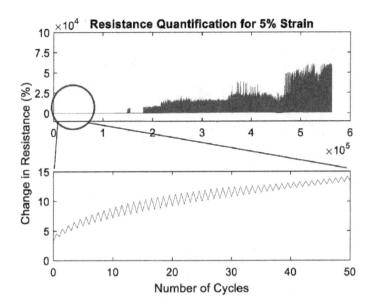

**FIGURE 13.53**
Fatigue stretch testing for 5% strain.

One point to be noted is that as the stretch level increases, there is difference in the zoomed-in versions of the graphs. That difference is due to the fact of keeping the frequency of 0.5 Hz same for all the levels. When the stretch level increases, the amount of travel increases. To keep the frequency same, the speed of the linear actuator was modified, which results in the difference.

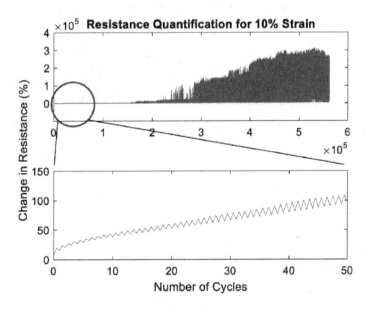

**FIGURE 13.54**
Fatigue stretch testing for 10% strain.

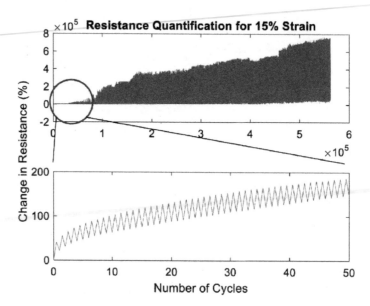

**FIGURE 13.55**
Fatigue stretch testing for 15% strain.

### 13.2.2.3 Resistance and Displacement Response

In this section, a displacement response against time is plotted in addition to the resistance to make sure that they are in phase with each other. Figures 13.57 and 13.58 show the response for two stretch levels of 5 and 10%. It is assumed that if these two levels are in phase with each other, the other two levels will be as well. From Figures 13.57 and 13.58,

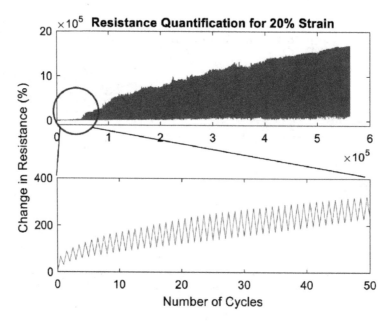

**FIGURE 13.56**
Fatigue stretch testing for 20% strain.

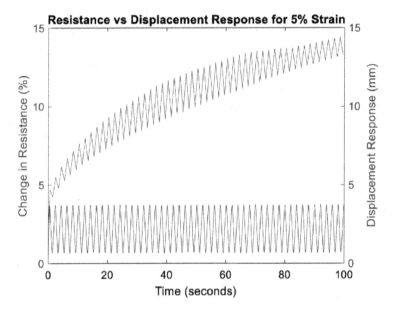

**FIGURE 13.57**
Resistance and displacement response 5% strain.

it can be seen that the crest of the displacement response is corresponding to the stretched state, which is shown as the crest in resistance response.

### 13.2.2.4 Comparison of Strain Levels

In this section, results presented in Section 13.2.2.3 are compared with each other. When the stretch level increases, we see a gradual increase in the resistance till the change plateaus

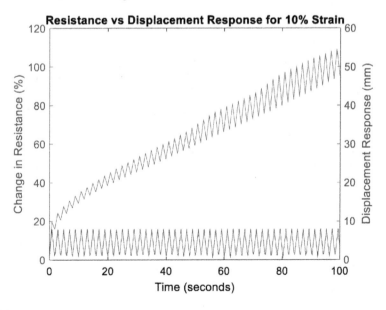

**FIGURE 13.58**
Resistance and displacement response for 10% strain.

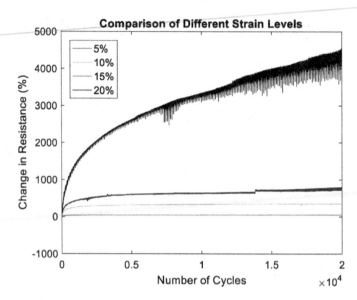

**FIGURE 13.59**
Comparison of fatigue stretch testing for different strain levels.

out and later depicting an unpredictable behavior. Figure 13.59 shows the comparison for about 20,000 cycles for each strain level 5, 10, 15, and 20%. The frequency of the testing was kept constant throughout the whole procedure of 0.5 Hz. However, to keep the frequency constant, other parameters needed to be changed such as actuator speed and travel distance.

In addition, since readings were taken for every stretch and unstretched state for a set strain level, different set of graphs are plotted to trend on each state for 50 cycles, shown in Figures 13.60 and 13.61. The unstretched and stretched state values vary quite differently

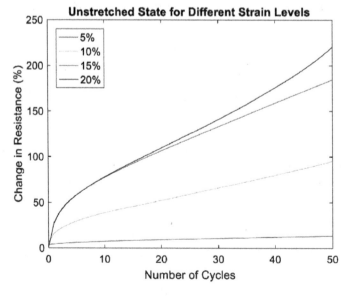

**FIGURE 13.60**
Unstretched state for different strain levels.

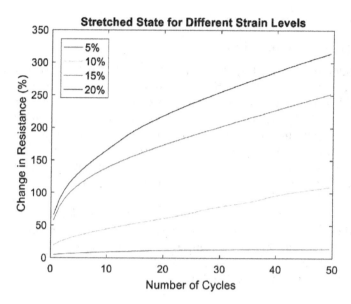

**FIGURE 13.61**
Stretched state for different strain levels.

for different strain levels; however, for 15 and 20% unstretched state, values are quite close to each other for initial number of cycles, but as the number of cycles increases, these tend to move farther apart, which can be seen in Figure 13.59.

Figure 13.62 shows the slope quantification of the percentage change in resistance for three different ranges of cycles: 0–1000, 1000–10,000, and 10,000–20,000. Inferences can be drawn from this such as the effect of higher strain levels is detrimental on the resistance values in as few as 1000 cycles, while it decreases as the number of cycles increases.

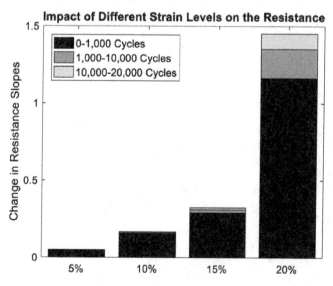

**FIGURE 13.62**
Slope quantification for different strain levels.

## 13.3 Summary and Conclusions

Flexible hybrid electronics may be subjected to flexing, folding, and stretching. Reliability results and failure modes for additively printed traces on a stretchable substrate were discussed subjected to mechanical flexing and stretching. Flexing data was discussed for flex-levels of 100°–80°. The stretching data is presented for multiple strain levels ranging from 5% to 20%, which are very common to occur in a wearable application. In addition, the effect of sintering temperature on the flexing reliability is presented. Data of resistance and shear load indicates that conductive traces printed with using silver nanoparticle have a lower susceptibility to flexure-failure with increase in sintering temperature up to temperature of 250°C. The results have been correlated with the microstructural changes in the trace geometry with the increase in sintering temperature. Reducing the cyclic amplitude by 20° makes a huge impact on the cyclic fatigue life. The samples can sustain more than 100,000 cycles at the lower angular amplitude as compared to 8000 cycles at the higher angular amplitude. Zigzag traces tend to fail early as compared to other trace geometries including straight traces and horseshoe traces.

## Acknowledgments

The project was sponsored by the NextFlex Manufacturing Institute under PC 2.5 Project titled – Mechanical Test Methods for Flexible Hybrid Electronics Materials and Devices. This material is based, in part, on research sponsored by Air Force Research Laboratory under agreement number FA8650-15-2-5401, as conducted through the flexible hybrid electronics manufacturing innovation institute, NextFlex. The U.S. Government is authorized to reproduce and distribute reprints for governmental purposes notwithstanding any copyright notation thereon. The views and conclusions contained herein are those of the authors and should not be interpreted as necessarily representing the official policies or endorsements, either expressed or implied, of Air Force Research Laboratory or the U.S. Government.

## References

[1] Xuyuan Tao, Vladan Koncar, Tzu-Hao Huang, Chien-Lung Shen, Ya-Chi Ko, and Gwo-Tsuen Jou, "How to make reliable, washable, and wearable textronic devices", Sensors, vol. 17, no. 4, pp. 673, March 2017.

[2] L. Dai, Y. Huang, H. Chen, X. Feng, and D. Fang, "Transition among failure modes of the bending system with a stiff film on a soft substrate", Applied Physics Letters, vol. 106, pp. 021905, 2015.

[3] T. Happonen, J. Häkkinen, and T. Fabritius, "Cyclic bending reliability of silk screen printed silver traces on plastic and paper substrates", in IEEE Transactions on Device and Materials Reliability, vol. 15, no. 3, pp. 394–401, Sept. 2015.

[4] I. Yakimets et al., "Polymer substrates for flexible electronics: Achievements and challenges, in Proceedings FuSeM, pp. 5–8, 2009.

[5] N. Lacerda Silva, L. M. Goncalves, and H. Carvalho, "Deposition of conductive materials on textile and polymeric substrates", Journal of Materials Science, vol. 24, no. 2, pp. 635–643, Feb. 2013.

[6] Hsien-Hsueh Lee, Kan-Sen Chou1 and Kuo-Cheng Huang, "Inkjet printing of nanosized silver colloids", Nanotechnology, vol. 16, no. 10, Published 2 September 2005.

[7] Jin-Woo Park, and Seong-Gu Baek, "Thermal behavior of direct-printed lines of silver nanoparticles", Scripta Materialia, vol. 55, no. 12, pp. 1139–1142, December 2006.

[8] Sunho Jeong, Dongjo Kim, Sul Lee, Bong-Kyun Park and Jooho Moon, "Fabrication of the organic thin-film transistors based on ink-jet printed silver electrodes", Molecular Crystals and Liquid Crystals, vol. 459, no. 1, pp. 35/[315]–43/[323], 2007.

[9] Julia R. Greer, and Robert A. Street, "Thermal cure effects on electrical performance of nanoparticle silver inks", Acta Materialia, vol. 55, no. 18, pp. 6345–6349, October 2007.

[10] Na-Rae Kim, Ji-Hoon Lee, Seol-Min Yi, and Young-Chang Joo, "Highly conductive ag nanoparticulate films induced by movable rapid thermal annealing applicable to roll-to-roll processing", Journal of The Electrochemical Society, vol. 158 no. 8, pp. K165–K169, 2011.

[11] Seol-Min Yia, Ji-Hoon Leea, Na-Rae Kima, Sungil Ohb, and Seonhee Jang, "Improvement of electrical and mechanical properties of ag nanoparticulate films by controlling the oxygen pressure", Journal of the Electrochemical Society, vol. 157, no. 12, pp. K254–K259, 2010.

[12] Cheol Kim, and Chung Hwan Kim. "Universal testing apparatus implementing various repetitive mechanical deformations to evaluate the reliability of flexible electronic devices", Micromachines, vol. 9, no. 10, pp. 492. 25 Sep. 2018, doi:10.3390/mi9100492.

[13] P. Lall, J. Narangaparambil, A. Abrol, B. Leever, and J. Marsh, Development of Test Protocols for the Flexible Substrates in Wearable Applications, in Proceedings of ITHERM, San Diego, CA, US, pp. 1120–1127, May 29–June 1, 2018.

# *Index*